Processing Series No. 6

Series Advisor: Professor B. L. Weiss

FABRICATION OF GaAs DEVICES

Other volumes in this Series:

Volume 1 **Silicon wafer bonding technology for VLSI and MEMS applications** S. Iyer and A. Auberton-Herve (Editors)

Volume 2 **Process technology for silicon carbide devices** C.M. Zetterling (Editor)

Volume 3 **MEMS packaging** T.R. Hsu (Editor)

Volume 4 **SIMOX** M. Anc (Editor)

Volume 5 **Silicide technology for integrated circuits** L.J. Chen (Editor)

FABRICATION OF GaAs DEVICES

by

Albert G. Baca and Carol I. H. Ashby

Sandia National Laboratories
Albuquerque, NM, USA

Published by: The Institution of Electrical Engineers, London, United Kingdom

The Institution of Electrical Engineers,
Michael Faraday House,
Six Hills Way, Stevenage
Herts., SG1 2AY, United Kingdom

British Library Cataloguing in Publication Data

Baca, A. G.
Fabrication of GaAs devices. - (EMIS processing series ; 6)
1. Gallium arsenide semiconductors
I. Title II. Ashby, Carol III. Institution of Electrical Engineers
621.3′8152

ISBN 0 86341 353 6

Typeset in India by Newgen Imaging Systems (P) Ltd., Chennai, India
Printed in the UK by MPG Books Limited, Bodmin, Cornwall

DEDICATION

To my wife, Beatriz, my children, Luis and Michelle, my mother, Nina, and my late father, Alfonso. A.G.B.

With my heart-felt appreciation to my husband, Jim, and my children, Paul and Lydia, for the many hours of my time that they sacrificed to the writing of this book. Et soli Deo gloria. C.I.H.A.

CONTENTS

ACKNOWLEDGMENT

We would like to acknowledge all who helped in the preparation and proofing of this book. Many thanks to Marcelino G. Armendariz, Matthew G. Blain, William G. Breiland, Ronald D. Briggs, Melissa Cavaliere, Jeffrey G. Cederberg, Ping-Chih (Pablo) Chang, Michael J. Cich, Kent M. Geib, Leonardo Griego, YoungMin Kim, Geraldine M. Lopez, Phil F. Marsh, Carolyn M. Matzke, Cedric Monier, Mark E. Overberg, Dennis J. Rieger, Carlos Sanchez, Charlie E. Sandoval, Randy J. Shul, Charles T. Sullivan, John P. Sullivan, Jeffrey Y. Tsao, and W. Graham Yelton.

Many thanks also to all of our colleagues and collaborators, past and present, whose interactions we greatly value, and who are too numerous to mention individually.

ABBREVIATIONS

Acronym	Chapter	Meaning
2DEG	1	two-dimensional electron gas
AES	3	Auger electron spectroscopy
ALE	2	atomic layer epitaxy
AR	3	anti-reflection
AUD	3	advanced unified defect (model)
BJT	1, 9	bipolar junction transistor
BOE	8	buffered-oxide etch
CAIBE	5	chemically assisted ion-beam etching
CBM	3	conduction band minimum
CMOS	1, 3, 8, 10	complementary metal oxide semiconductor
CRT	2	cathode-ray tube
C-V	2, 3, 7	current-voltage
CZ	1	Czochralski
DBR	10	distributed Bragg reflector
DC	2, 3, 5, 7–9	direct current
DFB	3, 4, 10	distributed feedback
DH	1	double heterostructure
DHBT	9	double heterojunction bipolar transistor
DI	3, 4	de-ionised
DIGS	3	disorder-induced gap state (model)
DLTS	2	deep level transient spectroscopy
EBIC	2	electron-beam-induced current
ECR	3, 5, 8, 9	electron-cyclotron resonance
EDS	2	energy-dispersive spectrometer
EDX	3, 9	energy-dispersive X-ray analysis
EMP	2	electron microprobe (analysis)
EW	1	electronic warfare
EWF	3	effective work function (model)
FET	1–4, 6–9	field effect transistor
FIB	2, 5	focused ion beam
FWHM	2	full width at half maximum

Acronym	Chapter	Meaning
GMR	3	giant magnetoresistance
GOI	10	GaAs-on-insulator
HBT	1–6, 8, 9	heterojunction bipolar transistor
HDP	3	high-density plasma
HDPE	5	high-density-plasma etching
HEMT	1, 2, 5, 7, 8	high electron mobility transistor
IBAE	5	ion-beam-assisted etching
IBE	5	ion-beam etching
IC	1	integrated circuit
ICP	3, 5, 9	inductively coupled plasma
IMPATT	1	impact avalanche and transit time
IR	6, 7	infrared
JFET	1	junction field effect transistor
LCR	2	inductance, capacitance, resistance
LD	1, 9	laser diode
LDD	7, 8	lightly doped drain
LEC	1, 2	liquid encapsulated Czochralski
LED	1, 2, 9	light-emitting diode
LEED	3	low-energy electron diffraction
LPE	1, 2	liquid phase epitaxy
MBE	1–3, 6, 8, 9	molecular beam epitaxy
MESFET	1, 5–8, 10	metal-semiconductor field effect transistor
MHEMT	8	metamorphic high electron mobility transistor
MIGS	3	metal-induced gap state (model)
MIS	3, 8–10	metal-insulator-semiconductor
MMIC	2	monolithic microwave integrated circuit
MOCVD	1–3, 5, 7–9	metal organic chemical vapour deposition
MODFET	4, 5	modulation doped field effect transistor
MOS	2, 3, 8	metal-oxide-semiconductor
MOSFET	1, 7, 10	metal-oxide-semiconductor field effect transistor
MTTF	9	median/mean time to failure
OEM	1, 2	original equipment manufacturer
OES	5	optical emission spectroscopy
PA	1	power amplifier
PE	5	plasma etching
PECVD	3, 7–9	plasma-enhanced chemical vapour deposition

Acronym	Chapter	Meaning
PHEMT	1, 7, 8	pseudomorphic high electron mobility transistor
PL	2, 3	photoluminescence
PLE	3	photoluminescence excitation
PMA	3	post-metallisation annealing
PR	5, 7	photoresist
QW	10	quantum well
RBIBE	5	radical-beam-ion-beam etching
RBS	2, 8	Rutherford backscattering spectroscopy
REDR	9	recombination-enhanced-defect-reaction
RF	1, 3, 5–9	radio frequency
RHEED	2, 3	reflection high-energy electron diffraction
RIBE	5	reactive-ion-beam etching
RIE	5, 7–9	reactive ion etching
RMS	5	root mean square
RTA	3, 6–8	rapid thermal annealer/annealing
SAN	1	storage area network
SBH	3	Schottky barrier height
SEM	2	scanning electron microscope/ microscopy
SIMS	2, 3, 6, 8	secondary-ion mass spectrometry
STEM	2	scanning transmission electron microscopy
STM	3	scanning tunnelling microscopy
TEM	2, 6–10	transmission electron microscope/ microscopy
TPD	3	temperature-programmed desorption
UDF	3	unified defect (model)
UHV	3	ultra-high vacuum
UV	1, 3, 4, 6–9	ultraviolet
VB	2	vertical Bridgman
VBM	3	valence band maximum
VCSEL	1, 10	vertical cavity surface emitting laser
VGF	2	vertical gradient freeze
VPE	1, 2	vapour phase epitaxy
WDM	1	wavelength division multiplexing
WDS	2	wavelength-dispersive spectrometer
XPS	2, 3	X-ray photoelectron spectroscopy

Chapter 1

Introduction to GaAs devices

1.1 SCOPE OF THIS BOOK

One might ask why another book about GaAs processing is needed. In every field, there are two bodies of knowledge required for true mastery. This is especially true in the field of device engineering. The first is related to understanding the ideal behaviour of devices predicted by fundamental laws of physics. This is the subject of many textbooks and courses and is readily accessible to every engineer. However, as we soon discover as we fabricate real-world devices, behaviour often deviates from that ideal. When that occurs, the questions "Why?" and "How can I fix it or avoid it next time?" must be answered or the same problems will continue to plague our devices.

The second body of knowledge helps us answer these challenging questions. It is that vast body of practical, even anecdotal, knowledge gradually accumulated by a practising engineer through everyday experiences with successes and failures in designing and fabricating devices. It is some of this second body of knowledge that we hope to provide in writing this book. While even the 30-year veteran will not have fully mastered this second body of knowledge, the aim of this book is to share many of the insights and solutions that we have learned during our combined years working with devices made of GaAs and related materials.

In addition, many books on GaAs devices contain sections or chapters on processing methods. However, most books provide only a part of the knowledge required to actually make high-quality devices and almost all of the previous books neglect some aspects of processing as specifically related to device operation and reliability.

Most books on GaAs devices are mainly concerned with a logical and educational presentation of the device operation, naturally focusing on the physics of the devices and the derivation of equations that describe device operation. Some good books on advanced processing techniques in general [1–5] and even on modern GaAs processing [6] are available. The GaAs processing book by Williams focuses on process techniques for GaAs electronic technology. The device-oriented books describe ideal operation and do not treat the impact of specific processing choices on device operation [7–17]. The process-oriented books address some impacts of processing choices but do not adequately relate these to subtleties in device performance and reliability. Especially lacking in any of these books is an adequate treatment of the critical importance of surfaces and interfaces in GaAs devices and their correlation with device performance.

This book attempts to address all aspects of GaAs processing that deal with GaAs free surfaces and with interfaces between GaAs and metal contacts or dielectrics. These subjects are intimately involved in virtually all aspects of device processing. We have attempted to provide both fundamental perspective and practical advice in the areas of cleaning and passivation (Chapter 3), wet etching and photolithography (Chapter 4), dry etching (Chapter 5) and GaAs-metal contacts (Chapters 6 and 7). Chapters 8 and 9 discuss device performance for HBTs and FETs and how this relates to processing choices. Chapter 10 deals with special processing issues such as wet oxidation that are especially important in optoelectronic devices. Chapter 2 provides a variety of background material about physical concepts, semiconductor materials and characterisation techniques so that the student does not have to consult multiple texts to gain perspective while using this book.

Our main goal is to provide both fundamental and practical information to aid both the beginning and the practising engineer in relating device performance, degradation and non-idealities with processing choices made during device fabrication. Above all, we have attempted to make this an "apprenticeship in a book" by presenting as much practical information as possible that we have learned through many years of processing of GaAs devices ourselves. To the extent that some of these goals are achieved, another book on GaAs processing is worthwhile.

1.2 GaAs MATERIALS

Soon after the discovery of the transistor in 1947, gallium arsenide and many other semiconductors were assessed as candidate materials for electronic devices. GaAs became recognised as a material with a large electron velocity that might be suitable for high-speed electronics. It took several decades before this "material of the future" made a significant commercial entry in the 1970s, much to the frustration and delayed ambitions of many of its practitioners. Now, GaAs is the basis of a several billion dollar worldwide industry for high-frequency and high-speed electronics, which places it at a few percent of the overall semiconductor market. Although by market share alone GaAs may not seem to deserve much attention as a semiconductor material, its importance lies in the applications it enables. As seen in Section 1.5.2, the applications that use GaAs electronics are some of the most exciting around.

Although GaAs materials properties are more easily understood after assimilating some of the topics in Chapter 2, a brief contrast of GaAs to other semiconductors is instructive here. One important classification is that of elemental versus compound semiconductors. The main elemental semiconductors are Si and Ge. These semiconductors are generally easier to grow, purify and process than compound semiconductors such as GaAs, so they were used for devices at an earlier date and continue to have important material-related advantages. Si, the dominant semiconductor commercially, is more defect free, has better mechanical properties, and is available as larger diameter wafers. Its commercial dominance in electronics is usually ascribed to two important advantages. The first is its unparalled surface passivation with SiO_2 that leads to unequalled transistors based on surface inversion, i.e. the MOSFET (metal-oxide-semiconductor field effect transistor). The second is the unrivalled small size and high packing density of its transistors and other integrated circuit components (e.g. trench capacitors). Part of this second factor is the sheer investment in tools for high-density, high-volume manufacturing that is, in principle, available to compound semiconductor technologies as well (but not in practice due to smaller wafers and smaller addressable markets necessary to pay for the capital investments). The foundational part of the high-density advantage for Si-based electronics is due to its superior material properties. High-density devices require small lateral and vertical dimensions in the device as well as low power density per device. Virtually every important dimension is easier to reduce in Si devices because of the unique advantages of its oxide and the processing advantages of an elemental semiconductor (i.e. shallower and more flexible doping).

Compound semiconductors find niches in certain applications due to other material properties. The most important material properties are the high mobility (carrier speed in response to an electric field), light-emission capability, and a large variety of bandgap-engineering approaches available. These properties derive in large part from the band structure of semiconductors (Section 2.2.1). In particular, GaAs has a direct bandgap (Section 2.2.1) that permits high-efficiency light emission as well as a conduction-band structure that leads to fast electron conduction. GaAs electronics find niches in small-sized circuits that require high speed or low power consumption. Soon after the semiconductor laser was discovered in 1962, the light emitting properties of GaAs gained a new importance and ultimately became a large commercial attribute. GaAs-based lasers and LEDs share a broad range of important applications with several other compound semiconductors. Interestingly, Si, the dominant electronic semiconductor, does not have a direct bandgap nor does it address light-emitting applications (except as part of the transmitting and receiving electronic functions).

1.3 TYPES OF GaAs DEVICES

GaAs devices are categorised broadly by whether they are electronic devices or light-based (photonic) devices. This categorisation is due to differences in specialisation ranging from understanding the physics of the device operation to the considerable differences in measurement techniques. Processing of photonic devices is remarkably similar to electronic device processing and many of the topics in this book are equally applied to both categories of devices.

In this section, the types of GaAs devices will be briefly defined and introduced for the purpose of referring to the types of applications of GaAs devices. This introduction will not include details of the physical principles or the operation of the devices, the subject of later chapters. Readers without a prior familiarity with semiconductor devices may find this chapter more useful after reading the rest of the book.

1.3.1 Electronic devices

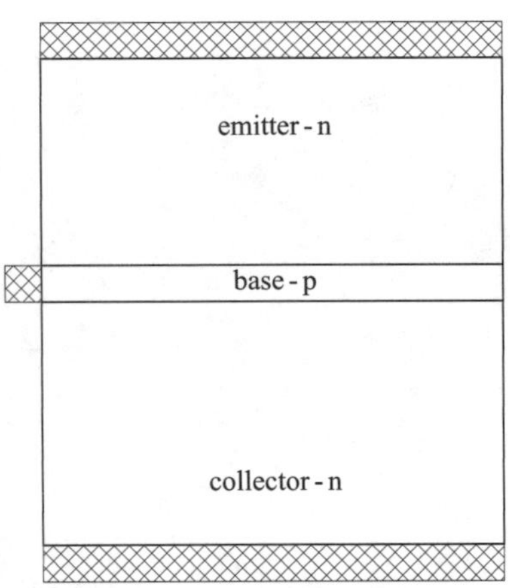

FIGURE 1.1 A schematic of a bipolar transistor with electrical contacts to the emitter, base and collector.

Electronic devices are of three basic types: bipolar transistors, field effect transistors and diodes. All three types have important commercial applications. The purpose of this introduction is to provide a level of understanding useful for a discussion of applications of GaAs devices (Section 1.5).

GaAs applications use heterojunction bipolar transistors (HBTs), field effect transistors (FETs) and diodes. The bipolar transistor is illustrated in FIGURE 1.1. The bipolar transistor can be considered to be two pn junction diodes connected together with a common middle region. The npn bipolar transistor is the most common and uses a common p-region. As shown in FIGURE 1.1, the bipolar transistor has an emitter, a base region and a collector region, and electrical contacts to each. A heterojunction transistor uses emitter material with larger bandgap (Section 2.2.1) than the base material. In an npn transistor, electrons flow from the emitter to the collector regions, passing through the base. The base is made very thin so that few electrons recombine in that region. The electron flow is modulated by an electrical signal applied to the base-emitter junction. A small change in the signal applied to this junction leads to a large current change at the collector. This amplification of the signal is what makes transistors have more uses than diodes. Details of the operation of bipolar transistors are found in Section 9.2.

Field effect transistors are illustrated in FIGURE 1.2. Like the bipolar transistor, the FET is a three-terminal device with metal contacts for the source, the gate and the drain. In the field effect transistor, electrons flow from the source to the drain when a drain voltage is applied. A voltage signal to the gate can cause a change in the drain current resulting in amplification. As with the bipolar transistor, amplification of an input signal occurs for a properly designed device.

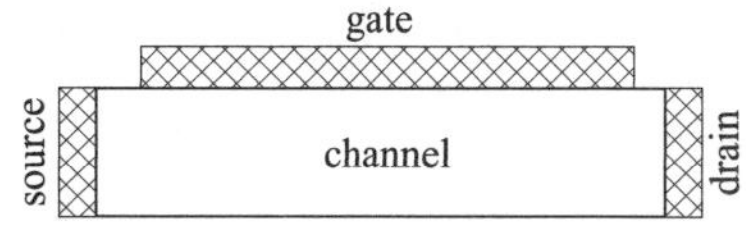

FIGURE 1.2 A schematic of a field effect transistor with electrical contacts to the source, gate and drain.

1.3.2 Photonic devices

Four broad categories of photonic devices will be introduced: light-emitting diodes (LEDs), laser diodes (LDs), photodetectors and waveguides. As for electronic devices, a rudimentary introduction of device types is useful for a discussion of applications of GaAs devices (Section 1.5).

Light-emitting diodes produce light after passing current through a pn junction in a semiconductor with a direct bandgap. A cross-section of an LED is shown in FIGURE 1.3. Electrical current flows between a top contact and a bottom contact, shown here at the bottom of the semiconductor. "Bottom" contacts may also be made to the front side of the wafer (but below the top contact). The structure of FIGURE 1.3 is generally rotated about the axis for a circular geometry, but other shapes are also possible. An applied electrical bias sends electrons from the negative region to the positive region and holes (Section 2.2.2) in the opposite direction. LEDs emit photons when electrons and holes in the same area of the semiconductor recombine and release the energy of recombination as a photon of light. The top contact is shown as a cross-section of opposite ends of a ring which conducts current from the edge of the cavity (central area between top contacts) to the bottom contact (and spreading in the process). Sometimes instead of circular ring contacts, planar transparent contacts are used. Because of the band structure of the semiconductor (Section 2.2.1), the light is generally monochromatic. GaAs-based LEDs generally emit in the red and yellow regions of the spectrum. Light is emitted in all directions in the LED. Many of the photons are absorbed by the semiconductor or by the package and never contribute to the light output. Researchers and applications engineers spend much effort

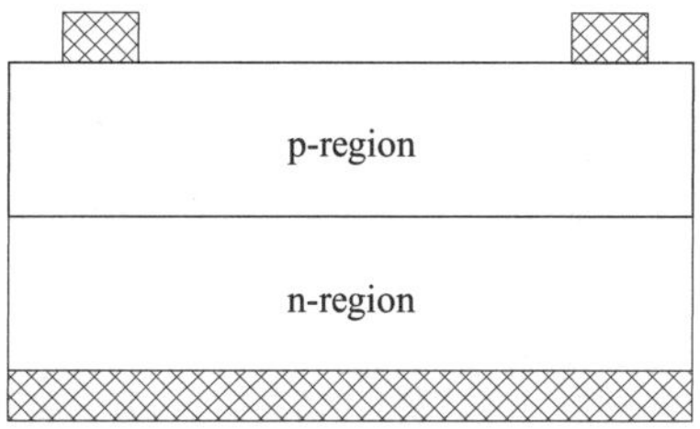

FIGURE 1.3 A schematic of a light-emitting diode with electrical contacts to the n- and p-regions.

trying to find ways to get as much light as possible out of the LED in order to make the devices more efficient and cost effective.

Laser diodes also have pn junction structures. As in the LED, photons are generated by passing the current through a pn junction. However, one establishes an optical cavity by creating mirrors in the semiconductor so that photons can make multiple passes through the cavity to enable the amplification required for lasing. An edge-emitting laser, as illustrated in FIGURE 1.4, will have cleaved edges to form the mirrors as reflecting edges of the semiconductor. A vertical-cavity laser, which is similar in cross-section to the LED of FIGURE 1.3, will have superlattice mirrors above and below the pn junction (Section 10.5.1 and FIGURE 10.13). Like the LED, laser diodes create electron-hole pairs by injecting minority carriers into an active region where they can recombine and emit light. Like any laser, the light is very directional and can have very narrow bandwidths for the emitted light. The wavelength of the light can be tuned by changing the bandgap or energy levels in the material of the active region.

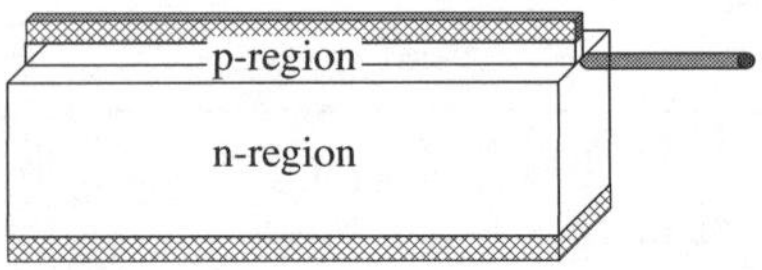

FIGURE 1.4 A schematic of an edge-emitting laser with electrical contacts to the n- and p-regions.

One type of photodetector, the p-i-n diode, may be considered to be an LED operated in reverse. However, instead of an abrupt pn junction, an undoped (or i for intrinsic) region with reasonable thickness is used for photon absorption. Some fraction of photons with energy above the bandgap of the i-region entering the detector are absorbed by the semiconductor and create electron-hole pairs. The electric field across the reverse-biased junction sweeps the charge out of the i-region to permit its detection.

1.4 A BRIEF HISTORY OF GaAs DEVICES

A brief history of GaAs devices will be presented here so that the reader can place the devices of interest within a historical context. For readers completely unfamiliar with GaAs devices, much of the historical context may be better understood by referring to this section as other material in the book is digested. The history of GaAs electronic devices is presented first (because electronic devices were researched earlier), followed by the history of GaAs photonic devices.

1.4.1 History of GaAs electronic devices

The transistor was invented in 1947 at Bell Laboratories. This transistor was called a point contact transistor but operated as a bipolar transistor. This and other early transistors used a germanium semiconductor. This invention opened the door for the exploration of many other semiconductor materials. Quickly germanium and silicon became the dominant semiconductors of the 1950s, with Si becoming dominant in the 1960s.

The field effect transistor was invented in 1952. However, most of the early transistors of any importance in the 1950s and 1960s were still bipolar transistors. Not surprisingly, the initial interest in GaAs devices centred on bipolar transistors that would be superior to germanium (and later silicon) transistors for high-frequency applications. The first GaAs bipolar transistor with RF performance superior to that of silicon bipolar transistors and approaching that of germanium devices was reported in 1961. Large efforts supported by US government funding aimed at developing GaAs bipolar transistors with superior high-frequency performance. Technological limitations forced these early devices to be based on homojunction analogues to Si and Ge devices, often with diffusion as the means of creating the pn junctions in bipolar devices. By 1980 a GaAs bipolar transistor with an operating frequency greater than 1 GHz (unity current gain, Section 8.2.1) had been achieved using ion implantation. These early attempts at making high-performance GaAs BJTs were not very successful and competitive GaAs bipolar transistors would have to wait for the development of heterostructure devices.

The field effect transistor was first analysed in the early 1950s by Shockley. In 1966, the first GaAs FETs were developed, both a metal-semiconductor FET (MESFET) using a Schottky barrier gate (Section 8.2.1) and a junction field effect transistor (JFET) with a Zn-diffused region for a pn junction gate. The first MESFET with GHz capability was demonstrated a year later. By 1970, MESFETs with 30 GHz performance (unity current gain) had been demonstrated with 1 μm gates. The 1970s saw a series of demonstrations of GaAs FET use in a variety of high-frequency analogue circuit applications such as low-noise amplifiers and power amplifiers. JFETs based initially on diffusion and later on ion implantation for pn junction gates never became as widely used as MESFETs because the pn junction gate has a parasitic capacitance that limits its performance. The majority of early MESFET and bipolar structures were grown by vapour phase epitaxy (VPE), which offered better active layer thickness control and higher surface quality.

In addition to GaAs transistors, there were two important microwave diode technologies discovered in the 1960s: the Gunn and the IMPATT (IMPact Avalanche and Transit Time) diodes. Apart from challenging physicists to explain the source of their microwave oscillations, these devices still play an important role as very-high-frequency power sources.

Two key ingredients of high-performance GaAs electronics are the semi-insulating substrate for low parasitics and the use of heterostructures in all GaAs devices. The first GaAs crystals grown by the Czochralski (CZ) method were produced in 1956. The liquid encapsulated Czochralski (LEC) technique (Section 2.3) was first applied to GaAs in 1965. High-purity crystal growth has been enabled by advances in this technique. By the early 1980s, round substrates grown by LEC in 2-inch and 3-inch sizes were commercially available. Prior to that GaAs crystals had odd shapes that were not conducive to the development of high-volume commercial applications.

A historically important development for semi-insulating substrates was the use of Cr doping, first reported in the early 1960s. Today Cr is no longer needed to produce semi-insulating crystals (Section 2.3), but in the early days of poor purity and stoichiometry control, it enabled high-frequency devices that would not otherwise have been possible.

Bringing performance to higher levels in the 1980s and beyond depended primarily on two factors, dimensional scaling and the emergence of new devices based on heterostructures. Dimensional scaling depended primarily on the development of better processing techniques and better lithography tools, which largely leveraged the investments made in Si technology. Devices based on heterostructures had been proposed long before the materials growth techniques were able to produce these structures.

In the 1960s, liquid phase epitaxy and vapour phase epitaxy were the main methods of creating heterostructures. Neither offered the type of control needed for thin active layers in heterostructure transistors. Dimensional control of thin AlGaAs and GaAs heterostructures was demonstrated by Cho in 1971 using a newly developed technique of MBE (Section 2.4). By 1975, high-quality electronic and optical devices were demonstrated.

The history of the heterostructure bipolar transistor is nearly as old as the transistor itself. In 1948, Shockley outlined the advantage of incorporating a heterostructure into a bipolar transistor. In 1957, Kroemer formulated the basic HBT theory. The first microwave HBT was reported in 1972. By 1982, the HBT was used in digital integrated circuits. By 1990, GaAs HBT operation was pushing past 100 GHz.

The high electron mobility transistor (HEMT), was first demonstrated in 1979 by Mimura and co-workers. It was preceded by the engineering of superlattice heterostructures of GaAs and AlGaAs by Dingle and co-workers and a single quantum-well "two-dimensional electron gas" (2DEG) structure. The 2DEG structure (FIGURE 1.5) uses electron dopant impurities placed in the higher bandgap material (AlGaAs). The band alignments of the adjacent GaAs and AlGaAs shifted the free electrons from the AlGaAs into the GaAs. Hence, electron mobility was improved

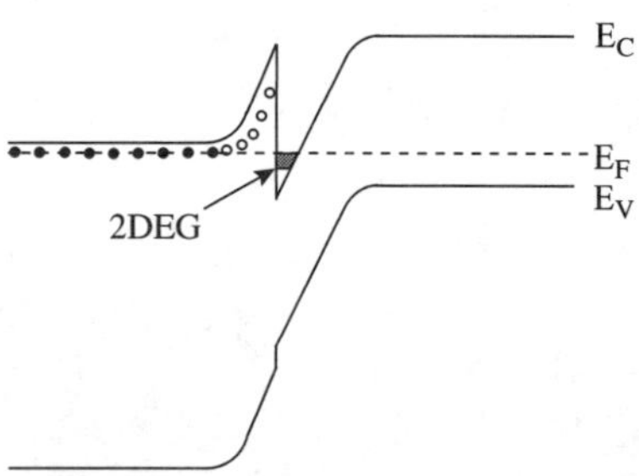

FIGURE 1.5 An illustration of a two-dimensional electron gas.

by transport in this pure, high-quality layer and the separation between impurity donors and the electron gave rise to the name "two-dimensional electron gas".

The next development in HEMT devices was the incorporation of a strained layer of InGaAs in the channel, termed a pseudomorphic HEMT (PHEMT). The higher transport of this strained layer pushed frequency performance towards 100 GHz by the late 1980s, although GaAs ceded the performance advantage to InP-based HEMTs by then. In spite of higher frequency operation for InP-based devices, GaAs devices' favourable combination of speed and breakdown characteristics led to the dominance of PHEMTs and HBTs in power amplifier applications (Section 1.5.2).

1.4.2 History of GaAs photonic devices

Prior to the invention of the solid-state laser in 1962, light emission from semiconductors was little more than a curiosity. Instantly, the potential for directed light sources of all types was recognised and laser research in semiconductors took off. Many early advances in GaAs lasers came out of the Ioffe Institute, Bell Laboratories and IBM laboratories [7,18]. GaAs homojunction lasers were demonstrated early on but suffered a number of problems relating to the lack of confinement. The double heterostructure laser was conceived theoretically well before the experimental means of achieving such a device was perfected. The double heterostructure (DH) was designed to inject electrons and holes into an active region while the potential barriers of the heterostructures serve to inhibit escape of the injected carriers and to enhance population inversion. The heterostructures are also engineered with a different dielectric constant to confine the light in the active region and prevent optical losses. Such a structure proved to be far superior to homojunction lasers, but was initially thought to be a paper curiosity. Little was known about the interface quality between candidate heterostructure materials and widespread skepticism existed about whether the ideal heterojunction with a non-defective interface could be fabricated [18]. Many early researchers considered the search for the ideal heterojunction very unlikely to succeed and few groups at that time attempted it.

Since GaAs was known from early physics and electronic investigations to show a favourable combination of properties, the search for a DH laser centred initially on GaAs for an active region. Its direct bandgap, low effective mass, high mobility at the conduction band minimum and sharp optical absorption were among its favourable properties. The search for compatible heterostructure materials was initially focused on GaP and AlAs. AlAs, while having a compatible lattice constant, was quickly found to be chemically unstable. Work then centred on GaAsP and the first double heterostructure GaAs laser was fabricated in 1966 [19]. However, the difficulty of achieving a lattice matched condition led to a search for other materials, and shortly thereafter AlGaAs was found to be a lattice matched material that exhibited the chemical stability that AlAs lacked. The first CW stripe laser using AlGaAs/GaAs was demonstrated in 1970. It exhibited a lower lasing threshold than earlier approaches. Thresholds were to continue improving throughout the 1970s. At this point the evolution of laser materials research turned to quaternaries. They offered the flexibility of independently varying the lattice constant and the bandgap.

These early laser developments were carried out mainly with liquid-phase epitaxy as the means of material growth. LPE produced heterostructures with adequate quality interfaces but had many shortcomings that would limit the further development of GaAs lasers. The lack of uniformity of the layers caused current crowding and optical self-focusing. Also, LPE is incapable of growing thin layers. The thick active regions grown by LPE caused high losses in the cavity. In the 1970s, two new growth techniques began to have a large impact on GaAs devices. Molecular beam epitaxy (MBE) and metal organic chemical vapour deposition (MOCVD) brought the ability to grow semiconductor films to monolayer accuracy with high purity. This control allowed the growth of thin, uniform active regions which ultimately led to quantum-well active regions. Quantum-well laser diodes were made with reduced threshold current and higher gain.

1.5 APPLICATIONS OF GaAs DEVICES

GaAs applications fall into two categories: those based on its light emitting properties (due to its direct bandgap – see Section 2.2.1) and those based on its transport properties. A complete discussion of all applications will not be possible, but we will present a wide enough range of applications to give a good overview. In the process, we will focus on the more established, high-volume, commercial technologies that represent real, not potential sales. In this section, the order of presenting electronic and photonic applications is reversed from the history of the devices (Section 1.4) because of their history; photonic devices were commercialised earlier than electronic devices while electronic devices were researched earlier than photonic devices.

Each of these categories may contain consumer, enterprise and military segments. In the consumer segment, it is understood that products must meet technical requirements, but only become truly successful by offering low cost. Every OEM (original equipment manufacturer) knows the competitive pressures of a consumer product all too well and plans to maintain profit margins by continually reducing manufacturing costs and receiving regular cost reductions from their suppliers. The enterprise segment, that which mainly consists of products for large companies, is also cost sensitive, but not nearly so much as the consumer segment, since the lack of extremely high volume will not necessarily attract the same range of potential competitors. The least competitive (but still competitive) segment of all is the military segment, since successful product introductions in this segment often require lengthy product development cycles and long-term customer-supplier relationships that are not easily upset by new entrants offering better prices. Each business category offers advantages and disadvantages for suppliers of GaAs devices. The consumer and enterprise categories offer high growth rates but are extremely competitive. The military category offers a stable pricing environment, but relatively low volume, and it is a business that can act as a buffer during economic downturns.

1.5.1 Photonic device applications

The biggest application of GaAs devices continues to be LEDs. LEDs had humble beginnings but have since followed a version of Moore's law for several decades (roughly ten times improvement per decade). Light output per watt, light output per lamp and cost per unit of light have shown a consistent exponential rate of improvement over three decades [20], although in some cases with more jumps and pauses than with Si technology. Cumulative improvements have reached 3–4 orders of magnitude over the first three decades since the introduction of red LEDs and similar rates over shorter periods for other colours. With continued innovation, LEDs can continue producing such gains for another decade or two. Practitioners of the LED art talk about a revolution in lighting that is under way and has three components. First is the recent addition of blue and green colours, which had proven elusive in traditional III–V materials but have recently come of age in the mid 1990s in GaN-based materials. Second is the recent introduction of white LEDs, which either combine three primary colours or use a near-UV LED with a phosphor. Third is continuing steady improvement in cost structure, luminous efficiency and total light output per device. Since the early 2000s, LEDs have become the preferred component for traffic lighting applications, portable electronic displays (cell phones, personal desk assistants, etc.), backlighting applications, automotive lighting and signage. Some of these applications are enabled by the high reliability of solid-state devices compared to traditional light sources (think traffic lights). Others are enabled because of the need for a single primary colour, which is very inefficient to produce with a white light source and a filter, and the light becomes too hot. Other applications are enabled by extending the battery life due to higher efficiency as well as the brightness of LEDs (such as portable electronics). Most other lighting applications are extremely cost sensitive and will adopt solid-state lighting solutions only when their cost comes down further. Ultimately, LEDs offer the potential to save billions of dollars in electricity if theoretical efficiencies can be practically attained.

Another photonic device is the edge-emitting laser. Edge-emitting lasers have applications as diverse as telecommunications, data storage, materials processing, medical surgeries, spectroscopy and sensors. In telecommunications, the laser is used to transmit encoded data along an optical fibre to carry telephone calls along with digital data, initially for long-haul communications in the mid 1980s, but recently for all areas of extremely high traffic or bandwidth such as metropolitan areas. Many telephone calls and other sources of data are combined electronically (multiplexed) and sent along an optical fibre using the output of a single laser.

The telecommunication industry is highly competitive and the GaAs (or other compound semiconductor) devices must enable a lower overall system cost in order to be widely adopted. Fibre-optic lines are more costly to put in place compared to twisted-pair copper lines (the legacy of voice technology), but they are cost effective if enough traffic is present. Their capacity is also more easily expanded by sending extra wavelengths (more lasers per fibre) of light down the same fibre at the same time by a technique called wavelength division multiplexing (WDM). To minimise cost, one should minimise the use of electronic repeaters as much as possible.

When sending light down an optical fibre two things happen that cause a loss of the signal. One is that some of the light gets lost along each length of fibre, mainly by absorption. The other problem arises because different wavelengths travel at different speeds (dispersion) through the optical fibre medium. This effect causes spatial separation of the different wavelengths along the length of the fibre that will eventually result in loss of the ability to decode the signal. Minimising both of these problems is largely the responsibility of the optical fibre makers, who try to reduce impurities in the fibre and engineer it for minimum dispersion. By sending light at the minimum in dispersion (1.3 μm), or at the minimum of loss (1.55 μm), maximum transmission distance can be obtained. GaAs-based lasers do not reach such long wavelengths (up to 1.1 μm is possible using InGaAs quantum wells strain-matched to GaAs) and initially had greater reliability problems, so the closely related InP-based lasers became the mainstay of fibre-optic communications. However, GaAs-based lasers acquired an important role in fibre-optic communications in the 1990s in all-optical repeaters.

When the optical signal gets weak enough, it must be amplified or regenerated. In the early days, this meant converting the light to an electrical signal, regenerating (reshaping pulses to their original form) and amplifying the signal, and driving a laser (or in very-high-frequency systems, driving a laser modulator) to again transmit the signal down the optical fibre. In the 1990s, a means of optical amplification was introduced where the weak optical signal was sent through a small section of Er-doped fibre along with a CW laser source to pump the Er-doped fibre (through Er absorption at 1400 or 1100 nm). After passing through the Er-doped fibre, the amplified signal is reintroduced into the standard fibre-optic line. InP-based lasers can be used to amplify at 1400 nm or GaAs-based lasers (with InGaAs active regions) can be used at 1100 nm. The GaAs-based lasers offer lower noise and have garnered a significant part of this market. Er-doped fibre amplifiers greatly reduced the cost of transmitting data over optical fibres and enabled the adoption of WDM techniques to greatly expand the capacity of fibre-optic systems without having to dig up streets and lay new cable lines.

Another large category of laser use is in materials processing in industrial applications [21]. These applications include precision cutting, welding, shaping and cleaning. The precision nature of the cutting as well as the flexibility in the laser placement have led to expanding usage of lasers in cutting and welding. Although the majority of these applications employ other high-power lasers such as CO_2 lasers, semiconductor lasers have gained a small part of this market as well and will continue to find new applications as the maximum output power levels of GaAs lasers are increased over time. One notable advance in solid-state lasers such as yttrium aluminium garnet (YAG) lasers has been the use of semiconductor diode lasers to pump the YAG laser rod. YAG lasers were first introduced with flashlamps for optical pumping. Since incorporating GaAs semiconductor lasers, they have become smaller and more reliable (flashlamps were notoriously short-lived). These diode-pumped solid-state lasers, which now operate in excess of 100 W and should continue to increase in power output with time, have gradually captured a significant part of the laser cutting and related markets.

GaAs edge-emitting lasers are also used in large quantities for CD/DVD players and data storage. Read-only applications have been around for nearly 20 years and use GaAs lasers at wavelengths 830 and 780 nm. Once reliable output power levels were pushed beyond 30 mW, read/write applications became feasible. Read/write applications initially used the 830 and 780 nm lasers and then later 650–680 nm to increase data storage densities. These applications will soon use blue GaN-based lasers.

The vertical-cavity surface-emitting laser (VCSEL) is another important photonic device. Since edge-emitting lasers emit parallel to the plane of the wafer along a cleaved edge, they are generally not testable until they have been packaged. This unfortunate reality makes edge-emitting lasers very costly to produce if the yields are not high because package and assembly costs are incurred in discarded parts. VCSELs, on the other hand, are testable at wafer level, since lasing can be characterised normal to the plane of the wafer. Bad dice are marked and discarded without incurring the cost of assembly and packaging. VCSEL processing is also simpler and cheaper than that for edge-emitting lasers. VCSELs have other desirable properties as well. They can be made into arrays, effectively building up speed (bandwidth) through combining parallel paths rather than building single, serial, high speed devices. Short-wavelength VCSELs (usually less than 1100 nm) are a less mature technology than edge emitters, but they are attractive for certain short-length data transmission applications. VCSELs may soon be suitable for replacing edge-emitting lasers in long-haul applications through progress in VCSEL research at 1.3 μm and longer wavelengths.

VCSELs are predominantly GaAs-based devices at present. In order to build high-reflectivity mirrors above and below the active region, one needs lattice-matched materials with high contrast in the index of refraction. This allows high-reflectivity mirrors (distributed Bragg reflectors) to be made from a large number of pairs of alternating planes made from the high- and low-index films. For GaAs-based devices, GaAs, AlAs and their alloys provide both a lattice match and a high-index contrast, but these are lacking for devices lattice matched to InP.

One of the early VCSEL applications that established their commercial viability was as the fibre channel standard for storage area networks (SAN), a method of allowing greater scalability in data storage by the addition of storage elements directly to a network rather than to particular servers. Data is transferred between the storage network elements by fibre-optic lines. More recently, VCSELs are being adopted for all kinds of short-to-intermediate-range computer and network connectivity applications.

Other photonic devices are also used in communications applications as well as a myriad of niche applications. These devices include photonic waveguides and related circuits, photodiodes and optical modulators.

1.5.2 Electronic device applications

GaAs electronic devices have historically served in a myriad of applications as well, but these have been winnowed down in recent years to a few very-high-volume applications. The applications of greatest importance today are cellular phone technology, fibre-optic communications, electronic test equipment and military applications.

The cell phone market comprises the cell phone handset as well as equipment infrastructure segments. Users with handsets originate calls within a calling area and the handset communicates with basestations built by the wireless service provider. An economic tradeoff exists in basestation placement versus the power requirements in the handset and its call quality. Obviously, fewer basestations built further apart will result in a less costly network, at least in the initial buildout. However, this scenario will result in greater requirements on the power-transmitting capability of the cell phone, which create a variety of difficulties, not the least of which is that its battery is small and discharges quickly. Cell phone OEMs like to differentiate their handsets on the basis of premium features and factors such as small size and length of talk time between battery charges. The cell phone power amplifier (PA) is the most power-hungry component in the handset because it is required to transmit up to several miles. GaAs technology is very competitive in meeting the technical requirements for cell phone PAs and has achieved a large market share in this technology by meeting cost targets for these parts (not always easy with demanding OEMs!). InP-based technology, in contrast, can achieve higher power-added efficiency PAs than GaAs, but it is not yet able to meet cost targets. Si and SiGe bipolar technology are generally less desirable because of lower power-added efficiency and are only attractive in the lowest cost segment of the market.

Apart from the PA, GaAs devices can be used in other parts of a wireless communications system, but they have less compelling advantages and must compete based on cost. All digital functions have always been more suitable for Si CMOS (e.g. digital signal processors). Even among other mixed-signal components, there are serious competitive pressures. Functions such as switches, receivers and mixers can often be implemented with better technical specifications in GaAs, but the differences compared with Si may not be compelling enough to gain design wins.

Besides cost advantages, Si offers increasing (mixed-signal) integration and even promotes the long-term idea of a system on a chip (digital/mixed-signal/analogue). GaAs competes by trying to integrate other functions such as switches into modules. For example, there are three worldwide standards for cell phones, each with a different frequency and requiring different PAs. One recent trend is to make cell phones that can be used on all three types of network (called tri-band phones) and integrate all three types of PA with GaAs devices that switch between each band.

Cell phone applications represent the most prominent but not the only type of application of wireless communications systems. The frequency spectrum is allocated for different purposes by the Federal Communications Commission (FCC) in the United States and other regulatory bodies in other countries. Cell phone communications have been allocated frequency bands in the 1–2 GHz range for different wireless protocols that are implemented by wireless communications providers. We will discuss some of the applications that have specific portions of the spectrum allocated (examples are from the United States; other countries have different ways of allocating the frequency spectrum) as examples of other addressable markets. One important frequency band is at 2.4 GHz. Below a certain power level that does not interfere with the licensed use, anyone can use this part of the spectrum. Wireless industry groups have developed standards for high-speed, local-area data transmission. These standards have resulted in the adoption of wireless networking cards for computers that are now widely used for wireless broadband home networks and wireless "hot spots" at certain commercial locations such as airports, hotels and coffee shops. The same frequency band is used for a "bluetooth" standard that allows almost any instrument or appliance to receive and send wireless communications. These wireless technologies are similar in many ways to cell phone technology, with one main difference being the much lower power levels at which they transmit. Other examples of commercial wireless spectrum allocations are the wireless local-area-network band near 5 GHz, and the wireless point-to-point band near 30 GHz. The former is similar to the 2.4 GHz wireless data transmission, while the latter was conceived as an alternative way of providing telecommunications carriers "last mile" access to the end user and spurring competition among incumbent wireline telephone companies. Not all of these technologies have developed or will develop in the way they were envisioned when the spectrum was initially allocated. These technologies represent opportunities for GaAs technologies, particularly with co-integration of multiple applications (the cell phone-PDA-satellite radio-computer), but they are more cost-sensitive than cell phone PA technology.

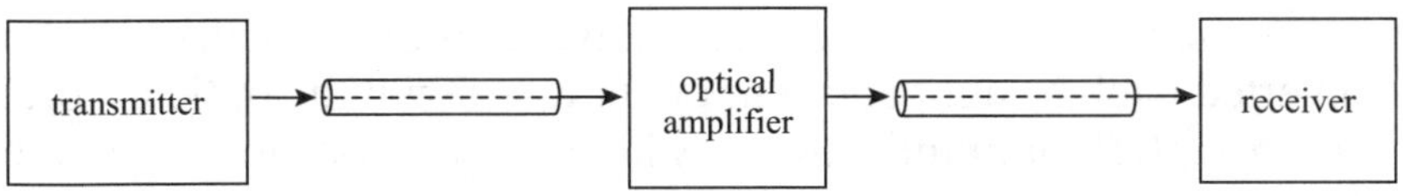

FIGURE 1.6 A schematic of a fibre-optic communications system.

The cell phone infrastructure market uses GaAs PAs for slightly different reasons than the handset market. Here, linearity is the key. In order to process as many calls as possible within a given frequency band, each call is assigned a certain part of the frequency spectrum. The details of how this is done and by what modulation scheme is beyond the scope of this book, but suffice it to say that the specifications of the PA become more stringent. The base-station broadcasts calls within its area. It is important that any distortion in the signal (such as intermodulation distortion) not be interpreted as data from another frequency band. In order to keep intermodulation distortion to an acceptable level, extremely high linearity is required of the power amplifiers. Again, GaAs devices compete well in terms of price and technical specifications, but a silicon device called LDMOS also performs well. Future competition from GaN PAs is also likely.

Both of these wireless technology segments are highly competitive. GaAs technology, although superior in many technical respects, is continually challenged by Si, SiGe and Si LDMOS technologies, both in terms of pricing and in attempts to close the technology gap.

Fibre-optic communications utilise GaAs devices for some of the components used in high-bit-rate electronics. A fibre-optic system schematic is shown in FIGURE 1.6. A communication signal is originated in the transmitter and converted from an electrical signal to a stream of bits through a series of electronic circuits ending with the high-speed multiplexer. The stream of bits is sent (within the transmitter) to the laser driver, which modulates the laser. In cases where the laser is not fast enough to be modulated directly, a modulator driver is used. The modulated light travels down an optical fibre until the signal becomes attenuated and needs amplification or regeneration (see Section 1.5.1 for fibre-optic amplification). After a few rounds of amplification without regeneration, the signal must be regenerated. The signal exits the fibre and is passed into a photodiode of the receiver. It is then amplified with a transimpedance amplifier. After the transimpedance amplifier, the signal will be reconstructed with a clock and data recovery circuit. If the signal is at its final destination, it will then go to a demultiplexer and be converted from a stream of bits to the electrical signal representing the communication. If it is not at its final destination, it will go to a laser driver or modulator driver to be converted back to light and sent on its way down the optical fibre.

Fibre-optic communications systems were first fielded in the 1980s for terrestrial long-haul and transoceanic communications. They were then gradually adopted for many other applications where high capacity or high bandwidth was required. InP-based devices have long been used for the lasers and photodiodes. Until recently, the highest bit rate optical communications systems used GaAs devices for multiplexers, demultiplexers, transimpedance amplifiers, clock and data recovery circuits, and laser (and modulator) drivers. Si ICs were used for other digital functions and lower-bit-rate systems. However, SiGe bipolar and even CMOS circuits are now readily available to build most of these circuits even in the high-bit-rate systems and GaAs devices have been relegated to a minor role in all but laser and modulator driver circuits. The driver circuits appear likely to remain the domain of GaAs-based or InP-based circuits in high-bit-rate systems because they require higher operating voltages than are achievable with CMOS or SiGe when clock rates are at 10 Gbit/s or above. Likewise, many other applications of high-speed digital logic circuits were once the domain of GaAs technology (even up to 100 k gates of logic). Now the speeds of Si technology have caught up with GaAs. Future, higher speed applications may evolve towards InP devices.

Military applications utilise a wide range of GaAs circuits at a wide range of frequencies. These include radars, communications systems, electronic warfare (EW) circuits and others. Because military systems operate worldwide and on irregular assignments, they do not deal with a well-defined spectrum allocation the way commercial communications systems do. Therefore, military systems design and build their radar systems around a broad range of frequencies from below 1 GHz at the low end to low mm-wave frequencies (near 30 GHz) at the high end. Radar and EW applications have higher bandwidth requirements than wireless communications systems. They are also qualitatively different since radars are an imaging technology. The wide-ranging frequencies, high bandwidth requirements and challenging performance specifications have led to widespread adoption of analogue and microwave GaAs and InP-based technologies in preference to Si devices. This trend is likely to remain for the foreseeable future.

Other applications of GaAs (or InP) devices include collision-avoidance radar, other commercial radars, space solar cells, radiation-hard electronics, space satellite uplinks and downlinks, space communications systems, ultra-low-noise receivers and certain niche communications components such as oscillators.

REFERENCES

[1] S. Mahajan, K.S. Sree Harsa [*Principles of Growth and Processing of Semiconductors* (WCB McGraw-Hill, Boston, MA, USA, 1999)]

[2] M. Madau [*Fundamentals of Microfabrication* (CRC Press, Boca Raton, FL, USA, 1997)]

[3] S.K. Ghandi [*VLSI Fabrication Principles: Silicon and Gallium Arsenide, Second Edition* (John Wiley and Sons, New York, 1994)]

[4] S. Wolf, R.N. Tauber [*Silicon Processing for the VLSI Era, Volume 1 – Process Technology* (Lattice Press, Sunset Beach, CA, USA, 1986)]

[5] S.J. Pearton, C.R. Abernathy, F. Ren [*Topics in Growth and Device Processing of III-V Semiconductors* (World Scientific, Singapore, 1996)]

[6] R. Williams [*Modern GaAs Processing Methods, Second Edition* (Artech House, Boston, 1990)]

[7] H.C. Casey Jr., M.B. Panish [*Heterostructure Lasers* (Academic Press, London, 1978)]

[8] S.M. Sze [*Semiconductor Devices: Physics and Technology* (John Wiley and Sons, New York, 1985)]

[9] W. Lui [*Fundamentals of III-V Devices: HBTs, MESFETs, and HFETs/HEMTs* (John Wiley and Sons, New York, 1999)]

[10] S.L. Chuang [*Physics of Optoelectronic Devices* (John Wiley and Sons, New York, 1995)]

[11] S.M. Sze (Ed.) [*High-Speed Semiconductor Devices* (John Wiley and Sons, New York, 1990)]

[12] M. Shur [*GaAs Devices and Circuits* (Plenum Press, New York, 1987)]

[13] C.Y. Chang, F. Kai [*GaAs High-Speed Devices: Physics, Technology, and Circuit Applications* (John Wiley and Sons, New York, 1994)]

[14] L.A. Coldren, S.W. Corozine [*Diode Lasers and Photonic Integrated Circuits* (John Wiley and Sons, New York, 1995)]

[15] J.T. Verdeyen [*Laser Electronics* (Prentice-Hall, New Jersey, 1981)]

[16] W.R. Wisseman, W.R. Frensley [*in VLSI Electronics Microstructure Science vol.11, GaAs Microelectronics* Eds. N.G. Einspruch, W.R. Wissman (Academic Press, Orlando, FL, USA, 1985) p.1]

[17] G.P. Agrawal, N.K. Dutta [*Semiconductor Lasers* (Kluwer Academic Publishers, 1993)]

[18] Z.I. Alferov [*Semiconductors (USA)* vol.32 (1998) p.1]

[19] Z.I. Alferov et al. [*Lov. Phys. Solid Atate* vol.9 (1967) p.208]

[20] A. Bergh, G. Craford, A. Duggal, R. Haitz [*Phys. Today (USA)* vol.54 (2001) p.42]

[21] D.F. Welch [*IEEE J. Sel. Top. Quantum. Electron. (USA)* vol.6 (2000) p.1470]

Chapter 2

Semiconductor properties, growth, characterisation and processing techniques

2.1 CHAPTER SCOPE

The purpose of this chapter is to provide a concise overview of a lot of different subject matter that forms the basis of understanding semiconductors, semiconductor devices and start-to-end product realisation using them. Much of this material is considered a prerequisite to understanding devices in traditional learning settings. To master the prerequisite material the reader will need formal courses or in-depth texts for self study. The authors recognise that not all readers may have become well schooled in these subjects prior to learning GaAs processing techniques, in part because of their breadth of scope and of the theoretical character. Hopefully, these readers will benefit at this time from simplified explanations of a number of broad subject areas and from the references to sources with more complete coverage. These “layman's” introductions should help readers place the processing techniques and the necessary multidisciplinary topics quickly into proper context within the more fundamental subjects.

The basics of semiconductors and their properties [1–5] most useful in semiconductor devices are given in Section 2.2. Section 2.3 introduces the basic techniques for GaAs crystal growth. Section 2.4 presents the main methods of epitaxial growth, metal-organic chemical vapour deposition and molecular beam epitaxy. Section 2.5 introduces some of the more important semiconductor characterisation techniques. In Section 2.6, some processing techniques that are common to many of the later chapters are presented. Specialised processing techniques and equipment are introduced in later chapters as the need arises.

2.2 SEMICONDUCTOR PROPERTIES

The electrical properties of solids allow their classification as conductors, insulators or semiconductors. The electrical properties of a solid are derived from the physical equations governing electrical charges through the interactions of protons and electrons in the solid. Solid materials come in three types of atomic ordering: amorphous, poly-crystalline and single crystal. Useful semiconductor properties are most often obtained from single crystal or polycrystalline solids, though amorphous Si is very important in solar cell applications and display technology. The energy states of many crystalline solids have been determined both theoretically and experimentally. These energy states, the crystalline analogues to energy levels in individual atoms, are commonly referred to as the band structure of the material. The details of the band structure of solids and its relation to the details of the crystalline lattice is the subject of solid state physics; we have listed both introductory [6] and advanced references [7]. In this section, some important semiconductor properties, concepts and equations will be introduced at a very basic level. The concepts chosen are those that will lead to a greater understanding of semiconductor contacts, the semiconductor surface, and devices such as the field effect transistor and the heterojunction bipolar transistor and issues related to their interaction with processing techniques. No attempt will be made to derive important equations and any attempt to explain these equations or certain physical concepts is only with the intent of later introducing related device concepts. A minimum of topics will be presented to enable a better understanding of semiconductor materials, useful characterisation techniques, electronic devices and common processing techniques. Our goal is not to provide a comprehensive discussion but rather to introduce the reader to relevant topics that may be pursued further using other resources, if necessary.

2.2.1 Energy levels and band structure in semiconductors

Using the electron energy levels of atoms as a starting point, the bonding of molecules will be extended to the case of solids. The electrons are attracted to the protons in a nucleus due to electrostatic forces but they are repelled from other electrons due to these same forces. The energy level of an electron state is the energy required to remove the electron to a position far away from the atom (such that the electrostatic attraction with the nucleus is negligible) and this energy state can be calculated from the quantum-mechanical equations governing the system. From quantum mechanics, it is found that the energy levels of electrons in the atoms are organised into shells, beginning with those closest to the nucleus of the atom. The first shell can accommodate up to two electrons in the 1s state. The second shell can accommodate up to eight electrons, which are called the 2s and 2p states. The third shell accommodates up to two 3s electrons, six 3p electrons and ten 3d electrons. Heavier atoms have additional shells. The actual number of electrons in a neutral atom is equal to the number of protons in the nucleus. In the neutral atom's ground (or lowest energy) state, the electrons fill each shell in sequence, with the outermost shell being only partially filled up to the number of available electrons. The difference between electron energy states is much greater between shells than within a shell. Atomic energy states are illustrated in FIGURE 2.1 for a three-shell atom.

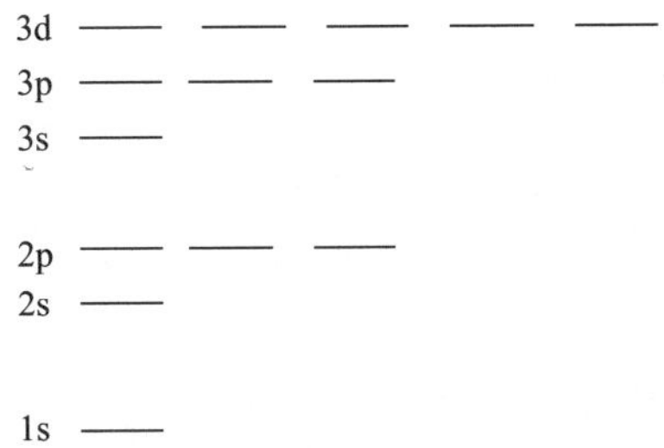

FIGURE 2.1 An illustration of electron energy states in an atom.

Molecules are built up from atoms by quantum mechanically combining their atomic orbitals that correspond to the atomic energy levels to produce molecular orbitals with different energy levels. A chemical reaction accompanies this process leading to pairing electrons in unfilled shells to lower the overall energy state. Atoms with completely filled shells cannot lower their energy state in this way and are generally unreactive. The process of forming a molecule is illustrated in FIGURE 2.2. Paired electrons in bonding orbitals have a lower energy state than the original electrons in the atomic orbitals. Energy levels of the molecule are calculated by similar methods as for atoms.

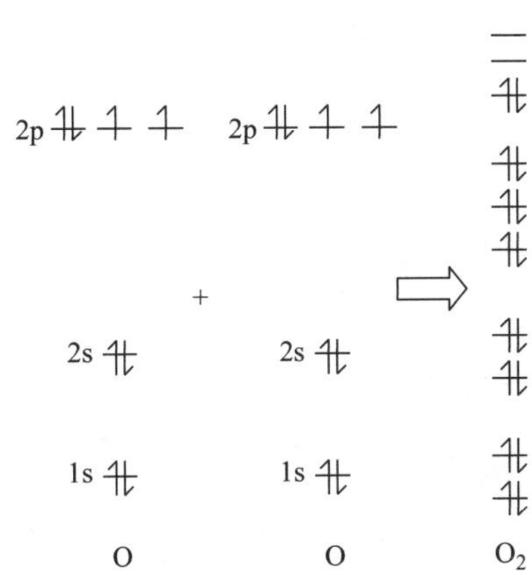

FIGURE 2.2 Energy levels of a molecule are derived from the outer atomic levels.

A solid can be thought of as an extremely large molecule, but it is not useful to think of it that way because calculating and keeping track of upward of 10^{23} valence energy states per cubic centimetre is very unwieldy. In the case of ordered solids, the calculation of energy states is greatly simplified by crystalline symmetry. Generally speaking, the properties of a unit cell (with several to dozens of atoms) is sufficient to describe the whole crystalline solid, since many of the properties can be accounted for by translation. Such a calculation will give a "band structure" for a solid, so named because large numbers of energy states are naturally grouped into bands that can be related back to the atomic energy levels. Rather than accounting for each paired electron energy state as is done in molecular-energy-state descriptions, the energy states in an ordered solid are classified according to the number of states in a given energy range, which is termed the density of states. An example of a band structure is shown in FIGURE 2.3.

Similar to the case for atomic energy or molecular states, the lower lying energy "bands" describe the core electrons that are localised to specific atoms (although this is not always obvious from band structure terminology). The upper lying energy bands describe electrons (valence electrons) that form the chemical bonds that hold the solid together. However, rather than specific electrons being associated with specific chemical bonds, many electrons in a given "band" are associated with many similar chemical bonds. For example, in a two-atom molecule such

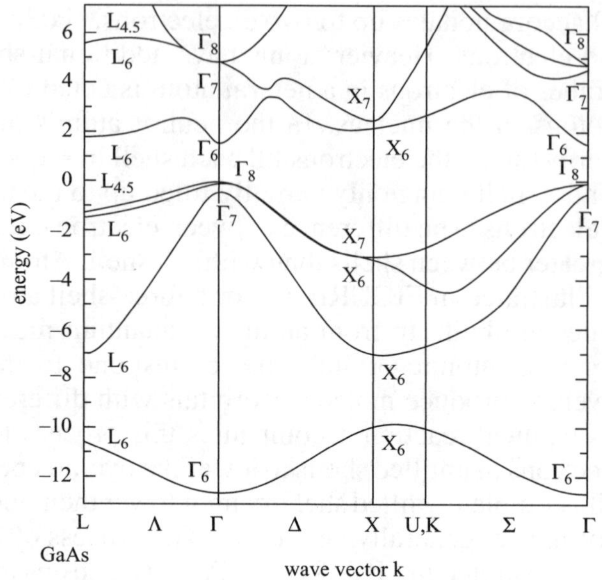

FIGURE 2.3 Band structure of GaAs. (*Properties of GaAs*, IEE, 1996, reprinted with permission.)

as hydrogen fluoride, the outermost valence electron of fluorine occupies an energy state shared with the hydrogen electron to form a localised chemical bond. In a Cu solid, the outermost valence electron of many Cu atoms combine to form an energy band that represents Cu-Cu chemical bonds. Each Cu bond can be thought of as having localised character, but the band describing those bonding electrons does not specify any particular origin of an electron occupying a particular state. Without getting into a philosophical discussion of the implications of the band-structure model for bonding, we will merely state that this concept will have important implications when the band structures of insulators and conductors are contrasted later in this section.

As is the case for atoms and molecules, electrons from the solid occupy the available states starting with the lowest energies available. The states are shown in FIGURE 2.3 as energy states as a function of k, which is a solid state analogue of the electron momentum and has units of inverse length. Separately from the energy states shown in FIGURE 2.3, one must account for the number or density of states at a given energy, since electrons are allowed to occupy a state only if it is "empty". The state-filling process (a thought process only, since electrons in reality do not fill the states sequentially in time) proceeds until all available electrons in the solid have filled the available states. These are the occupied bands. At this point the electrical properties of the solid are determined by the character of the remaining unfilled bands and their density of states.

Before describing the electrical implications of a given band structure, a few qualitative comments about the band structure are in order. In the band structure of a solid, as in atomic and molecular states, transitions may occur from the occupied to the unoccupied bands. These may occur in all of the usual ways: thermally assisted (according to the equations governing the probability of a thermal transfer with that energy separation), by absorption of radiation or by inelastic charged-particle scattering. The occupied bands are termed the valence bands, while the unoccupied bands are called the conduction band, for reasons that will become apparent shortly. At zero temperature and in the absence of transitions to higher levels, the bands are filled strictly according to energy levels from the lowest to the highest. The remaining unfilled levels may be continuous with the filled levels or they may have an energy gap separating filled and unfilled states. In the latter situation, no electrical conduction can occur in the presence of an electric field. The reason for this situation is that an electron's properties are governed by its energy state and these states represent chemical bonds whose attractive forces with atomic nuclei are greater than an externally applied electric field. At finite temperatures some electrons have a reasonable probability of occupying upper states. Due to finite temperature or the presence of transitions to higher energy bands through photonic excitation or scattering, electrons in these higher states may move in response to an external electric field. The reasons are two-fold. First, the non-valence states are not associated with bonding (they are analogues of non-bonding, or antibonding orbitals). Second, and perhaps more importantly, the states in ordered solids tend to be delocalised over many atoms and readily allow carriers to respond to an external field. Rather than dwell on this point or attempt to be more precise, it will be taken as a given that electrons in upper states (higher than the ground state) can respond to an electric field. These states are called conduction bands in the terminology of solid-state physics, electrical engineering and materials science.

Unfilled valence band states can also conduct electrical current in response to an electric field, as will become apparent in the discussion of "hole" conduction.

Ordered solids are grouped into conductors, insulators and semiconductors based on the energy separation of the valence and conduction bands. Materials with no energy separation between the valence and conduction bands are conductors (usually metals). The energy bands form a single continuum in energy. Materials with large separation between the valence and conduction bands are insulators. Those with intermediate energy separation

TABLE 2.1 Bandgaps at 300 K.

Semiconductor	Bandgap (eV)	Semiconductor	Bandgap (eV)
AlN	6.2	ZnS	3.8
AlP	2.43	ZnSe	2.58
AlAs	2.16	ZnTe	2.28
AlSb	1.6	CdS	2.53
GaP	2.25	CdSe	1.74
GaAs	1.42	CdTe	1.50
GaN	3.35	Si	1.12
GaSb	0.69	Ge	0.67
InAs	0.36	Ga_2O_3	4.3
InN	1.89	Al_2O_3	8–10
InP	1.28	SiO_2	9
InSb	0.17		

are called semiconductors. The definition of the energy separation of the valence and conduction bands is called the "bandgap". As a general guideline, conductors have bandgaps less than 0.2 eV, semiconductors have bandgaps from 0.2 to about 4 eV, and insulators have higher bandgaps. TABLE 2.1 lists some of the more common semiconductors and their bandgaps. In practice some insulators can be grown in the same crystal as semiconductors for "bandgap engineering" and are then considered semiconductors. An example of the latter is the aluminium nitride/gallium nitride combination. AlN with a bandgap of 6.2 eV is considered a ceramic in many applications, but is also a semiconductor when combined with GaN. Partly, these designations came from the early days of electronics where one's imagination as to how to use semiconductors was limited by prevailing materials technology. The reality nowadays is that many insulators can be engineered to have semiconducting properties.

Two more comments about semiconductor band structures (such as that of FIGURE 2.3) are in order. First is the concept of direct and indirect semiconductors. When the conduction minimum and the valence-band maximum line up at the $k = 0$ (inverse length units) momentum point, termed the Γ point, electrons and holes can recombine directly with the emission of light and the semiconductor is termed direct, as is the case for GaAs and a number of other III–V semiconductors. When the conduction band minimum does not line up in k with the valence band maximum, the semiconductor is termed indirect and electrons at the conduction band minimum recombine with holes in the valence band through the emission of phonons (lattice vibrations) rather than light. Such materials are indirect semiconductors. Second, the curvature of the bands (the dE/dk derivative) is inversely proportional to the effective mass of the electron for the conduction band and of the hole for the valence band. Higher curvature in a band results in better transport properties (Section 2.2.3).

Another important semiconductor concept is that of the Fermi level. Whereas the density of states tells us how many states exist at a given energy, the Fermi function f(E) specifies how many of the existing states at the energy E will be occupied. More precisely, f(E) specifies the probability that an available state at an energy E will be occupied by an electron and is given by the relation

$$f(E) = (1 + \exp((E - E_F)/kT))^{-1} \qquad (2.1)$$

where E_F is the Fermi level, k is the Boltzmann constant and T is the absolute temperature. As the temperature goes to 0, f(E) goes to 1 for $E < E_F$ and to 0 for $E > E_F$, i.e. the Fermi level defines the energy below which electron states are occupied and above which they are unoccupied. At non-zero temperatures, the Fermi function f(E) gives the probability of electrons occupying higher energy states. The Fermi function plot is given in FIGURE 2.4, where the x-axis is plotted in units of kT. The Fermi function is 0.5 at $E = E_F$, approaches 1 for $(E - E_F) < 3kT$, and approaches 0 for $(E - E_F) > 3kT$. At room temperature, kT is approximately 0.026 eV and 3kT is approximately 0.078 eV. Compared to the bandgap of GaAs (1.42 eV), these are small numbers and very few carriers will be in the conduction band. In order to calculate the carrier distributions (electron and hole occupancy versus energy), the Fermi function is multiplied by the density of states.

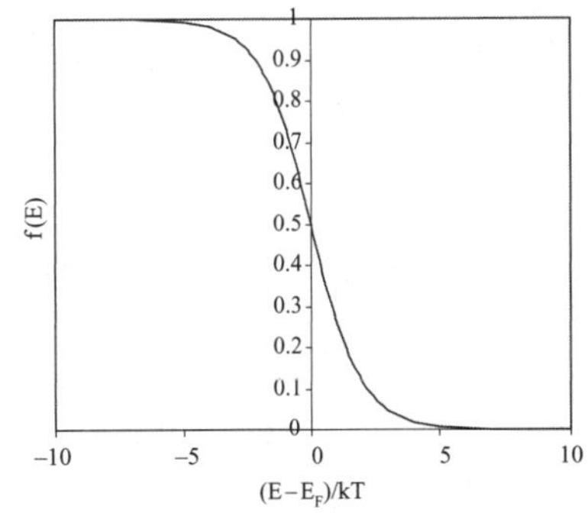

FIGURE 2.4 A plot of the Fermi function.

2.2.2 Charged carriers in semiconductors

Next the conducting properties of semiconductors will be discussed. A pure semiconductor such as Si with a bandgap of 1.12 eV can conduct electrons if they reach the conduction band. An interesting situation arises if the electron's thermal energy, which at room temperature is 0.026 eV, allows it to reach the conduction band. Although the thermal energy is only a small fraction of the bandgap, some electrons nevertheless reach the conduction band because their numbers are so large. A Si atom has 14 electrons, 4 of which are valence electrons and a Si crystal has more than 10^{22} atoms per cubic centimetre. The equation governing the equilibrium (meaning the semiconductor has had plenty of time to respond to a perturbation such as a temperature change and has reached its lowest energy state) electron concentration in the conduction band is given by

$$n_i = (N_V N_C)^{1/2} \exp(-E_G/2kT) \qquad (2.2)$$

where n_i is the intrinsic (meaning for a pure semiconductor) carrier concentration, N_V is the density of states of the valence band, N_C is the density of states of the conduction band, E_G is the bandgap, k is the Boltzmann constant and T is the absolute temperature. The density of states is calculated by theoreticians and reasonable estimates are available. For Si, n_i is about $10^{10}\ cm^{-3}$ and for GaAs, n_i is about $10^6\ cm^{-3}$ at room temperature. Although these numbers seem large, in terms of current they are actually quite small due to the unit of charge of an electron (1.6×10^{-19} C) in relation to an ampere (coulomb/s). It is seen that pure semiconductors with bandgaps greater than 1 eV are excellent insulators at room temperature, but that their conductivity will rise exponentially with temperature. At 300°C, for example, n_i rises to about $10^{14}\ cm^{-3}$ for Si and $10^{12}\ cm^{-3}$ for GaAs.

Pure semiconductors can be manipulated by adding atomic impurities to change their electron concentrations. First one considers a pure silicon crystal with four-fold bonding symmetry (one bond for each unpaired valence electron). Rather than the band structure description of the bonding (which is accurate but hopelessly complicated for a brief discussion), simple chemical pictures will be used. The ideal bonding configuration for four-fold symmetry is the tetrahedron, in which four points equally spaced on a sphere represent the nucleus of the four atoms bonded to the centre Si. Each bond comprises one valence electron from each silicon atom. Next, consider the introduction of a P impurity. Phosphorus occupies a tetrahedral site in place of a Si atom. Instead of four valence electrons it has five, so four of these are used in chemical bonding and a fifth is left over and allows P to be an electron "donor" to the crystal. The lattice has an extra proton which stays fixed at the P atom. The leftover electron is free to move over the whole crystal, subject to relevant physical effects including the (weak) attractive force from the fixed positive charge that is left with the phosphorus. In the case of P and several other dopants, the weak attractive force to the donor ion is easily overcome by thermal energy and the electron freely conducts in response to an electric field. This situation is illustrated in FIGURE 2.5, where for convenience, the tetrahedral three-dimensional bonds are shown as if in a single plane. If, instead of phosphorus, boron is used to dope the silicon, only three valence electrons are added to the crystal at the place where boron replaced a Si. One of the silicon bonds will be missing a valence electron. In such a case the valence band can participate in electron conduction. The unpaired silicon bond can take an electron from its neighbour in response to an electric field leaving a fixed negative charge with the boron and a mobile positively charged "hole" in the neighbouring Si atom. The electron "hopping" process can repeat itself leading to "hole"

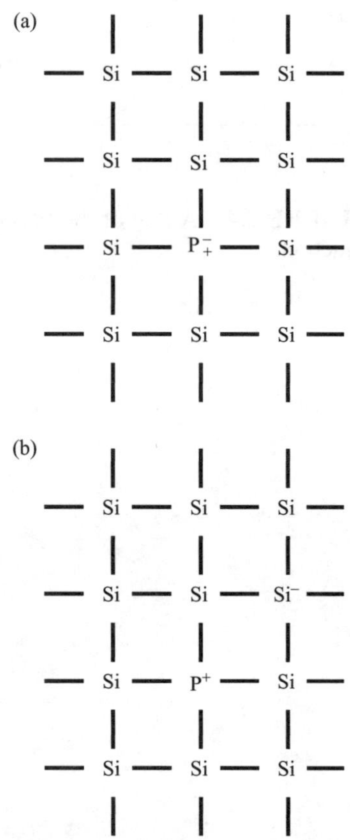

FIGURE 2.5 Impurity doping in Si. The donor electron of P is free to move about the crystal, while the P remains with a fixed positive charge.

mobility in response to an electric field. Rather than call this an electron conduction mechanism through the valence band, it is more useful to think of this transport and its energetics in terms of particles called holes. GaAs is doped by an analogous method using a wide variety of dopants. For example, Si with four valence electrons provides electrons to the crystal if it replaces Ga and provides holes if it replaces As.

The preceding discussion can be summarised as follows. Semiconductors can be made more conductive by introducing impurities that occupy lattice sites normally occupied by the elements of the original semiconductor. Those that add an extra electron to the lattice contribute to electron conduction and those that are missing an electron contribute to hole conduction. Because an electron is negatively charged by convention, electron-adding impurities are called n-type dopants. Likewise, impurities that add holes are called p-type dopants. Whereas pure semiconductors with a bandgap greater than 1 eV have no more free carriers than $10^{10}\,cm^{-3}$ at room temperature, it is fairly common to dope semiconductors with donor or acceptor impurities to levels greater than $10^{17}\,cm^{-3}$, leading to a 10^7 or greater increase in conductivity.

In fact, the situation of a doped semiconductor is not quite so simple because the electron does not necessarily go into the conduction band of a pure semiconductor (because of the attraction to the fixed ion), but the states of an n-type semiconductor are recalculated with the perturbation of an impurity dopant and there may be slight changes to the band structure. Any impurity atom and other things like crystal defects will usually add energy states to the bandgap of the semiconductor. The case of the impurity dopant, for example, can be modelled most simply as a hydrogen-like atom in a dielectric matrix. The atomic hydrogen states are known from free-space calculations:

$$E_n = m_e{}^* q^4 / 2(4\pi n \varepsilon \hbar)^2 \qquad (2.3)$$

where m_e is the electron mass, q is the electron charge, ε is the dielectric constant and $\hbar$ is Planck's constant divided by 2π. The dielectric constant of GaAs is 13 times greater than the free-space dielectric constant and the electron mass in GaAs is about 13 times smaller than the free-space electron mass. E_n in free space is 13.6 eV for $n = 1$, representing the ground-state ionisation energy of hydrogen. In GaAs, the hydrogen-like model gives E_n of approximately 6 meV. Coupled with the band structure model, the donor state is placed 6 meV below the conduction band and the donor electron is said to have a binding energy of 6 meV. More sophisticated calculations are, of course, used for this and other states involving impurities and defects in semiconductors when accuracy is needed. The hydrogen-like structure is called a bound exciton, to describe the fixed ion and the loosely bound electron. Unbound excitons describe the situation where mobile electron-hole pairs are loosely bound in such an arrangement.

If we recall the interpretation of the Fermi function of EQN (2.1), and its plot in FIGURE 2.4, we can see that hydrogen-like donor binding energies fall well within 3kT of the conduction band at room temperature, indicating that this type of donor impurity gives up its electrons to the conduction band at room temperature. In fact many common donors and acceptors in GaAs have binding energies less than 30 meV and are assumed to be fully ionised at room temperature.

EQN (2.2) gives the free-electron concentration for a pure semiconductor. Similar equations can be generalised for pure or doped semiconductors. Some of the most useful results are quoted here:

$$n = (N_C)^{1/2} \exp(E_F - E_C/kT) \tag{2.4}$$

$$p = (N_V)^{1/2} \exp(E_V - E_F/kT) \tag{2.5}$$

$$np = (N_C N_V)^{1/2} \exp(-E_g/kT) = n_i^2 \tag{2.6}$$

where n and p are the free electron and free hole concentrations and the other terms have been defined previously. EQNS (2.4)–(2.6) are derived under the assumption that E_F lies at least 3kT below the conduction band and at least 3kT above the valence band, i.e.

$$E_V + 3kT < E_F < E_C - 3kT \tag{2.7}$$

If these assumptions are met, the semiconductor is said to be non-degenerate and if they are not met, the semiconductor is said to be degenerate. EQNS (2.4)–(2.5) have two unknowns, E_F and n (or p). The reader is directed to the references for methods of solving these equations [1,6,7]. However, some general comments are in order. E_F is roughly in the middle of the energy gap for an intrinsic semiconductor (its actual position depends on the values of N_C and N_V). E_F moves logarithmically towards E_C as n increases from the intrinsic value towards values determined by the number of ionised donors. E_F reaches (E_C − 3kT) at about 10^{18} cm^{-3} for GaAs, which makes the doping degenerate, or more commonly, the GaAs is said to have n^+ doping. In a similar way, E_F moves logarithmically towards E_V with increasing p-type doping and GaAs is said to be degenerate for p^+ hole concentrations generally above 10^{18} cm^{-3}. One last comment relates to unintentional doping. The intrinsic carrier concentration of GaAs is near 10^6 cm^{-3} at room temperature, but GaAs cannot be made pure enough to realise this condition. Today's highest purity GaAs will have n- or p-type donors in the low $10^{14}\,cm^{-3}$ range which come from impurities incorporated during the growth process (Sections 2.3 and 2.4).

Other energy levels that exist in the bandgap of GaAs and other semiconductors are related to crystal imperfections or substitutional incorporation of complex impurities such as metals. These energy levels are often termed "deep levels" to contrast their energies with those of shallow donors and acceptors. Potentially large numbers of defects can fall into this category and their study keeps material scientists busy. The importance of these types of deep levels stems from their ability to trap mobile carriers. Consider a semiconductor with shallow donors and acceptors in the presence of a deep level, as illustrated in FIGURE 2.6. Four processes can occur: electron capture (a), electron emission (b), hole capture (c) and hole emission (d). Processes (b) and (d) are also referred to as carrier generation events. The carrier capture events (a) and (c) depend on factors such as the density of the deep levels, spatial overlap of carriers and deep levels, and electric fields in the semiconductor. The carrier capture events are often referred to as "trapping" of mobile carriers. The term "trap" relates to the fact that thermal emission of electrons or holes occupying the deep level is improbable due to the energy separation from the conduction or valence bands. Because of the importance of trapping effects in active devices, much effort has gone into the study and characterisation of deep levels in GaAs and other semiconductors, including specific types of growth (Section 2.3), measurement (Section 2.5), and device modelling (Section 8.2.2).

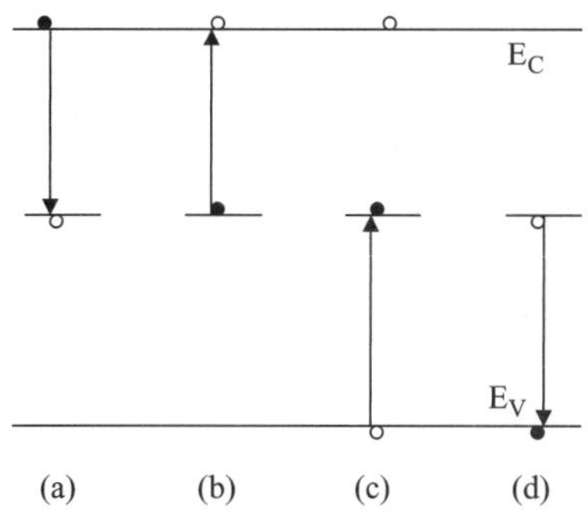

FIGURE 2.6 Four processes that can occur in the presence of a deep level.

As an example, let's consider the most important defect in GaAs, EL2. EL2 is identified by its energy location 0.76 eV below the conduction band edge and is generally believed to be related to an antisite defect, an As atom on a Ga site, perhaps complexed with a nearby vacancy (empty lattice site). Its concentration in most types of bulk crystal growth (Section 2.3) is near $10^{16}\,cm^{-3}$, although it is much lower than that in epitaxial growth (Section 2.4). EL2 and other deep levels are thought to compensate impurity donors and acceptors to make GaAs semi-insulating and insensitive to impurities at levels below $10^{16}\,cm^{-3}$. EL2 may also trap electrons in GaAs devices when they are accelerated into the semi-insulating regions of the device and may create regions of fixed trapped charge that can be detrimental to devices (Section 8.8). Its presence and that of other deep levels must be accounted for in device design and operation to minimise detrimental effects. Detrapping of electrons (or holes) may also occur and generally happens on a timescale that is long compared to device transit times that correspond to high-frequency operation.

2.2.3 Carrier transport and continuity equations

Under normal operating conditions there are two types of carrier transport in semiconductors: drift, which is the response of a charged particle to an electric field, and diffusion, which is a response to a concentration gradient. In a semiconductor, a charged particle in an electric field, E, experiences a force which accelerates the particle, but it will not accelerate indefinitely because collisions with the lattice can redirect its energy and velocity. At any given electric field, an average velocity will result, which represents the net average of the response to the electric field, collisions with the lattice, and random thermal motion. The electron drift mobility is a parameter that relates the electron or hole velocity to the applied electric field,

$$v_d = \mu_d E \tag{2.8}$$

where μ_d is the drift mobility, E is the electric field and v_d is the drift velocity. From the drift velocity, a current density is given by

$$J_n = qn\mu_d E \tag{2.9}$$

where J_n is the electron current density, and the other parameters were previously defined. An equivalent expression for hole current can be used. EQNS (2.8) and (2.9) imply that velocity and current density can increase without limit as E is increased. On the contrary, as the particles get more energetic, the effects of scattering and electric field acceleration balance and offset each other and velocity saturation occurs. The electrons can also scatter more easily into higher bands. An electron, for example, can scatter from the Γ band of FIGURE 2.3 to the L band and its transport degrades due to the higher effective mass in the L band. Thus, at higher electric fields, the electron velocity is an average of those in the Γ and L bands and actually decreases at higher electric fields. In general, the electron velocity for GaAs looks something like that of FIGURE 2.7. At low electric field the electron velocity is linear with electric field and is well described by its drift mobility. At intermediate electric field, the electron velocity peaks and overshoots its high field value. The saturated electron velocity is given by the high-field value. The important transport properties depend on the specifics of the device design, which determine electric fields for the operating conditions of that design (more in Sections 8.2 and 9.2), and in particular, whether low-field mobility, the saturated velocity or the velocity overshoot is more important.

FIGURE 2.7 The electron velocity versus electric field for GaAs.

A second type of carrier transport is diffusion, or response to a concentration gradient, which may occur in a semiconductor device as a result of how its fixed carriers are distributed. In one dimension, for simplicity, this current is given by

$$J_n = qD_n(dn/dx) \quad (2.10)$$

where D_n is the diffusion coefficient for electrons and the other parameters are as before. A similar equation describes the diffusion current for holes. In a pn junction, for example, this type of current will continue until an electric field is set up to oppose any further current flow. The carrier transport equations for both electrons and holes are given by:

$$\begin{aligned} J_n &= qn\mu_{dn}E + qD_n(dn/dx) \\ J_p &= qp\mu_{dp}E + qD_p(dp/dx) \end{aligned} \quad (2.11)$$

Continuity equations are used to solve for transport in semiconductor devices. In many cases of commonly studied devices, these equations are solved in semiconductor device books, as examples, or in seminal research papers. Most readers will have no need to solve these equations, but should be aware of their use, not only in advanced textbooks, but also in commercially available device simulation packages. These equations are solved simultaneously with any other equations, such as the Poisson equation, that describe the physics of the device. In one dimension, the continuity equations are

$$\begin{aligned} (dn/dt) &= G_n - U_n + q^{-1}(d(J_n)/dx) \\ (dp/dt) &= G_p - U_p + q^{-1}(d(J_p)/dx) \end{aligned} \quad (2.12)$$

where G_n (G_p) is the generation rate for electrons (holes), U_n (U_p) is the recombination rate for electrons (holes), and the other symbols are as previously defined.

2.3 BULK CRYSTAL GROWTH

Semiconductor crystals, including GaAs, are grown from a liquid "melt" with a small single crystal "seed" used as the starting material. An ingot or boule of material is grown from the seed as it is pulled from the melt and the interface between the melt and the crystal forms a growing surface. The melt is confined in an enclosed area constrained in ways defined by the specific method of growth and is usually grown as a boule that is one to several times longer than the diameter of the wafer desired. After growth, the boule is removed from the crucible (container), ground to the desired diameter, and sliced into wafers which are lapped and polished. Electronic devices require high-resistivity semi-insulating GaAs ($>10^7$ Ω-cm) in contrast to many photonic devices, which generally require doped substrates. Ideally, GaAs wafers are extremely pure and free of defects. In practice, GaAs wafers do have reasonably high purity (7–8 orders of magnitude lower impurity concentration than the effective GaAs "concentration"). However, temperature non-uniformities that exist both at the growth front and during the cooling process cause stresses in the GaAs crystal, which lead to the formation of dislocations. Most of the art in GaAs growth is aimed at producing economical, high-yield substrates with both radial and axial uniformity while minimising dislocations and achieving uniform high resistivity. Other parameters of interest for the sliced wafer include flatness and surface preparation for epitaxial growth (Section 2.4). As if these were not challenging enough requirements, manufacturing technology progresses relentlessly towards larger diameter substrates, which bring additional challenges for temperature uniformities and other manufacturing issues in boule growth.

Section 2.3.1 gives an overview of the major types of crystal growth in use today. A great amount of flexibility exists in the choice of technology available to meet the requirements of the device engineer. Section 2.3.1 will present a discussion on the tradeoffs in substrate properties, substrate technology and device requirements.

The basics of the two main growth methods are presented in the rest of this section. As the details of each method are introduced, their influence on substrate quality is discussed.

2.3.1 Methods of crystal growth

The main methods of crystal growth in use nowadays fall into two categories: Czochralski methods and Bridgman methods. Historically, liquid encapsulated Czochralski (LEC, FIGURE 2.8(a)) and horizontal Bridgman were commercially available earlier than today's vertical Bridgman methods and produced low-cost wafers. The early horizontal Bridgman method did not compete well with LEC because it produced D-shaped ingots, which required oblique slicing and its associated material waste to obtain round wafers.

In the Czochralski method, a seed crystal is placed over a GaAs melt contained in a heated crucible that is held at a temperature close to the liquid-solid phase transition. The seed crystal is moved upwards under continuous rotation. The Bridgman method involves the directed solidification of a melt starting from the seed. The solidification direction can be vertical or horizontal. The movement of the crystal interface with the liquid can be

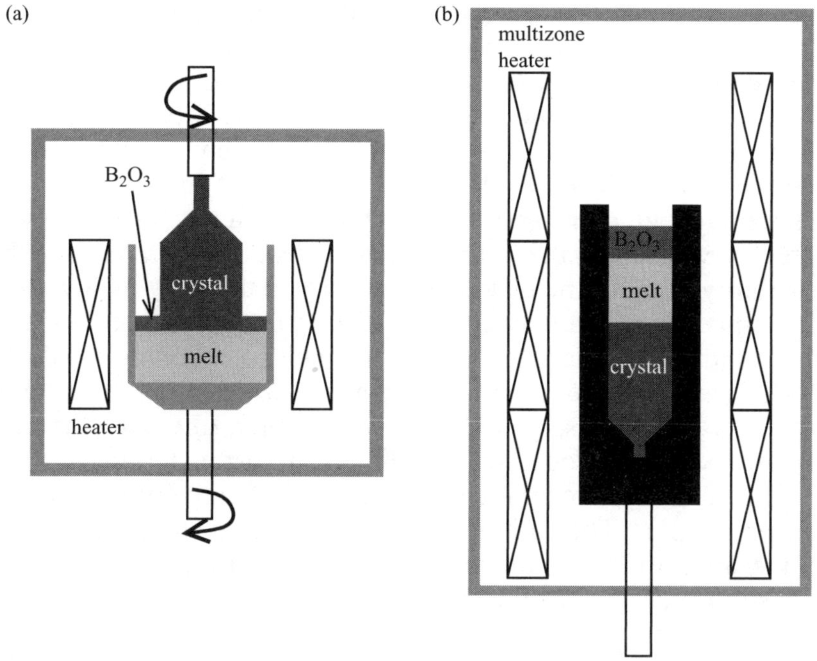

FIGURE 2.8 Schematic illustrations of crystal growth chambers (a) LEC and (b) VGF.

carried out either by moving the crucible in a static temperature profile or by moving the temperature profile within a static crucible. Typical Bridgman processes are the vertical Bridgman (VB) technique and its well-known variant, the vertical gradient freeze (VGF, FIGURE 2.8(b)) technique. Actual processes used by commercial companies are mostly proprietary technologies based on modifications of the classical crystal growth techniques. Both techniques offer wide process latitude to tailor crystal properties, such as dislocation density.

A necessary factor in crystal growth technology is control of the stoichiometry of the GaAs melt in order to maintain its stability. This is done by maintaining an equilibrium vapour pressure of As, which is commonly near 1 atmosphere. Two possibilities of achieving the As pressure are available: the use of a closed hot wall system for the crystal growth or encapsulation of the melt with liquid B_2O_3. In the case of the Czochralski method, B_2O_3 encapsulation is the best method, resulting in the well-known LEC process (FIGURE 2.8(a)). For the Bridgman process, both methods can be used.

Ideally, GaAs crystals are free of dislocations. However, temperature gradients (the main factor) along the growth front lead to differences in the GaAs lattice constant, which allow strains to be built up during growth. The strains are ultimately relieved by forming dislocations. Accordingly, it is highly desirable for the growth method to minimise temperature gradients to minimise dislocations. Unfortunately, this ideal is only approached if the growth rate is made uneconomically slow. When the molten GaAs crystallises at the growth front, heat is generated due to the thermodynamic heat of solidification. This heat contributes to the temperature gradient and is more problematic for faster growth rates. Nevertheless, the Bridgman methods of growth do have the flexibility to tailor the dislocation density in this way by lowering the temperature gradients. In the LEC growths, the presence of the liquid B_2O_3 layer at the top of the melt surrounding the GaAs crystal sets limits for the minimum achievable temperature gradient close to this phase boundary. Consequently, LEC growths have usually been used for faster growth, lower cost material.

The thermodynamic heat of solidification releases heat to the crystal interface as the crystal grows. This heat can lead to temperature gradients which need time to equalize, and therefore, a faster growth rate unavoidably leads to higher temperature gradients, all else being equal.

Crystal growth techniques are still being actively pursued as research topics. The variations on each technique that are pursued in research often blur the distinctions generalised in this section. Consult Reference [8] for more detail.

2.3.2 Substrate properties and device requirements

Material properties important for semi-insulating GaAs substrates include dislocation density, high electrical resistivity, low susceptibility to breakage, substrate flatness and spatial uniformity of all these properties. Dislocation density and the influence of crystal growth techniques on dislocations were discussed in some detail in Section 2.3.1. The resistivity is often required to be greater than 10^7 Ω-cm. This may be achieved in a variety of ways, including ultra-low impurity levels, careful compensation of residual donors and acceptors and compensation by deep levels. The low impurity levels can be hard to consistently achieve for meeting the resistivity specification. Accordingly, Cr doping was often used in the early days of GaAs technology as a deliberately introduced deep impurity level in order to increase both resistivity and its uniformity at high resistivity values. Since the early 1980s, however, the use of Cr doping has largely given way to another method of compensation using a native GaAs defect commonly identified as EL2. The structural origin of EL2 has been the subject of many studies and debates. Two competing structural descriptions of EL2 define it as either an As_{Ga} antisite defect or an As_{Ga} antisite defect associated with an As interstitial. More detail on the properties of EL2 can be found in [5]. In contrast to its structural uncertainty, much is known about its electrical characteristics. It can be described as a deep double donor with energy levels 0.75 and 0.54 eV above the valence band, corresponding to neutral and positively charged states. An important empirical fact is that EL2 is controlled by the stoichiometry of the melt. As-rich melts are associated with high EL2 levels, balanced stoichiometry with stable EL2 levels, and Ga-rich melts with low EL2 levels. Crystal growth using EL2 works especially well with growth conditions where the residual impurities are p-type. In modern commercial crystal growth systems, C is the major impurity and growth conditions are optimised to achieve EL2 concentrations where it compensates the expected levels of carbon (and other impurities). Because EL2 and other deep levels have some level of conductivity that is greater than the intrinsic pure semiconductor, variations in EL2 across a wafer and from tail to seed end of the boule lead to variations in resistivity. Another interesting fact regarding EL2 is that its spatial distribution closely resembles that of dislocations, perhaps because excess As drives both types of defects. Since all of these types of deep levels can influence the transient characteristics of GaAs field effect transistors, substrate details such as this can be important in analysing device anomalies (Section 8.2.2).

Apart from the resistivity and the dislocation density of the substrate, the uniformity of these properties is of great importance. Both properties have signature variations across a boule and from seed to tail, as has been noted. The presence of excess As and the distribution of electrically active complexes like EL2 are greatly influenced by post-growth annealing. The perfect situation would be an extremely narrow resistivity distribution across the axial and longitudinal extent of the boule. One annealing condition that produces material close to this desired ideal result consists in a rather fast cooling of the crystal from the maximum temperature down to room temperature. However, fast cooling results in a situation similar to the case of rapid growth of the original crystal. The thermal gradients associated with the rapid cooling result in the formation of dislocations. High-dislocation-density material, therefore, allows for a faster cooling speed than the low-dislocation-density material, as the strain-reducing dislocations have already been generated. Generally, high-dislocation material, whether Czochralski or Bridgman, can be made with more uniform properties. Assuming that Bridgman growth methods (VGF/VB) are utilised primarily for low-dislocation-density substrates (Section 2.3.1), each growth method has one major advantage and consequently one major shortcoming: VGF/VB material has lower dislocation density and LEC material offers greater uniformity of electrical properties, principally resistivity but possibly also uniformity of trap-related electrical anomalies (Sections 8.2.2 and 8.8).

FETs and HBTs are the subject of Chapters 8 and 9. Some brief comments about performance requirements will be made in this section in order to complete the discussion of substrate specifications. While manufacturers of GaAs FETs and HBTs specify many properties of GaAs substrates similarly, they generally specify substrates somewhat differently for dislocation density and resistivity. Similar substrate specifications might include the need for epi-ready substrates (Section 2.4) with high-quality surfaces and well characterised and consistent oxide layers for HEMTs and HBTs. For HEMTs, the resistivity of the substrate can influence the turn-on voltage of the device. As a result, fluctuations in the resistivity, either across the wafer or from wafer to wafer, can cause variations in the performance of the devices. HEMT manufacturers, therefore, specify high uniformity of substrate resistance. Uniform substrates are also important for HBTs, but to a lesser extent. For HBTs, the most important substrate requirement is that the leakage through the substrate between different devices must be extremely low. Low leakage can be accomplished by using substrates with very high resistivity. Consequently, HBT manufacturers may place higher importance on the minimum substrate-resistivity specification than on its overall variation. This is but one example showing why HBT manufacturers often prefer VGF/VB substrates and HEMT manufacturers prefer LEC material.

Other material properties that are not linked to dislocation density are similar for Czochralski and Bridgman methods of growth. Impurity concentrations, for example, are determined by the equipment quality and skill of the people staffing the equipment.

Much of the art in crystal growth is related to the scale-up from one generation of substrate diameters to the next. In 2003, the market is using a mixture of 100 and 150 mm (4 and 6 inch) diameter wafers, with most major users at or moving to 150 mm. In the future, a need for even larger diameter wafers will develop. The growth of 200 mm wafers has already been demonstrated. Preliminary research at this level indicates that the main ideas presented in this section remain valid at 200 mm. GaAs electronics suppliers can continue to expect to have choices between LEC and VGF/VB substrates that they can specify based on the needs of future generations of FETs and HBTs. As with any significant scale-up, the processing technology required to produce 200 mm diameter wafers will again be very challenging for the substrate suppliers.

2.4 EPITAXY

Epitaxial growth techniques involve the growth of the active (those that are part of the device) semiconductor films, usually less than 1% of the total substrate thickness, for a device of interest on a suitable substrate. The use of epitaxial techniques allows more flexibility in the growth and placement of semiconducting layers and dopants than is achieved by other techniques such as bulk crystal growth, ion implantation and diffusion. Bandgap engineering becomes possible by choosing an arrangement of different semiconductor layers so that the bandgap along the growth direction is tailored to achieve certain desired properties. The substrate provides a template, usually of the same crystal structure and lattice constant (atomic spacing), for high-quality crystalline growth. For example, GaAs on GaAs, InGaAs on InP and SiGe on Si are all common types of epitaxy along with many other variants. Elemental constituents of the film to be grown are brought into contact with the substrate surface at elevated temperatures under conditions optimised for the particular growth technique. Impurity dopants may be introduced with the elemental constituents to dope particular layers. Generally, the quality of the epitaxial layers is much higher than that of the underlying substrate. Although crystal defects in the substrate can propagate into an epitaxial layer, they tend to diminish in number as the growth continues. For this reason, active layers of a device are normally grown only after growth of a suitable thickness of a "buffer" layer that filters out defects.

Epitaxial growth involves the optimisation of layer quality by manipulation of the experimental conditions available with a particular technique. It is very much an art because direct links between experimental techniques and the final result are often difficult and costly to prove. The starting surface condition of the substrate is critically important (Chapter 3). Surfaces are often terminated by a different structural arrangement of atoms than in bulk crystals because dangling bonds (due to a missing atom at the other end of the bond) have higher energy levels. To lower their energy, surfaces are terminated by reconstruction or incorporation of impurities, most often those commonly in the atmosphere (oxygen and carbon). Ex-situ and in-situ cleaning (often high-temperature anneals to desorb several monolayers from the surface) are utilised to make a reproducible starting condition for repeatable, high-quality growth. Today, most of the ex-situ surface treatments have become the responsibility of the substrate suppliers, who provide "epi-ready" substrates.

Epitaxial growth is often performed on a substrate that is cut so that the crystal surface is several degrees away from a major crystal plane. On GaAs, for example, a 2° miscut from the [100] plane is used to give better morphology.

Most of the methods of epitaxial optimisation depend on the particular growth technique used. Molecular beam epitaxy (MBE) and metal-organic chemical vapour deposition (MOCVD) are the major techniques in use today. Older techniques have been largely phased out (liquid-phase epitaxy or LPE) or morphed into other techniques. Chloride vapour-phase epitaxy (VPE) has given way to MOCVD for most GaAs growth.

2.4.1 Molecular beam epitaxy

MBE is an ultra-high-vacuum technique (base pressure near 10^{-10} torr). In its classic form, it involves equipment that sends a flux of atoms to a heated substrate where epitaxial growth occurs. The atom fluxes typically come from heated crucibles that evaporate solid sources in a vacuum. The flux of atoms is directed to a sample by an orifice in the source crucible which can be switched on and off by a shutter. Each atom flux is controlled by the temperature of the crucible through often painstaking calibrations. An MBE growth chamber is shown in FIGURE 2.9.

The atoms reach the heated substrate and most of them stick on the surface, where they diffuse around until they find a lattice site or the edge of a two-dimensional "island" where they can incorporate into the lattice. The key to good growth is to heat the substrate sufficiently such that good atomic mobility occurs on the surface without excessive desorption of the atoms. Of course, every rule has its exception, and sometimes one desires growth

FIGURE 2.9 A picture of a research MBE growth chamber.

under conditions of low surface mobility (an example will be given later in this section).

Generally the growth rate is defined by the flux of one or two of the atoms and the other(s) are kept under excess flux conditions. Gallium and arsenic atoms are asymmetric in the sense that As is much more easily desorbed than Ga and one must maintain a high V/III ratio during growth. In general, the As flux must be maintained during growth interruptions or during initial surface treatments so that the lattice does not begin to decompose. The same is true of the P-based semiconductors.

During the growth, progress is monitored by a variety of diagnostic tools. Electron diffraction (by a technique called reflection high-energy electron diffraction or RHEED) can be used during growth to monitor the periodicity of the two-dimensional surface as a quality metric. When the periodicity is high, the diffraction spots are bright. The order observed in the RHEED pattern can indicate a variety of surface conditions, especially during the initial substrate cleaning step, when surface reconstructions and contaminants can be numerous and complex. MBE growth proceeds ideally as a layer by layer growth. During the completion of a layer, the RHEED spots are brightest and during an approximate half layer completion, they are the dimmest. RHEED intensity oscillations are used to "count" layers during growth and provide either real-time growth rate feedback or calibration.

Another common in-situ diagnostic is optical reflectance. The intensity of near-normal-incidence reflected light can be analysed for periodic structures that are common in optical devices but reflectance can be used during special calibration runs for electronic devices as well. Real-time growth-rate information about individual layers comes out of this analysis. An MBE chamber will have other diagnostics as well. See references [9,10] for more details. Samples are also analysed ex-situ (Section 2.5).

Because MBE is an ultra-high-vacuum technique and due to the sheer size of its components, it is fairly time consuming to initially bring the chamber to the desired vacuum. A bake-out near 200°C is required to "outgas" the interior surfaces of the chamber so that they will not be sources of gases at the 10^{-7} to 10^{-8} torr level. This step can take around a week. In addition, source material that has been vented to atmosphere must be conditioned so that impurities adsorbed on the source material surfaces do not become incorporated into and degrade the semiconductor. Source conditioning is extremely time consuming as well. In order to keep vent time to a minimum, MBE machines have fairly large sources that take months or years to be consumed. Since venting and preparing a chamber is so time consuming, MBE machines must be very reliable to minimise the number of times they are vented, particularly those used in production.

An MBE chamber may have 8–12 fairly large sources (visible on the left side of FIGURE 2.9), for flexibility in research machines and for redundancy in production machines. The source material is gradually depleted under continual use and the source operating conditions to attain a given flux will drift over time and require fairly regular calibrations. The methods of performing the calibrations have become routine and good growers can maintain excellent repeatability over long periods of time.

It is sometimes assumed, implicitly, that MBE growth is an equilibrium process. However, some of the strength of the technique is that certain, non-equilibrium methods can be employed to yield useful results, which will be illustrated here with a single example. Strained layer growths are important for HEMT structures (Section 8.2.2), such as the AlGaAs/InGaAs HEMT. InAs has a larger lattice constant than GaAs and In-containing alloys of GaAs grow with residual strain on GaAs substrates. When the InGaAs layer exceeds a critical thickness, thermodynamic equilibrium will require the InGaAs to relax to a lower energy state with misfit point defects created in the InGaAs. By using non-equilibrium conditions for the growth, the InGaAs layer can exceed the critical thickness and more useful devices have resulted.

The advantage of using MBE is that very precise control of the structure can be obtained. Structures with sharp interfaces are routine and high compositional precision and uniformity are achieved. Purity and control of dopants are also generally very high due to the high vacuum and the ready availability of high-purity source materials. Since its inception in the 1970s, MBE has been considered the technique of choice for the highest quality epitaxial structures. In recent years MOCVD has narrowed or eliminated most of the advantages that MBE has enjoyed, as will be described in Section 2.4.2. Disadvantages of MBE include the high cost of the equipment, its complexity, and the relatively low growth rate compared with MOCVD. Modern, large diameter, multi-wafer machines, in particular, are enormous beyond imagination, even by modern semiconductor manufacturing standards.

The preceding material has been a presentation of conventional MBE. Variants, of course, are available as alternative techniques. These may or may not prove permanently useful.

The most common variation is to use gaseous sources in place of solid source material. In this way, the replenishment of source material can occur without breaking vacuum. For example, triethyl gallium can be the source material for Ga and either arsine or tertiary butyl arsine can be the source material for As, depending on the cost of complying with safety standards for the use of arsine. The growth mechanism can be considered to be a low-pressure variant of MOCVD (Section 2.4.2).

Another variation of MBE is atomic layer epitaxy (ALE), a variation on the MBE process in which the group III and group V atoms are sequentially deposited on the wafer by alternating the shutter openings on the Ga and As sources. In conventional MBE, both shutters are open at the same time. The group III atom has more time to migrate over the surface to yield a more perfect lattice structure. In ALE the intent is to form one atomic layer of each species at a time, to attain this more perfect structure. The potential advantage of this technique is that the first group III layer deposited forms chemical bonds to the crystal and the extra atoms form weaker group III with group III bonds. By switching to the group V atoms, the physisorbed atoms are easily displaced and a complete atomic layer is easily grown without requiring precise delivery of the right number of each type of atom.

2.4.2 Metal-organic chemical vapour deposition

Whereas MBE is a growth technique utilising a directed flux of mainly atomic source material, MOCVD is related to vapour-phase epitaxy with a non-directed flux of source material. Vapour-phase epitaxy is an older technique which also involves a non-directed flux of source material; its essential difference from MOCVD is in the choice of source materials and their implications for reactor design. VPE uses solid Ga and $AsCl_3$, which react in the presence of H_2 to form GaAs and HCl. Some limitations of the technique are that HCl is corrosive in the chamber and that Ga is consumed rapidly, often requiring replenishment after a few runs. Consequently, with the appearance of metal-organic sources in the 1980s, modern day MOCVD reactors were designed and rapidly gained popularity. Metal organic sources use various Ga, Al and In compounds with hydrocarbon substituents. Typical reactions are of the following types:

$$\mathrm{Ga(CH_3)_3 + AsH_3 = GaAs + 3CH_4} \tag{2.13}$$

$$\mathrm{Al(CH_3)_3 + AsH_3 = AlAs + 3CH_4} \tag{2.14}$$

$$\mathrm{In(CH_3)_3 + AsH_3 = InAs + 3CH_4} \tag{2.15}$$

where the group III$(CH_3)_3$ is named trimethyl (Ga, Al or In) for the methyl ligand organometallics. The gases are introduced into a chamber where they are mixed efficiently so that they flow uniformly and are then reacted with a heated compound semiconductor substrate such as GaAs. Because MOCVD machines do not require a directed flux but rather can mix large volumes of source gases, they can, in principle, scale-up more easily to handle a greater number of wafers simultaneously. This ease of scalability often leads to a manufacturing advantage for MOCVD, which in turn has led to its wide adoption for epitaxy of GaAs HEMTs and HBTs.

Arsine and the organometallic gases are introduced into the chamber where they are directed onto the hot substrate. Like MBE, the reactants must stick to the semiconductor surface to diffuse around and react. However, the reactants must first "crack", i.e. dissociate the alkyl group (or H) and allow the group III or group V elements to stick to the surface. Typically, this process leaves a complex variety of surface species, such as CH_3, $Ga(CH_3)_2$, AsH_2, etc. Ultimately, all of the groups III and V atoms must become atomic species in order to diffuse and form the (Ga, Al, In)As. These complex chemical reactions have the potential to incorporate undesirable impurities into the semiconductor, namely C and H. Organometallic precursors are chosen for their ability to allow the hydrocarbon radicals to recombine with H to desorb the resulting hydrocarbons and H_2. Fortunately, hydrocarbon radicals prefer to recombine with H atoms and desorb rather than incorporate into GaAs-like semiconductors with surprisingly little C incorporation. High-purity growth results with metal-organic gases as long as these can be provided with electronic grade purity.

Typically, the hardware required for MOCVD growth (FIGURE 2.10) would be much less complex and costly than for MBE, were it not for the safety features required to safely handle the highly toxic arsine. Key hardware includes a gas handling system capable of greatly varying flow conditions under a wide variety of pressures, a sample introduction system (usually with loadlock), a method of sample rotation (for gas flow dynamics, as well as better uniformity), a sample heater and a pumping system for moderate vacuum and high gas flows. As alluded to, arsine has such high toxicity that fail-safe and highly redundant safety features have to be built in. Because of the safety concerns associated with AsH_3, a large body of work has been carried out with group V organometallics in the hope of eliminating the need for arsine. While much progress has been made with some of these approaches, many MOCVD growers continue to use arsine. With safety features built in, MOCVD and MBE systems can often be similarly priced, though the details depend on the features of the system.

MOCVD is attractive relative to MBE for at least two reasons. First, it does not require ultra-high vacuum and can usually be

FIGURE 2.10 A picture of a research MOCVD growth chamber.

vented for service and brought back into operation more quickly and easily than MBE. Second, MOCVD is easier and perhaps more economical to scale up to larger usable area depositions. Both types of system lend themselves well to automation.

As is the case with MBE, growth conditions are chosen carefully to optimise crystal quality, layer uniformity and the abruptness of the transitions between layers. The adsorbed group III and group V elements must have some degree of surface mobility, as in MBE, in order to find active sites for crystal growth. Temperatures are carefully optimised so that an appropriate balance exists between the temperature required for dissociating the precursors at the (AlGaIn)As surface (while preventing gas phase reactions, ideally), providing adequate surface mobility and preventing excess desorption. In general, GaAs and AlGaAs growth occurs at slightly higher temperatures in MOCVD than MBE. Optimum conditions of the V/III ratio are also needed (generally excess group V) as well as optimisation of flows and pressure. MOCVD growth is generally more complex than MBE, not only because of the greater number of types of reactions, but also because of the non-directed nature of the flux, which has the potential to carry out gas-phase reactions and change the desired chemistry at the semiconductor surface.

Many diagnostics are used to characterise and monitor the growth, both in-situ and ex-situ. Some of the in-situ techniques will be described here and other ex-situ techniques are described in Section 2.5. As with MBE, in-situ reflectance is used to characterise the growth rate, both for calibrations and for real-time growths. A typical set-up may have a near-normal-incidence visible light source and a detector of the reflected light. The reflected signal may vary due to many factors, including the surface roughness and reflections from heteroepitaxial interfaces. A well-controlled homoepitaxial growth will show a very flat signal and reflectance is not an especially helpful diagnostic. Buried heteroepitaxial layers, such as in the growth of alternating layers of GaAs and AlGaAs, will have different optical constants and reflect the light differently than for a simple GaAs growth. The reflectance will depend on the actual compositions and thicknesses. In addition, phase changes between the surface reflected signal and the buried interface will modulate the intensity of the reflected signal as the growth of the top layer progresses. As long as the thicknesses of the layers are reasonable fractions of the light's wavelength, good modulation of the reflected signal will result. The expected reflectance can be modelled before actual growth and used to guide the process in real time [10]. In GaAs-based HEMT or HBT growth, in-situ reflectance is of use for controlling the thickness and composition of InGaAs, AlGaAs and InGaP layers. Because an actual device structure (Sections 8.2.3 and 9.2.3) will normally incorporate relatively thin wide bandgap layers, reflectance is more useful for calibrations than a real-time growth. Nevertheless, an abnormal reflectance signal during a real-time growth will alert the grower to possible serious problems, such as a roughening surface. For a calibration run, a structure with a series of layers will be grown, which will address alloy compositions needed in the device structures but using layer thicknesses designed to give good reflectance modulation.

Today, MOCVD can generally grow the same types of structures as MBE, with similar quality. At one time, MBE was considered to have the advantage in purity (because of higher vacuum and source materials) and interface abruptness. Today, a slight advantage may exist for MBE in the quality of thin, abrupt layers (measurable by high resolution imaging techniques such as TEM, Section 2.5.2), giving an advantage in some types of structures (e.g. low-temperature mobilities of special types of growths), but these do not appear to be big differentiators for practical HEMT and HBT structures.

2.5 MATERIAL CHARACTERISATION

The purpose of this section is to give an introduction to methods of qualifying substrates, epitaxial samples for device fabrication, and often in-process or finished devices. One goal is to identify bad samples to discard them before the high cost of doing epitaxial growth and processing them into devices is incurred. Another goal is to identify problems with epitaxial growths as soon as possible so that they can be corrected before large amounts of material must be scrapped. Lastly, many of these techniques are used in failure analysis for devices that either did not meet specifications or have failed in operation before their expected lifetime. The process engineer should have a rudimentary knowledge of these techniques so that interactions between different groups of engineers for the purpose of quality improvement or failure analysis can be fruitful in determining the often complex interactions between materials, processing and assembly. For greater detail than is available in this section, refer to references [11,12].

2.5.1 Light-based techniques

Optical techniques presented in this section include optical microscopy, ellipsometry, photoluminescence, X-ray photoelectron spectroscopy (XPS) and X-ray diffraction.

The most obvious and widely useful characterisation technique is optical microscopy. Optical microscopes take collimated light that is reflected (usually) off a sample and imaged through a high-quality lens system to eyepieces or to a video camera. Objectives with different magnification can be switched in and out to allow the user a range typically between 50- and 500-fold magnification. Microscopes are marvellous instruments that make complex optics transparent to the user.

Ellipsometry is a technique where changes in the polarisation state (the amplitude and phase) of light reflected from a surface are measured. It can be used to measure the optical constants of surfaces or thin films (usually transparent) on highly absorbing surfaces, from which thickness and refractive index can be derived. While complex analysis can be brought to bear on complex research problems, this discussion is limited to the capabilities and common uses of commercial desktop equipment. When light is reflected from a single surface, its amplitude and phase will generally be changed. For multiple reflecting surfaces, the various reflecting beams will interact further and give maxima and minima as a function of wavelength or incident angle. Equipment makers generally automate the measurement and hide the complexity of the calculations needed for easy turnkey operation. Under certain combinations of incidence angle, film thickness and wavelength, complete destructive interference of the reflected light can occur, resulting in null points that provide no useful information when the sample is measured. Recent upgrades to basic ellipsometry equipment include variable-angle and variable-wavelength features that can largely avoid the null point issue, as well as enable measurement of thicknesses much smaller than previously possible. However, multiples of a given path length (due to thicker films) can give the same optical constants. Ellipsometry therefore requires independent determination of the sample thickness relative to the full cycle thickness.

In photoluminescence (PL), a semiconductor sample is illuminated with an optical source (usually a laser) and the light emission of the sample is detected with a spectrometer. Electron-hole pairs are generated by the absorption of light after which the carriers recombine to emit at the wavelength near the semiconductor bandgap and sometimes at other wavelengths. Some of the transitions are illustrated in FIGURE 2.11. An emission signal can arise from band-to-band recombination (FIGURE 2.11(a)), free excitons (FIGURE 2.11(b)), bound excitons (FIGURES 2.11(c), 2.11(d)), donor-acceptor transitions (FIGURE 2.11(e)) or deep level emissions (FIGURE 2.11(f)). Shallow donors and acceptors are readily identified in PL. The radiative-emission intensity is proportional to the impurity density. Impurity identification by PL is very precise because the wavelength resolution of the technique can be very high. However, PL is not as useful for quantitative analysis because it is not easy to draw a correlation between the intensity of an impurity emission and the density of that impurity. The PL signal can vary greatly from sample to sample due to the presence of non-radiative recombination centres, either from bulk deep levels or the surface (Section 3.2). Because PL is a rapid and non-destructive technique, it is often used as a screening method for substrates and epitaxial growths. More details on the proper use of PL are provided in Section 3.3.1.3.

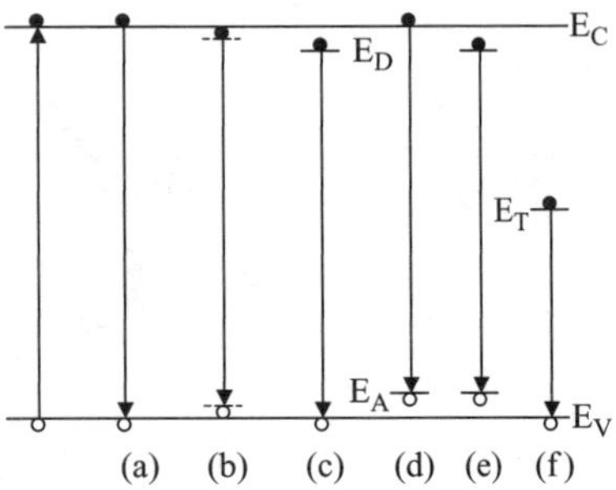

FIGURE 2.11 Some of the transitions possible in photoluminescence.

XPS is a chemical analysis method that uses X-rays to illuminate samples and the emitted electrons are detected as a function of energy. The energy of the electron identifies the atomic species absorbing the X-ray through the inferred binding energy of a (usually) core electron. The energy of the X-ray peak is slightly modified by chemical bonds of the atom, so the chemical environment can also be inferred. Glancing incidence can give surface sensitivity to the technique. XPS is sometimes used to analyse changes in surface chemical composition due to etches or growth steps.

X-ray diffraction is used to determine the crystal structure of a semiconductor. A collimated, monochromatic beam is incident on a sample and the scattered beam is collected and analysed for intensity and angle. Perpendicular to the plane defined by the incident and reflected beam, planes of the semiconductor can diffract the X-ray. At a certain angle, the diffracted beams constructively interfere according to the equation

$$\theta_B = \sin^{-1}(\lambda/2d) \tag{2.16}$$

where θ_B is the Bragg angle, λ is the wavelength of the X-ray and d is the spacing between crystal planes. In addition to determining the crystal structure, the full width at half maximum (FWHM) of the X-ray peak can be used as a measure of the quality of crystal growth. When a crystal incorporates many defects, impurities or dislocations, they affect the long-range periodicity of the crystal and degrade the FWHM.

2.5.2 Electron-beam techniques

Electron-beam techniques are powerful methods of imaging semiconductors because the wavelength of electrons is considerably smaller than that of photons. As De Broglie discovered in the 1920s, all particles can behave as waves. The electron wavelength is given by the relation

$$\lambda_e = h/mv = h/(2qmV)^{1/2} \tag{2.17}$$

where λ_e is the wavelength of the electron, h is Planck's constant, m is the mass of the electron, v is the electron velocity, q is the electron charge and V is the accelerating voltage applied to the electrons. Generally, the accelerating voltage in an SEM is 10–30 kV and λ_e is a fraction of a nanometre compared to 400–700 nm for visible light. When sample charging is an issue, lower voltages of the order of 1 kV are used. Even in this case, λ_e is sub nanometre, giving electron beam techniques much higher potential resolution than optical microscopes. Incident electrons are absorbed, reflected, transmitted or involved in the emission of other electrons. They can also cause light or X-ray emission. Electron beam techniques discussed in this section are scanning electron microscopy, transmission electron microscopy, cathodoluminescence and Auger spectroscopy.

Scanning electron microscopy (SEM) utilises an electron beam to image a sample located in a vacuum chamber. An incident electron beam is focused on a sample, the beam is scanned across the sample, and secondary or backscattered electrons are imaged by a detector. The signal at the detector is amplified and imaged on a cathode-ray terminal (CRT) display. A one-to-one correspondence is established between each point on the sample and the display. Magnification up to 100,000 is possible in an SEM, or about 100 times better than optical microscopes. The contrast in SEM depends on a number of factors including atomic number, surface conditions, local electric fields, and topography. A flat sample with no variation in composition or topography will show a uniform image in the SEM from uniform secondary electron emission. Topography aids in contrast due to the fact that electrons have a mean free path in the solid and tilting the angle of emission from normal increases the effective depth of sampling compared to normal emission by $1/\cos(\theta)$.

The SEM instrumentation includes a vacuum chamber, electron gun and electron optics, a sample holder, an electron detector and electronics that amplify the detector signal and image it on a CRT. The electron gun consists of a source (tungsten filament, LaB_6 filament for higher brightness, or field emission source) and electron optics for focusing and rastering the beam on the sample. The instrument console has knobs and gauges for focusing and adjusting the contrast and image quality as well as controls for the placement and tilting of the sample. The emitted secondary electrons are accelerated and collected at the detector, often based on a scintillation material that emits light when struck by energetic electrons and allows use of a high-gain photocathode to drive the CRT.

Insulating surfaces can charge up and interfere with the trajectory of the incident and emitted electrons, effectively degrading the image. Surface charging is eliminated by coating the sample with a thin conducting film such as Au or by operating the SEM at low enough voltage so that the number of primary electrons is approximately equal to the number of secondary and backscattered electrons. The energy for this balance is around 1 keV. Devices that can be damaged by high-energy electrons are also candidates for low-energy SEM. Low-energy beams suffer from reduced signal-to-noise that can degrade image quality and are not used if higher energy beams are acceptable.

SEM imaging is extremely useful for inspecting transistor circuits during and after processing and for failure analysis. Images can be taken of the surface of the devices. In many cases, it is desirable to view the device in cross-section to determine device dimensions, for example. SEMs are available both as research tools and for production lines where they are often used for inspection and line-width measurement.

Several electron-beam techniques are associated with SEMs. Electron-microprobe analysis (EMP) is a technique where incident electrons bombard a sample in order to produce X-ray emission that can be analysed. The incident electrons ionise core electrons. The core vacancies can be filled by other atomic electrons accompanied by emission of an X-ray whose energy is the difference of the two atomic energy levels. The characteristic energy of the X-ray can be used to identify the element. The electron beam must have high enough energy to ionise core levels, usually 10–30 keV. Its mean free path can be several microns in semiconductors. The emitted X-rays largely escape the crystal at these depths and can be energy-analysed to identify the atomic elements present in a sample. Addition of an X-ray detector to an SEM is a common method of gaining EMP analysis. Two different types of detectors are used: energy-dispersive spectrometers (EDS) and wavelength-dispersive spectrometers (WDS). EDS is commonly used for rapid sample analysis with low resolution and WDS is used for high-resolution measurements. EDS resolution is sufficient for atomic analysis, while WDS resolution can be used to infer the chemical environment of an analysed element. EDS typically uses a reverse-biased semiconductor as a detector, while WDS uses an analysing crystal to spatially separate the wavelengths by Bragg diffraction (Section 2.5.1).

Electron-microprobe analysis with EDS is often used in failure analysis for quick spatial maps of elements. EMP is not a trace analysis method because of poor sensitivity. The sensitivity of the technique depends on several factors, including the atomic number of the element. X-ray emission increases with the atomic number of the element and EMP is relatively insensitive to low atomic-number elements. EMP complements Auger spectroscopy as an analysis tool.

Another electron-beam technique is transmission electron microscopy (TEM). In TEM, a focused electron beam, typically 100–400 keV in energy, passes through a thin sample that is to be imaged and is detected from the backside of the sample. Some electrons are backscattered as well. The sample is made thin in order to achieve high spatial resolution by limiting beam spreading that occurs when secondary electrons are created. Another reason for keeping the sample thin is to limit the number of electrons that are absorbed, which can raise the sample temperature and change the lattice constant. The transmitted and forward scattered electrons can diffract and can be imaged with appropriate electron optics. This diffraction pattern contains a wealth of structural information. The principal attractiveness of TEM and related techniques is in the resolution of the techniques and the ability to gain atomic-level structural information. The optimisation of heterointerfaces has been greatly aided by TEM.

A bright-field image is formed when the transmitted electrons are imaged. A dark-field image is formed when a specific diffracted beam is imaged. Neglecting diffraction effects for the moment (as in amorphous samples), image contrast is affected by the atomic number of the atoms, as higher atomic number results in more scattering and less brightness in the bright-field image, and mass contrast (a void, for example). Including diffraction effects, the image contrast depends also on diffraction contrast and phase contrast. Resolution in a TEM can be as good as 0.15–0.25 nm, or atomic scale.

Conventional (stationary) TEM utilises a coherent, collimated incident electron beam. In contrast, scanning TEM (STEM) utilises a tightly focused (0.5 nm) electron beam which is rastered across the sample. The STEM image is constructed from the positional information of the scanned electron beam. The dwell time at each position in STEM must be sufficient to gather enough signal-to-noise, making the technique slower than conventional TEM. To make image collection times reasonable, dwell times are limited and STEM images are usually of lower resolution than TEM. However, the power of the technique is realised by the extra capability that becomes available for position-specific analysis within the sample. Electron energy-loss spectroscopy is one such technique that can be used with STEM at specific positions identified in an image to identify chemical species by their characteristic energy absorption from the electron beam. For example, one can determine whether certain impurities segregate near defects.

TEM images are obtained by specialised sample preparation techniques. Samples are mounted on an appropriate fixture and thinned by a combination of lapping, polishing, and ion milling down to a thickness of tens of nanometres. Traditionally, these techniques have been very time consuming and have required a great level of skill. In addition, specific areas of devices are extremely difficult to precisely locate and image. This situation has improved greatly with the recent combination of TEM with focused ion beam (FIB) techniques.

An FIB is an instrument similar in construction to an SEM, but which accelerates ions to the sample rather than electrons. The FIB can be used to ion mill a hole in a sample so that an SEM image can be taken of the cross-section. Alternatively, the FIB can be used to drill down the opposite sides as well, so that the piece can be removed from the FIB with a thickness appropriate for TEM for imaging. The use of FIB with TEM has greatly increased the flexibility of the technique and reduced the complexity of sample preparation. Free standing films can be cut in 30 min with greater precision and reliability compared to many hours for the traditional methods of lapping and ion milling. The main drawback to using FIB is the possibility that the energetic ion beam perturbs the sample. The surface of the sectioned film may become amorphous. Nevertheless, the power to image precise areas of a device (such as across the active area) has led to great advances in device optimisation and failure analysis.

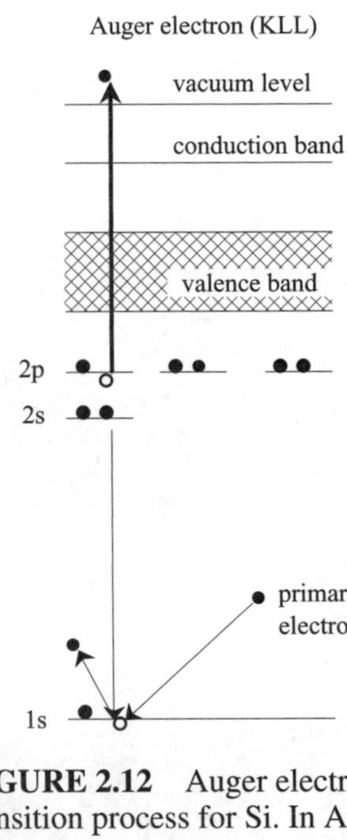

FIGURE 2.12 Auger electron transition process for Si. In Auger terminology, the K shell represents 1s electrons, the L shell represents 2s and 2p electrons, etc.

Another electron beam technique in widespread use is Auger analysis. In the Auger effect, an incident electron is sufficiently energetic to eject a core electron from an atom in the sample (FIGURE 2.12). The core electron vacancy (1s or K shell in FIGURE 2.12) may be filled by a higher energy core or valence electron (2p or L shell in FIGURE 2.12). The energy released by filling the ionised core state with another electron in the atom results in emission of an X-ray of the proper energy or the ejection of another electron. The latter process is called the Auger process and the energy of the ejected electron is characteristic mainly of the atom involved but also includes small shifts in this energy due to the chemical environment. X-ray emission (as in EMP) and Auger emission are competing processes for filling atomic core vacancies. Generally, the Auger process is favoured for low atomic number elements and X-ray emission dominates for high atomic number. The Auger electron will typically have energy of hundreds of eV and a mean free path in the semiconductor of tens of angstroms, making Auger spectroscopy a surface-sensitive technique. Auger spectroscopy finds its greatest utility in analysing semiconductor surfaces after processing or cleaning steps. It is particularly sensitive to carbon and oxygen contamination and can detect sub-monolayers of a surface film.

Auger spectroscopy is most often performed in stand-alone high-vacuum (or ultra-high) chambers but also as an in-situ module for MBE growth chambers or cluster tools. Its instrumentation consists of an electron gun, electron-beam control, an electron-energy analyser and data-analysis electronics. The incident electron energy is typically 1–5 keV. Auger instruments can be set up for two-dimensional scanned analysis, if desired, or in combination with sputter depth profiling. The latter method, which uses an ion beam to sputter away the material to expose buried regions, is often used for profiling ohmic-contact reactions (Section 6.2.3). Auger analysis will be contrasted with ion-based analytical techniques in the next section.

Several other electron-beam analysis techniques find important uses but will not be described in detail here. These techniques include electron-beam-induced current (EBIC) and electroluminescence. These techniques can be used in conjunction with an SEM.

2.5.3 Ionic techniques

Ion-beam techniques of analysis and characterisation are high-vacuum electrostatic techniques like their electron beam counterparts. Incident ions are absorbed, emitted, scattered, or reflected from a sample and are analysed for ion scattering, energy and mass, among other properties. Incident ions can also initiate photon or electron emission and these can also be analysed.

Secondary-ion mass spectrometry (SIMS) is perhaps the most widely used ion-beam technique. SIMS instruments are usually stand-alone high-vacuum chambers specifically designed for elemental analysis. In SIMS, an ion beam is focused at a sample with sufficient energy to sputter it away (tens of keV ion energies for ion penetration of several monolayers). The energy and mass transfer of the ion is sufficient to break surface bonds and displace atoms in the sample and to implant the incident ion. These atoms or fragments may react, rearrange themselves, or desorb as ions or neutrals. It is the desorbed ions (not neutrals) that can be accelerated and focused towards a mass spectrometer and detected by it. Although only a small fraction of ejected particles are ionised, these are efficiently detected with high discrimination against other possible elements by the mass spectrometer. The sample is analysed as it is sputtered and elemental analysis as a function of depth is provided with reasonably high depth resolution. Typically, a list of elements is requested for analysis. Some of these may be dopants and the rest are usually impurities of concern. Impurities may be detected at the parts per billion level, or lower in favourable cases. Detection limits in SIMS range between 10^{14} and 10^{16} cm^{-3} for many trace elements.

The ion beam is chosen based on its electronegativity and ability to ionise certain elements. In SIMS, it is the total desorbed ion yield of elements of interest that enhances the sensitivity of detection. Primary (sputtering) ions such as O_2^+, Cs^+, O^-, and Ar^+ are chosen based on their ability to ionise a particular element or group of elements for analysis. For example, electronegative oxygen (O_2^+) enhances the ionisation of electropositive elements such as Mg, Ga and other group III elements. Electropositive Cs^+ enhances the ionisation of electronegative elements such as O, As, and other group V elements. Whereas the electronegative property of oxygen itself explains the ionisation of electropositive elements, the electropositive effect of Cs^+ ions is explained by work functions that are reduced by the implantation of Cs in the near surface region of a sample.

The secondary-ion yield varies greatly depending on the primary ion chosen and the chemical nature of the sample. Two important practical effects arise from this situation. First, calibration using known standards is a necessary part of SIMS analysis. Selection of a standard that does not sputter at the same rate as the sample will lead to errors in the supposed depth being sampled and give erroneous composition-versus-depth information. Second, the ion yields at surfaces are highly unpredictable. For example, an oxidised surface may enhance the ion yields by factors of 1000. Typically, the first few monolayers of analysis must be ignored in most cases. Calibration samples can be prepared by ion implantation, since the total implanted dose has an uncertainty of less than 5% in most cases.

SIMS instrumentation is extremely complex. For the purpose of this section, a few general descriptions will suffice. Ultra-high vacuum is necessary to avoid contamination. Within the vacuum chamber, incident ions are accelerated and focused at the sample with electrostatic ion optics. The secondary ions require high mass resolution to distinguish chemically different but similar mass ions such as ^{31}P and ^{30}SiH. The mass spectrometer will consist of electrostatic and magnetic sector analysers in tandem.

Impurity and dopant profiling are the most important applications for SIMS in semiconductor manufacturing. For GaAs devices, the characterisation of heterostructure or other semiconductor interfaces is important as well. For example, the growth initiation procedures of epitaxial growth on bulk substrates by MBE and MOCVD are studied by SIMS to ascertain the effectiveness of cleaning procedures. SIMS is widely used to identify impurity incorporation during the growth and processing of GaAs devices.

Rutherford backscattering spectroscopy (RBS) is another important ion-beam analysis technique. In this technique, the energy of backscattered ions is measured and related to the sample under study. In this case the incident ion energies (typically for He) usually exceed 1 MeV. Incident ions are backscattered through collisions with the nuclei of atoms in the sample. In RBS, the following equation holds due to conservation of energy and momentum:

$$E_1/E_o = 1 - 2R(1 - \cos\theta)/(1 + R^2) \qquad (2.18)$$

where $R = m_1/m_2$, the mass ratio of the ion and the sample's scattering atom, E_1 is the backscattered ion energy, E_o is the incident ion energy and θ is the scattering angle. EQN (2.18) is derived assuming that $R \ll 1$ (for He ions this is usually a good assumption) and θ is close to 180 degrees (near total backscattering). From this relation, the mass of the scattering atom can be determined by measuring the energy of the backscattered He ion.

In addition to the mass identification in a sample (and correspondingly, element identification), concentration and thickness can be determined as well. The relation between energy loss of ions in solids is well established and can be used for thickness determination of a layer in a sample. EQN (2.18) determines the energy cutoff condition for RBS. Below the surface of the sample, the incident ion loses energy as a function of depth to the solid and EQN (2.18) would be modified for

$$E_o^* = E_o - \Delta E \qquad (2.19)$$

where E_o^* is the incident energy modified by the energy loss in the solid up to the depth d. Energy losses are slightly energy dependent and are listed in tables of stopping power. The concentration is determined by the intensity of the backscattered signal, generally without need for calibration. The thickness resolution of a thin film can be as good as 1 nm.

An RBS system consists of a high-vacuum chamber that contains a He-ion generator, an accelerator capable of achieving 1 MeV ion energies, the sample holder and the detector. Most of the ions pass through the sample and perhaps one out of 10^6 is backscattered. The backscattered ions are detected by a semiconductor detector that operates much like the X-ray detector described for EDS. The ions pass through the semiconductor detector, creating many electron-hole pairs that are detected as current pulses that are amplified and analysed for pulse height, which is proportional to the ion energy.

RBS is useful in semiconductor manufacturing as a means of quantifying composition of thin films. Due to the cost and complexity of the technique, it has not achieved routine use, but rather is employed as a method of calibrating other analytical techniques for determining composition of thin films or as a research tool.

Another ion-beam technique is not strictly used for analysis but rather for sample preparation in conjunction with other analysis techniques. Focused ion beams (FIB) can operate within an SEM or similar instrument to ion-mill in specific areas of devices that are imaged by the SEM and create a cross-section of the area. The cross-section can be imaged directly within the FIB instrument or SEM or TEM. The cross-section can also be cut out of the sample and transferred to another TEM. FIB-assisted techniques have become extremely useful in failure analysis of GaAs devices and in relating ohmic-alloy reactions (Section 6.2.3) to electrical properties, to mention two examples.

2.5.4 Electrical characterisation

Electrical characterisation techniques can either be somewhat specialised (and usually designed for materials scientists) or can be based on measurements of two and three terminal devices such as diodes and transistors.

Resistivity measurements are an important part of characterising thin semiconductor films, whether produced by ion implantation or epitaxy. Resistivity measurements are also made for thin conducting films. Some techniques used are the four-point-probe method and contactless methods. The techniques described here measure resistance. If a sample has uniform resistivity in a known thickness, the resistivity can be calculated from the measured resistance.

In a two-probe resistance measurement technique, current is measured as a function of voltage and the resistance is calculated according to Ohm's law. The resistance will include a contribution from the probes and the probe contact resistance. A four-point-probe method (FIGURE 2.13) is a means of eliminating the extraneous sources of resistance by using two extra probes to measure the voltage across a section of the current path and calculating its resistance across a known distance using a known current, forced by the first two probes. The term "four point probe" is used to describe an instrument that carries out such measurements in a calibrated way. Such an instrument typically uses a spring loaded mechanism that pushes small diameter metal probes against the sample and is calibrated for the spacings in the probe (usually changeable for different resistivity ranges and materials). Four point probes are widely used for quick measurements of thin metal films but also on occasion for GaAs wafers if an appropriate probe is attached (with finer tips for better contact). The four point probes can also be used to map wafers; this is widely done in Si manufacturing. However, because of the possibility of chipping the wafer with the probe, contactless resistivity measurements are preferred for GaAs.

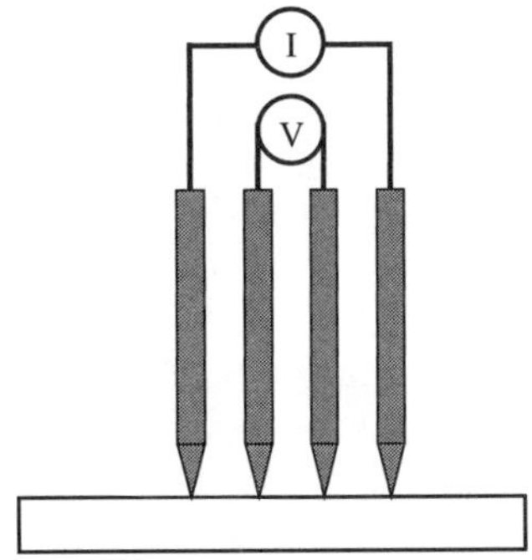

FIGURE 2.13 Illustration of the four-point-probe method. The two outer probes are used to force a current, I. The resistance is calculated from I and the voltage measured across the two inner probes.

Contactless resistivity usually refers to a measurement method based on eddy currents. A sample is placed between two sets of coils. Each coil is embedded within a ferrite material. A current in the upper coil will induce a current in the lower coil. When a conductive sample is placed between the coils, the Q (quality factor) of the circuit is changed and with it, the induced current in the lower coil. By using samples of known conductivity, the instrument can be calibrated to measure arbitrary samples. The measurement is quick and can be instrumented to step across a wafer and generate wafer maps of resistance. Such instruments are popular for mapping the resistivity of epitaxial or ion implanted layers created on semi-insulating substrates.

Carrier mobility measurements are important to growth and process engineers as a measure of the quality of the semiconductor. Mobility can be of two types, drift (mediated by an electric field) or Hall (mediated by a magnetic field). Although the former more directly relates to operation in GaAs devices, the two are often not very different and Hall mobility measurements are more quickly carried out. In addition, both methods can assess material quality. Mobility is reduced by physical factors, including temperature. More importantly, from a material quality perspective, impurities reduce mobility. For a more complete description of the differences between the two techniques, see [1].

Capacitance-voltage (C-V) profiling can provide a wealth of information. It is most well known for assessing the quality of the oxide in MOS devices, but it is also widely used for determining dopant profiles in semiconductors. The latter technique relies on the fact that the width of a depletion, or space charge region, of a semiconductor junction can be modulated by an applied voltage. For GaAs devices, Schottky junctions and pn junctions can be probed. The former are important for field effect transistors (Chapter 8) and for dopant profiling, while the latter are important for heterojunction bipolar transistors (Chapter 9), laser diodes and LEDs. We consider a Schottky-barrier diode with doping of N_D. A DC bias is applied to the Schottky contact. The differential or small signal capacitance is given by

$$C = dQ/dV = \varepsilon A/w \tag{2.20}$$

where dQ is the incremental charge induced by a small change in the voltage dV, ε is the dielectric constant of the semiconductor, A is the area of the Schottky contact and w is the width of the depletion region. The equation (derived using the depletion approximation – mobile carriers are assumed to be zero in the depleted space-charge region) relating C to the doping is given by

$$d(1/C^2)/dV = 2/q\varepsilon A^2 N_D \tag{2.21}$$

By measuring C as a function of applied voltage and plotting $(1/C^2)$ as a function of V, it is seen that the doping concentration N_D is obtained from the slope of the $(1/C^2)$ versus V plot as long as the area of the Schottky contact is known (Section 7.2.1 and FIGURE 7.2).

Capacitance-voltage measurements can be made with whatever Schottky contacts are available, often deposited metal contacts. However, the quickest measurement uses a temporary Schottky contact with a mercury-probe instrument. In this technique, a small capillary is immersed in a mercury reservoir at one end and is placed within a millimetre-sized enclosure at the other end. When a semiconductor is placed over the enclosure, a vacuum seal results once the volume is pumped down. The vacuum causes the mercury to rise until it reaches the top of the capillary. At that point, it would spill over the capillary and get pumped out, were it not for the small dimensions involved and the surface tension of the mercury. The mercury acts as a temporary Schottky contact that will be used to measure the capacitance as a function of voltage, whereupon the vacuum is released and the mercury is returned to its reservoir. The second electrode for the measurement may be a second, larger mercury dot, or a plate pushed into contact against the backside of the semiconductor. The surface tension of the mercury keeps the probe operating properly as long as the mercury is kept clean.

Dirty mercury will lower the surface tension and cause the loss of mercury to the vacuum line. Mercury probes are inexpensive and useful tools.

Current-voltage (I-V) techniques are available for measuring the electrical properties of diodes and transistors. The details and interpretation for these measurements will be left to the chapters relating to the particular devices of interest (Chapters 6–10). However, the hardware and procedures for the apparatus will be described here. One generally uses a probe station, which is mounted on a vibration-isolation table and is connected to vacuum lines for wafer and probe clamping. The probe station, or prober, typically contains a mounted microscope for viewing the probes with the device of interest. The sample is placed on a moveable (often motorised) stage and clamped with vacuum suction. Probe positioners are likewise clamped with vacuum suction or magnetic bases. Probe tips are mounted on the positioners and are moveable within an area covering several millimetres and can be lowered to contact metallised probe pads connected to the device of interest. If more than a few probes are needed, special probe cards can be built to probe multiple pads on the sample. Special coaxial probes are available for high-frequency measurements. The positioners are connected in a convenient way to the test equipment. The probers often have heated stages that can be enclosed in light-tight and electrically-shielded enclosures. Software is available to interface with test instruments, collect data, auto-step to multiple die on a wafer and map the results. A multitude of test instruments can interface to the probers, including semiconductor parameter analysers and other instruments that offer voltage sources, current sources, voltage probes and current probes as well as oscilloscopes, pulse generators and LCR (inductance, capacitance, resistance) meters.

Several considerations are crucial for getting consistent results with electrical testing. The contact resistance of the probes must be minimised by repeatable methods of placing them down. Proper levelling of the stage and the use of overdrive sensors, and proper probe pressure are necessary for automatic probing. Also, GaAs and other high-gain devices are prone to oscillation and the extended wiring often found in probe stations can often make the ideal external circuit to support low-frequency oscillations. If these are a problem, precautions such as the use of shielded cable or use of bypass circuitry (such as ferrite beads to create an LCR circuit) at each probe to bypass low-frequency oscillations may be necessary.

Another electrical technique of interest will be briefly mentioned for familiarity with its existence only. Deep-level transient spectroscopy (DLTS) is a method of probing the traps in a semiconductor material. Transient measurements, I-V or C-V, are made as a function of temperature and information relating to traps can often be deduced.

2.6 PROCESSING TECHNIQUES

Semiconductor processing techniques consist of three basic steps plus several specialised procedures. The basic steps are sample patterning, thin-film deposition and etching (a material removal technique). Specialised techniques include cleaning procedures, dopant introduction methods such as ion implantation and semiconductor regrowth. Excellent texts have been available for many years offering comprehensive coverage of processing techniques as well as equipment and materials used to implement them [13–17]. It is not our intent to reproduce that material in this book; rather we discuss processing techniques in terms of how they affect GaAs device performance and reliability, with special attention to the surfaces and interfaces in these devices. We present critical processing details for a wide variety of situations where the process can affect device performance and reliability but leave comprehensive coverage of most process techniques to other texts.

Processing techniques referred to in this book are introduced closest to the subject matter where they are first required. For example, photolithography is presented in Chapter 4 and rapid thermal annealing is introduced in Chapter 6 and both are referred to in later chapters as well. The reason for this type of presentation is that processing subtleties and interactions occur for most combinations of two or more process steps and examples of these interactions are best given in the appropriate chapter where the complexity becomes apparent.

The remaining processing techniques presented in this chapter are those of back end processing, which are introduced in the next section.

2.7 BACK END PROCESSING AND ANALYSIS

No presentation of active device processing is complete without context for how these steps fit in within the overall product realisation process. The steps for die separation, packaging, reliability and failure analysis are generic steps that are necessary for the manufacture of many semiconductor devices, not just GaAs devices. With the exception of reliability, the execution of these steps is more similar than different compared to Si devices. Those aspects of reliability testing that have unique bearing on processing GaAs devices will be elaborated on in succeeding chapters where such a discussion is useful and relevant. In this section, a general presentation of the basics of die separation, packaging, reliability and failure analysis is briefly introduced.

2.7.1 Backside processing, die separation and packaging

Once frontside processing is completed, a complete electrical test of all die is usually performed to identify devices that meet electrical specifications. During this process, devices that fail one or more specifications are sometimes "inked" to visually identify them as bad devices. Even though they are tracked through a database from the output of the electrical test, the ink is often used because it makes any mishandling less likely.

The wafer can now be mounted to another substrate for backside operations if the application calls for wafer thinning (as most do). A thinned GaAs wafer is extremely difficult to handle, will deform so it does not remain flat, and therefore requires the use of a host substrate. This supporting substrate is typically glass, quartz or sapphire, whose transparency allows the use of infrared aligners for backside alignment. An adhesive such as wax is used to bond the GaAs to the host substrate. These adhesives are available in spinnable versions for uniform application to the substrate. The frontside of the GaAs may be coated with resist to protect it before applying adhesive. The mounting process itself requires bonding of the frontside of the GaAs to the substrate, usually with the application of pressure within an appropriate fixture.

After bonding of the wafer to the host substrate comes wafer thinning. The wafer/substrate is mounted on a fixture that fits in a lapping machine. The fixture places the GaAs wafer upside down on a large rotating plate coated with a mixture of water, lubricant and grit. The grit is a specific size of particles of a hard material such as alumina, silicon carbide or diamond. Plates are made of a material that is harder than GaAs, so they do not wear away readily, but they do eventually wear and deviate from flatness. They then will need resurfacing or replacing. During lapping, one generally starts with a large-sized grit (15–30 μm) for fast lapping and works down to a fine grit (1 μm or less) as the lapping nears completion. With care, the lapping process can leave the GaAs wafer with excellent flatness.

The final thickness of the wafer depends on the application. Low-frequency and low-power applications for discrete transistors may not require any wafer thinning for electrical or thermal reasons. However, these wafers are often lapped to some extent to make the chips fit properly in their package and to make them easier to scribe and break. Substrates for low-frequency or low-power applications are often lapped to a thickness of 250 μm. High-frequency or high-power applications require thinner wafers. Power amplifiers must have a means of dissipating the heat generated without causing too great an increase in the junction temperature. The best means of accomplishing this goal is to thin the wafer to make a low thermal resistance path to a heat sink in the package. For thermal reasons, thinner is better, but somewhere in the range of 50–100 μm is the practical lower limit for subsequent handling of the dice. Electrical reasons for thinning wafers are related to the design of monolithic microwave integrated circuits (MMIC). The physical size of the transmission lines is proportional to the thickness of the wafer. Circuit size can be minimised by thinning wafers. Also, via holes are impractical to etch in overly thick wafers.

Once lapping is completed, the wafer can be polished. The purpose of this step is to remove mechanical damage and to restore an optical finish for subsequent lithography steps. The polish can use a combination of a very fine grit impregnated in a polishing pad and an etchant. This final step is relatively brief compared to the rest of the wafer thinning process.

Next the wafer is cleaned and further processed. For high-frequency devices and MMICs, perhaps 10 GHz and above, a low-inductance method of grounding a transistor is needed. Wire bonds from a frontside groundpad to a backside ground plane have significant and variable inductance that is problematic for designing microwave circuits. For this reason, via holes to the backside of the wafer are made to provide a direct, low-inductance ground. The lapped wafer, still mounted to a carrier substrate, is patterned with the via hole mask using photolithography. The GaAs is then etched through the substrate until the metal pad on the frontside of the wafer is reached. At this point, the resist is stripped from the backside of the wafer and the wafer is prepared for backside metallisation.

Backside metallisation is necessary to make electrical contact to the frontside of the wafer through the via holes. A good backside metal also allows the finished chips to be attached to a package or carrier with a metal-based die-attach process. An epoxy die attach may be used for low-power and low-cost processes. It is necessary to uniformly coat the backside of the wafer, the sides of the via holes and the bottom with a low-resistance metal in order to ensure good electrical contact to the frontside of the wafer, and sputter deposition is usually a good way to accomplish this goal. After deposition, the metal is usually thickened by electroplating to increase the conductivity of the via hole. In order to facilitate die separation, the wafer is often patterned in order to avoid electroplating in the streets between chips (die).

Once the backside metallisation is complete, the wafer is ready for die separation. Three choices are available for die separation: dicing, sawing and etching. Etching is not often used for die separation and won't be discussed further. Dicing is accomplished by dragging a diamond stylus along the surface of the wafer parallel to one of the cleavage planes. The scratches created by this process initiate fractures along the scribe lines. The scribing process requires dismounting the wafer to finish the cleaving process. The sawing process utilises high-speed circular saws to cut through the GaAs. The same host substrate used in backside processing can be used for sawing. If this choice is made, the chips are separated from the host substrate using solvents to dissolve the adhesive used in the mounting process. This process leaves a pile of chips that need to be sorted.

Instead of sawing with the wafer still mounted to the host substrate, one can remove it first, as with the dicing process. A vacuum wand the size and shape of the wafer holds the wafer from the backside as the host substrate is heated. Then the wafer is pulled sideways to slide the wafer off. After cleaning off the wafer, it is mounted backside down (using another vacuum wand transfer) on sticky, stretchable tape held taut in a frame (for handling). The wafer is then diced or sawed. After scribing, the chips are separated by moving rollers (specially sized metal cylinders, usually) over the wafer and the gentle pressure breaks the dice. This whole process is automated in modern scribe and break machines. Once either sawed or diced wafers are separated on the tape, the tape can be stretched. The stretched tape allows individual die to be removed from the tape. The tape stickiness is sufficient for holding down the wafer, but will allow individual die to be removed with a small force such as the pull of a vacuum wand.

Once the chips are separated, they are ready for packaging. The good die were preferably identified at wafer level test and now can be picked from the stretched tape and placed in a package. In high-volume factories, pick-and-place machines are available to do this job quickly and accurately. Otherwise, tweezers and hand labour are appropriate. The chips are bonded to the package with epoxy, solder or metal-based die attach. The latter two are more appropriate for high-power devices because a metal bond has better thermal conductivity. The solder is available in "preforms", appropriately sized for the die being packaged. The package is then heated to complete the eutectic or solder die attach process.

2.7.2 Reliability

Finished devices must test well and meet all specifications of the end application. They must then meet the customer requirements for ambient operating conditions over a specified lifetime. It is the purpose of reliability analysis to ensure that the devices meet these requirements and to uncover and study degradation mechanisms. Although reliability analysis is a field in itself, reliability studies are not done in isolation because it is not enough to evaluate device longevity and degradation mechanisms. One must also be able to uncover the causes for unacceptable degradation and have the ability to control or eliminate them. Process and material engineers must be intimately involved in the reliability process to engineer robust processes.

One difficult aspect of reliability analysis is the long times required to evaluate a batch of devices. Customers will typically specify operating lifetimes at a maximum operating temperature and bias conditions, often for ten years or more with low failure rates. Manufacturers don't have the luxury of waiting that long to test the parts so they look for ways to accelerate the testing. Acceleration is usually done by using some form of extra stressing of the device, typically using elevated temperature or bias stress. It is often found that device lifetime follows the Arrhenius equation:

$$F = F_o \exp(E_a/kT) \tag{2.22}$$

where F is the median time to failure, F_o is a constant, E_a is the activation energy, k is the Boltzman constant and T is the temperature. The Arrhenius equation applies if a single degradation mechanism is responsible and a single value of E_a represents the activation energy of the degradation process. When EQN (2.22) applies, the failure rate will be linear with the logarithm of T and data from several temperatures can be taken to extract the activation energy. Once the activation energy is known, the median time to failure at the maximum operating temperature can be extrapolated.

Typically three temperatures are chosen to do the study. One chooses a set of samples for each temperature, often with as few as 5–9 samples for initial studies, and conducts the study until half of the samples have failed. For larger populations of samples, a log normal plot of percentage failed versus time is plotted. If the population follows a straight line (the population is said to follow a log normal distribution), one can have greater confidence extrapolating the median failure time without proceeding until half of the devices fail, thus saving precious test time, but at the expense of using more devices and tying up more equipment.

The art and skill of reliability analysis is in dealing with the non-idealities of the real world. Even if the assumption of a single activation energy is valid, one still faces the challenge of finding the right choice of temperatures to use. If one is too conservative with low temperatures, the test may take a very long time. Being too aggressive has its pitfalls as well, since too high a choice of temperature may accelerate a failure mechanism that does not apply at operating temperatures. In order to gain insight into these difficult tradeoffs, a reliability engineer will often resort to a step stress study prior to undertaking a full reliability study. The stressing condition will be stepped at regular (but short) time intervals for a population of devices and the failures and degradations at each time interval are noted. The cumulative failure rate is plotted versus time with regular steps in the stress condition. One may also characterise devices after each stress condition to analyse samples for material degradation, electrical degradation or catastrophic failure. From such analysis, one often gains insight for designing better studies.

Although we have illustrated the principles of reliability using temperature as the stressor, electric field or current density can be equally important stressors for FETs and HBTs (Chapters 8 and 9). In general, one can keep the junction temperature constant by careful manipulation of the ambient temperature and the device self-heating.

The process engineer uses the results of reliability studies to engineer better devices and to get a better idea of which epitaxy, which ohmic contact (Chapter 6), which Schottky contact (Chapter 7) or which process of interest is most robust. Chapters 8 and 9 give some examples of the main reliability issues for GaAs-based FETs and HBTs.

Eliminating changes in device operation during long-term operation is the goal for designing high-reliability devices. Sometimes, if this goal is not totally achieved, another procedure called burn-in can be acceptable. For burn-in, all devices are electrically biased for a short period (up to several hours) of time. The purpose of this procedure is twofold: to allow weak devices to fail quickly and to stabilise devices that drift upon initial operation.

We'll conclude this section with a warning that reliability data is a powerful marketing tool and can be subject to manipulation. Device manufacturers and their customers, original equipment manufacturers (OEMs), engage in fierce negotiations over reliability specifications. The OEM wants to have very few product returns (perhaps specified in returns per thousand) and sometimes places unrealistic demands on its suppliers. The supplier feels this intense pressure to not only produce great reliability, but to do so early in a technology cycle (read new GaAs device) with never enough time to do careful studies and with intense competition with other potential suppliers. Further, the supplier may be in a better position than the OEM to understand the technology tradeoffs and may conclude (or gamble) that many causes of early product returns will be unrelated to the GaAs devices and will agree to impossibly aggressive reliability specifications for GaAs devices. Often the result is an early unrealistic reliability study presented at a conference that generates interest and excitement. The study will be likely to have been accelerated with too high temperatures that give an unrealistically high activation energy and a wonderful extrapolation to lower temperatures showing million hour or better "reliability". The GaAs process engineer should look with a critical eye at all assumptions or lack of details concerning them and learn to understand the difference between reliability hype and careful studies.

2.7.3 Failure analysis

When devices have failed, it is often of interest to determine the failure mechanism, particularly if one has gained some special insight into the device through prior specialised testing, through reliability testing or through a field return of a product. By identifying failures, one can highlight aspects of the process that can be modified for improvement. Many failure analysis techniques were described in Section 2.5. Some of the more useful ones include FIB, SEM, TEM and SIMS. Many specialised techniques have also been developed for failure analysis.

2.8 CONCLUSION

Background material for GaAs processing was presented in this chapter for easy reference and to give an overview of GaAs technology. The material presented includes semiconductor physics, GaAs material growth technology, methods of characterisation and an overview of the entire product realisation process. The explanations and descriptions were simplified and condensed to fit in one chapter. They are not a substitute for in depth study of any particular subject area, but rather intended as a concise reference for material in later chapters or a starting point for further study. The authors also believe that a reader should not have to interrupt reading material to consult detailed reference books on specific subject areas when concise summaries are sufficient for the purpose at hand.

REFERENCES

[1] S.M. Sze [*Physics of Semiconductor Devices* (John Wiley and Sons, New York, 1981)]
[2] S.L. Chuang [*Physics of Optoelectronic Devices* (John Wiley and Sons, New York, 1995)]
[3] M. Shur [*Physics of Semiconductor Devices* (Prentice Hall, Englewood Cliffs, NJ, USA, 1990)]
[4] C.M. Wolfe, N. Holonyak Jr., G.W. Stillman [*Physical Properties of Semiconductors* (Prentice Hall, Englewood Cliffs, NJ, USA, 1989)]
[5] M.R. Brozel, G.E. Stillman (Eds) [*Properties of GaAs, Third Edition* (IEE, London, UK, 1996)]
[6] N.W. Ashcroft, N.D. Mermin [*Solid State Physics* (Holt, Rinehart, Winston, New York, 1976)]
[7] P. Philips [*Advanced Solid State Physics* (Westview Press, Cambridge, MA, 2003)]
[8] P. Rudolph, M. Jurisch [*J. Cryst. Growth (Netherlands)* vol.198/199 (1999) p.325]
[9] J.Y. Tsao [*Materials Fundamentals of Molecular Beam Epitaxy* (Academic Press, Boston, 1993)]
[10] W.G. Breiland, K.P. Killeen [*J. Appl. Phys. (USA)* vol.78 (1995) p.6726]
[11] D.R. Brundle, C.A. Evans Jr., S. Wilson (Eds) [*Encyclopedia of Materials Characterization* (Butterworth-Heinemann, Boston, MA, USA, 1992)]
[12] D.K. Schroder [*Semiconductor Material and Device Characterization, Second Edition* (John Wiley and Sons, New York, 1998)]
[13] R. Williams [*Modern GaAs Processing Methods, Second Edition* (Artech House, Boston, 1990)]
[14] S.J. Pearton, C.R. Abernathy, F. Ren [*Topics in Growth and Device Processing of III-V Semiconductors* (World Scientific, Singapore, 1996)]
[15] S. Mahajan, K.S. Sree Harsa [*Principles of Growth and Processing of Semiconductors* (WCB McGraw-Hill, Boston, MA, USA, 1999)]
[16] M. Madau [*Fundamentals of Microfabrication* (CRC Press, Boca Raton, FL, USA, 1997)]
[17] S.K. Ghandi [*VLSI Fabrication Principles: Silicon and Gallium Arsenide, Second Edition* (John Wiley and Sons, New York, 1994)]

Chapter 3

Cleaning and passivation of GaAs and related alloys

3.1 CHAPTER SCOPE

The surface quality of GaAs is very important in the reproducible fabrication of devices. There are two rather distinct aspects of surface quality: cleanliness and electronic passivation. Both aspects are the topics of this chapter. We begin with a practical discussion of surface treatments aimed at the removal of organic contaminants and native oxides from the semiconductor surfaces. We will then proceed to a review of the electronic properties of the gallium arsenide surface, especially those of its native oxide, and discuss approaches to improve those properties, i.e. to passivate the surface to improve device performance.

3.2 CLEANING AND NATIVE OXIDE REMOVAL

The surface cleaning of GaAs and other semiconductors involves two different aspects. The first is the removal of contaminants, such as organic compounds and metal ions. The second is the removal of the native oxide to expose the bare semiconductor for subsequent processing such as metal contact deposition.

3.2.1 Removal of organic and metal ion contaminants

The surface of a semiconductor wafer can pick up organic contaminants in much more subtle ways than by being handled incautiously so as to allow it to come into direct contact with something greasy (including fingers). There are always some organic vapours in the air of a clean room, and a wafer that has been exposed to them for even a moderate amount of time can become contaminated.

An organic solvent cleaning protocol can be used to remove serious organic contamination. Some protocols are especially suitable for preparing a wafer surface for growth. One such protocol for gross degreasing consists of the following sequences

Degreasing

1) Trichloroethane (TCE)
2) Acetone
3) Methanol

4a) Isopropanol (IPA)

4b) DI Water

of treatments with different organic solvents: 1) boil for 10 min in 1,1,1-trichloroethane (TCE), 2) boil for 10 min in acetone, 3) soak for 10 min in methanol and 4) soak for 10 min in isopropanol (IPA or isopropyl alcohol). This works well for GaAs and InP. For antimonides such as GaSb, which form chemically resistant oxides that do not readily desorb at high temperature, a 10 s oxide removal step with 1 : 1 HCl/H_2O precedes the organic solvent cleans. When the surface is not heavily contaminated by organics, a 10 min boil in acetone, a 5 min boil in methanol and a DI water rinse can be sufficient. Note that heating organic solvents can be a fire hazard. This is especially true with acetone, which has a very low flash-point, so extreme caution should be employed and any potential sources of electrical sparks should be kept well away. Boiling may well not be necessary to achieve sufficient cleanliness; avoid the hazard whenever possible. Sonication using ultrasound can often be used as an alternative to boiling to enhance cleaning efficiency.

For optoelectronic devices, a semiconductor surface may have been patterned by etching prior to a second insertion into an MBE or MOCVD growth chamber for regrowth of additional semiconductor layers. It is especially important in these cases to obtain a clean and oxide-free surface before beginning regrowth or an unacceptable number of defects may exist in the final wafer. While the preceding process removes organic contaminants, the surface layer is still a native oxide rather than the simple binary or ternary semiconductor that one wishes to use as the growth template. If oxide is present on the surface as growth is initiated, a highly defected interface between the former and latter grown materials will be present. In general, this oxide is removed by heating in the growth chamber in the presence of a column V source gas. The specific heating protocol will depend on the growth technique, the semiconductor material, and the composition of material that is to be regrown on the surface; good deoxidation recipes are available in the literature for a variety of materials. An example of an oxide removal process for MOCVD growth on GaAs involves heating under AsH_3 flow at 750°C for 15 min before initiating growth. Similarly, InP can be prepared for growth by heating at 650°C for 15 min under a flow of the P-source gas. Heating over 600°C in the absence of a column V source gas will lead to excess elemental Ga on the surface. Heating under the column V source gas is also effective for MBE growth.

Thermal deoxidation

1) Degrease surface.
2) Heat under column V source gas with no oxygen present.

If oxide formation on the surface is not an issue or if the oxide is to be removed before later processing, atomic oxygen can be used to remove traces of residual organics. Ozone is a good source of atomic oxygen that produces no atomic displacement damage at the semiconductor surface. Commercial ozone cleaning tools are readily available. A very-low-energy oxygen plasma can also be used. This plasma cleaning approach is discussed in greater detail in Section 5.5. The tool employed for this approach is often referred to as an "asher" or barrel etcher. The sample is immersed in the plasma so that the energy of the bombarding ions is less than 10 eV, well below the atomic displacement threshold of approximately 40 eV for compound semiconductors.

For removal of metal ion contaminants, rinsing with chemicals that can form complexes with a wide variety of metal ions can be quite effective. Ammonium hydroxide solutions are an excellent choice for this. NH_4OH is formed by dissolving NH_3 in water: $NH_3+H_2O \leftrightarrow NH_4OH$. The ammonium ion is in equilibrium with ammonia (NH_3) in solution and ammonia is known to form soluble metal complexes, $M(NH_3)_x^{y+}$, with many transition metals. This is highly desirable since transition metals can degrade electronic properties. Two additional advantages of NH_4OH solutions are their ability to remove the native oxide of GaAs and their saponification capability. Saponification is the process of hydrolysing a fat (a fatty acid ester) with OH^- to form glycerol and a fatty acid salt (soap), both of which are highly soluble. This ability to solubilise oils and greases is shared by other hydroxides, such as NaOH and KOH, but these two bases do not form metal ion complexes, and Na-containing solutions are themselves a source of metal ion contamination. It is worth noting that "pure" de-ionised (DI) water (15 MΩ or higher resistance) will also readily dissolve metal ion contaminants.

Removal of Metal Ions
1) NH_4OH rinse.
2) DI water rinse.

3.2.2 Removal of native oxide

While an NH_4OH/H_2O solution will assist in the removal of organic and metallic impurities, its primary role as a surface treatment is in the removal of native oxide from the GaAs surface. Useful dilutions range from 1 : 10 to 1 : 20 NH_4OH/H_2O. Irrigation (sweeping the surface with a stream from a squeeze bottle) for 30 s generally works well. Dilute acidic solutions, such as 1 : 1 HCl/H_2O or dilute H_3PO_4 or H_2SO_4, can also be used. An understanding of the process of native oxide formation and the differences in the surface composition resulting from using different oxide removers can guide the selection of the most appropriate solution.

All the III–V semiconductors are readily oxidised by atmospheric oxygen, so the surface of GaAs and other compound semiconductors that have been exposed to air will be covered by a thin layer of native oxide. This is typically of the order of 1–2 nm after long-term air exposure. Immersion in either acidic or basic dilute solutions will dissolve the native oxide. If no oxidiser, such as H_2O_2, is present, there will be negligible dissolution of the actual III–As semiconductor in the oxide-etching solution. Phosphide semiconductors such as InP and InGaP can etch in HCl-based solutions without the addition of oxidiser. It is worth noting that the main difference between cleaning and etching solutions for III–V semiconductors is the addition of an oxidising agent to form an etching solution that continually produces fresh oxide for the acid or base to dissolve. Consequently, any wafer that has been etched may have an oxide on the surface that is at least as thick and possibly thicker than the native oxide.

Depending on the particular acid or base employed, the surface stoichiometry of the deoxided surface may be different. A surface may be either Ga-rich, As-rich or nearly stoichiometric (equal amounts of Ga and As), depending on whether one element is preferentially removed by the cleaning solution. Acidic cleaning solutions tend to leave an As-rich surface due to their preferential reaction with Ga. Surface studies of GaAs deoxided with HCl and H_3PO_4 show evidence for Ga-Cl and Ga-PO_4 bonding before the surfaces are rinsed with DI water [1]. Water rinsing effectively removes the anions. However, neither HCl nor H_3PO_4 react effectively with As; this results in a surface enrichment of elemental As. When these cleans are performed in a nitrogen atmosphere, the As remains unoxidised. When performed in air, the elemental As can oxidise upon removal from the solution. More As is seen with the HCl solution than with H_3PO_4. As will be discussed in Section 3.3 on passivation, a surface excess of As can be highly undesirable from an electronic perspective.

In contrast to its low solubility in acids, As dissolves readily in basic solutions. Consequently, NH_4OH solutions preserve a more nearly stoichiometric surface (Ga/As $=$ 0.99). The As surplus left by acidic solutions can be removed by a step-wise treatment with HCl, water, NH_4OH and then water. Changing the solvent system can affect the stoichiometric ratio. While HCl/H_2O leaves an As-rich surface, diluting the concentrated HCl with ethanol instead of DI water produces a surface closer to stoichiometric (Ga/As $=$ 0.93) [2]. A solution of $(NH_4)_2S$ will dissolve oxide, but it has the additional property of reacting with the surface to produce covalently bound sulphides. As discussed extensively in Section 3.3, this profoundly alters the electronic properties of the surface.

3.2.3 Regrowth of native oxide

Although the oxide may be effectively dissolved while in solution, a surface oxide begins to form again as soon as the surface is re-exposed to air. For highly reproducible device results, it is important to understand this reoxidation process and to take care to handle wafers in a manner that keeps it as reproducible as possible.

Because an oxide-free surface is hydrophobic, water beads up on the surface as a deoxided wafer emerges from an aqueous solution. Exposure to oxygen and reoxidation of the surface begins as soon as the water withdraws from a portion of the surface. In contrast to water, alcohols such as methanol and ethanol wet the deoxided surface. If an oxide-free surface is really important, one needs to deoxide the surface in an oxygen-free atmosphere. This can be easily accomplished by removing the oxide within a glove box or a glove bag filled with nitrogen or argon. The deoxided wafer should ideally be kept under nitrogen or argon to keep it oxide free. Unfortunately, the transfer of the wafer into a tool for subsequent processing, such as metallisation, will almost certainly re-expose it to some oxygen unless a nitrogen-purged load lock is part of the system. The base pressure of a typical electron beam metal evaporator is a few 10^{-7} torr. At this pressure, the time to adsorb a monolayer of gas is a few seconds. It also takes tens of minutes to pump the system down to that level. A monolayer can form in a few milliseconds at atmospheric pressure. Since most of these tools do not include load locks, some reoxidation is inevitable.

The reoxidation of a deoxided GaAs surface is characterised by at least two types of reaction sites [2]. One site is extremely reactive with even trace oxygen in a nitrogen atmosphere. Surface studies that monitor the O 1(s) signal are more sensitive to the total amount of oxygen present than are the more typical XPS (X-ray photoelectron spectroscopy) studies (FIGURE 3.1) that monitor the oxidation state of Ga and As species at the surface. Oxide removal under nitrogen using 1 : 20 NH_4OH/H_2O or using 1 : 10 conc. HCl/ethanol leave surfaces with 7 and 4 atomic %, respectively, as measured by the O 1(s) signal. Wafers cleaned in air exhibit 2–3 times more residual oxide signals than those cleaned under nitrogen gas originating as boil-off nitrogen from a liquid nitrogen tank (the most common type of nitrogen gas available in fabs and labs). The lower value achieved with HCl/ethanol is probably due to a reduced time of exposure to trace oxygen in the nitrogen gas while the ethanol wets the surface before loading in the UHV chamber for analysis.

Additional sites on the surface reoxidise more slowly over a period of minutes, as measured by changes in the photoluminescence intensity. Photoluminescence as a method for monitoring surface composition is discussed in Section 3.3.2. The most uniform device results may be expected when care is taken to perform subsequent process steps with the times between and the times for those subsequent steps being as uniform as possible following deoxidation. The presence of oxygen at the interface between a

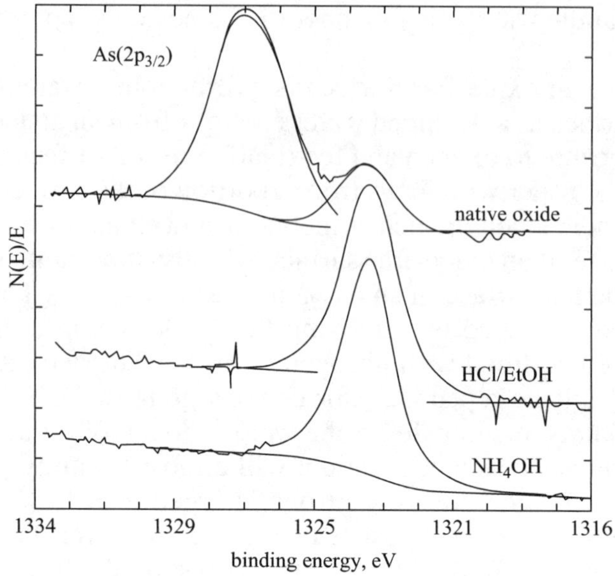

FIGURE 3.1 $As(2p_{3/2})$ XPS spectra showing chemical shift for As bound to O in native oxide versus "oxide-free" surfaces.

contact metal and GaAs can affect the process metallurgy and the formation of defects at the interface, so consistency in cleaning can be important in determining consistency in contact behaviour.

The structure of the surface oxide can depend on how it was formed. When an oxide is grown by thermal oxidation at 420°C to a thickness of about 4 nm, the oxide is Ga-rich with an As gradient that has more As-oxide present as one moves towards the free surface. In contrast, an ECR-plasma-grown oxide of comparable thickness is nearly stoichiometric (equal amounts of Ga and As) and contains a larger amount of As^{+5} versus As^{+3} than does the thermal oxide [3]. These differences may be due to the products of the thermal oxidation being determined by their relative thermodynamic stability while the plasma oxidation is controlled more by the kinetics of the reactions. XPS results suggest that the thermal oxide is somewhat closer to Ga_2O_3 while the ECR plasma oxide is closer to the stoichiometric $GaAsO_4$. In either case, the quality of the semiconductor-oxide surface is equally poor in terms of interface charges in C-V measurements after deposition of PECVD SiO_2 and Al metallisation. Composition, structure and electronic surface quality are the topics of Section 3.3.

3.3 PASSIVATION OF GaAs

The term "passivation" carries with it some rather distinct yet related connotations. The first deals with the alteration of the surface of the semiconductor to achieve more favourable electronic properties than those obtained with the native oxide. The second has to do with protection from reaction with the ambient over the lifetime of the device. Since reaction with the ambient, especially oxygen, can alter the electronic properties, even a device with excellent initial operating characteristics may gradually degrade over time. The third common use of the term "passivation" relates to the deactivation of dopants or defects through reaction with readily diffusing species such as H atoms. Surface treatments and especially plasma processing conditions can be a source of hydrogen that can diffuse deeply into the semiconductor before reacting with a dopant or a defect. While defect passivation is generally desirable, dopant passivation alters free carrier concentrations and can negatively affect device performance. This third type of "passivation" will be discussed in Section 5.7.2.

We will begin this section with a discussion of the physical and chemical issues that affect the important surface and interfacial electronic properties of GaAs. Since reaction with the ambient, especially oxygen, can alter critical electronic properties, we will conclude with a discussion of some methods that seek to isolate the surface from such effects.

3.3.1 Electronic properties of the GaAs surface

The bulk electronic properties of semiconductors in general and GaAs in particular were discussed in Section 2.2. Many of the concepts, such as band structure and bandgap, are derived assuming the three-dimensional translational symmetry of an infinite bulk crystal. However, the presence of a surface immediately introduces different boundary conditions that make the bulk description no longer adequate. The most significant effect of the surface is the introduction of additional energy levels that may lie within the forbidden energy gap between the valence and conduction bands. The number and energy of these levels, called surface states, will determine how important they are in controlling the electronic behaviour near the surface.

An abrupt termination of the three-dimensional crystal lattice would leave surface atoms coordinatively unsaturated with many unpaired electrons; this would produce an unacceptably high-energy state. In reality, this condition is avoided by either surface reconstruction or reaction with other chemical species that may be present.

In surface reconstruction, the surface atoms move from their hypothetical bulk positions into positions where they can bond with each other to reduce the total energy. There are a variety of different reconstructions that can occur, depending on the particular exposed crystal face and whether the surface is primarily terminated by Ga or As atoms. Since the GaAs(100) surface is of the greatest technological importance, we will restrict most of our discussion to that crystal face. However, any mesa-type device will have other crystal faces exposed on the sidewalls. The ideal GaAs(110) surface, which is a natural cleavage facet and can be obtained as vertical facets by crystallograpically selective etching (Section 4.3.2), is supposed to be free of surface states within the bandgap; however, real device facets can be expected to deviate somewhat from ideal since they are not perfectly planar. The kinks and steps found on surfaces that are not planar at the atomic level can introduce additional energy states. Many studies of the surface structure of GaAs(100) using LEED (low-energy electron diffraction) or RHEED (reflection high-energy electron diffraction) have been reported. From the LEED electron diffraction patterns, it is possible to determine the crystal symmetry of the surface layers; from the LEED geometry, the possible atomic arrangements are deduced. There are a number of reported reconstructions for GaAs(100) depending on the history of the surface, and they have different work functions, i.e. energies relative to the vacuum level. This is to be expected since they will have different surface electronic states that determine the work function. Some surface geometries are shown in FIGURE 3.2. While it is possible to measure structures of carefully prepared samples under ultra-high-vacuum (UHV) conditions ($<10^{-8}$ torr), it is important

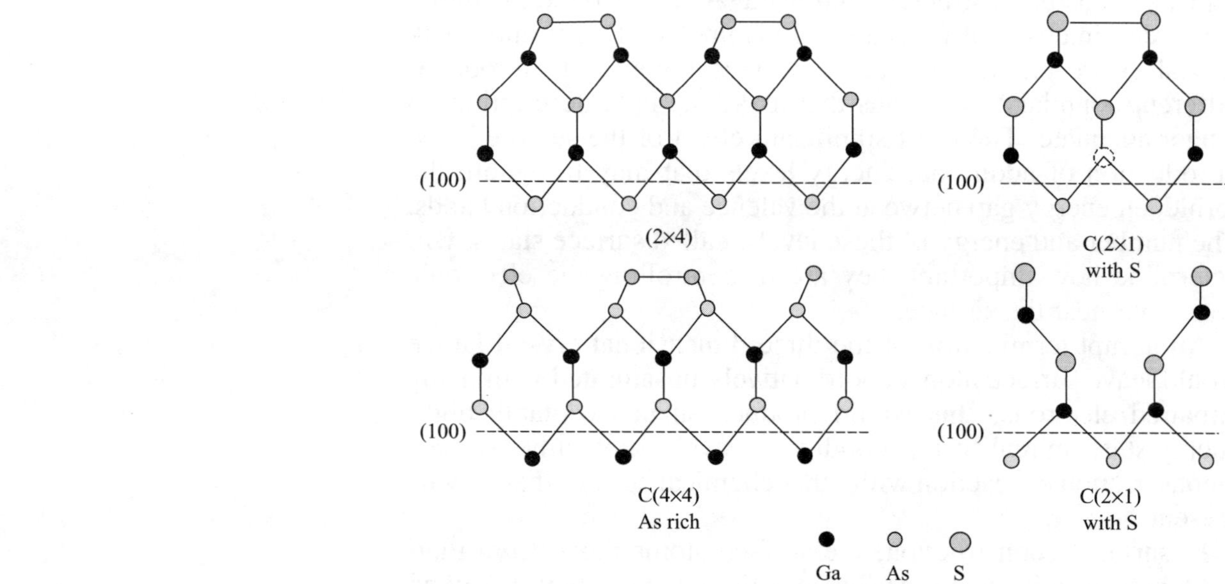

FIGURE 3.2 Side view of C(4×4) and (2×4)GaAs(100) and C(2×1)S of sulphided surface.

to remember that the surface of an actual device will not usually have a perfectly planar topography, exposing only one crystal plane, nor will it be in UHV. Therefore, great care must be taken in concluding which scientific studies will map appropriately onto the situation for a particular device.

The thermal history of the wafer can profoundly influence its surface structure. It is common to describe a surface geometry using $(A \times B)$ nomenclature. For example, the (2×4) surface reconstruction refers to a two-dimensional unit cell that is $2\times$ larger in one direction and $4\times$ larger in the other direction compared to the bulk unit cell. If a GaAs wafer capped with amorphous As in an MBE machine to prevent oxidation is heated in vacuum to 380°C, it assumes a $C(4 \times 4)$ geometric structure with 1 monolayer of adsorbed As. If it is further heated to 480°C, it assumes a (2×4) geometry. Further heating to 600°C without an As overpressure converts it to (4×1) with a slight excess of Ga on the surface [4]. Reactions with species such as oxygen or sulphur produce other changes in the surface geometry. While it is possible to obtain well-defined geometries under controlled vacuum conditions, great care must be taken in extrapolating the results to real-world wafers undergoing device fabrication. The character of the surface may be very different from that described in the literature, and very different behaviour may result when that is the case. Nevertheless, such studies can provide starting points for thinking about the condition of a particular wafer.

An alternative way to lower the surface energy is by reaction with the ambient, especially with oxygen, to form an oxide surface. This will generally be the condition of the GaAs surface unless deliberate effort has been taken to remove the oxide and replace it using a more favourable surface reaction. A major goal for many years has been the achievement of an insulator/III–V interface of the same quality as the nearly ideal SiO_2/Si interface that is the foundation of CMOS technology. Unfortunately, unlike the SiO_2/Si interface with its extremely low density of interface states (as low as $10^9/cm^2$) and a very high-quality oxide yielding trapped charge densities of about $10^{10}/cm^2$, the oxides of III–V semiconductors suffer from many interface states that are located within the bandgap. This is true whether the oxide is formed thermally or by plasma oxidation. For GaAs with its native oxide, the density of these interface states is in excess of $10^{13}/cm^2$, roughly one state for every 100 surface atoms. This high interface-state density results in Fermi level "pinning", as discussed below, and has so far precluded a successful metal-insulator-semiconductor (MIS) technology in the III–V systems. The majority of "passivations" involve reactions whose products are meant to be more electronically favourable than the native oxide.

Replacement of the native oxide with an alternative interfacial species is essential because the formation of the native oxide itself can generate defects that ensure poor electronic behaviour [5]. During the initial period of oxidation with atomic oxygen, even before a general disruption of the GaAs lattice or the formation of unstable arsenic oxides that can liberate elemental As (Section 10.2.1), oxygen atoms displace the top-layer As atoms and bind to the second-layer Ga atoms. Scanning tunnelling microscopy (STM) of GaAs (100) with the (2×4) surface structure has shown that the displaced As atoms localise in surface trenches and on top of surface As-dimer rows to place As atoms in what would normally be Ga sites, forming the donor As_{Ga} antisite defect and an acceptor state that may be associated with an undimerised As atom with a half-filled dangling bond. Increases in total O coverage lead to preferential formation of oxygen dimers on the surface with reduction of the undimerised As sites. Such dimer formation appears to be the natural step for oxidation with molecular oxygen. In addition, metallic clusters of As atoms can form as the oxygen reaction proceeds. With sufficient oxidation, the Fermi level is pinned for p-GaAs somewhere between 0.75 and 0.5 eV, which are the energy levels for singly and doubly ionised As_{Ga} antisite defects.

Vapour deposition of Ga_2O_3 oxide onto GaAs in vacuum can minimise the formation of As-oxides. This is significant because As_2O_3 is thermodynamically unstable in contact with GaAs, with the reaction

$$As_2O_3 + 2GaAs \rightarrow Ga_2O_3 + 4As$$

being thermodynamically favoured ($\Delta G^0 = -62$ kcal/mol). Vapour-deposited Ga-oxides [6] have shown some of the best electronic behaviour to date (FIGURE 3.9), but problems still remain (Section 3.3.5).

An important point to remember when considering surface reactions is that even a partial monolayer of reacted material can dominate the electronic structure of the surface or interface. The chemical bonding of the surface monolayer defines the energetics of the interface states and determines whether they lie within the bandgap (undesirable) or not (highly desirable). Therefore, the preparation and cleaning of the surface before any passivation process may totally dominate the results and great care must be taken to ensure reliable and reproducible results. When differences between literature predictions and personal fabrication results occur, they can often be traced to differences in the starting surface condition.

A word of caution is in order about relying simply on surface stoichiometry or geometry as a predictor of electronic properties.

The surface state energies and the density of states at each energy are not merely dependent on what chemical elements are at the surface. Details of the bonding configuration of those atoms will determine the energetics of the surface, and the same nominal composition can display quite different bonding. The standard methods of identifying surface chemical composition, like Auger electron spectroscopy (AES), X-ray photoelectron spectroscopy (XPS), secondary ion mass spectrometry (SIMS) or energy-dispersive X-ray analysis (EDS, EDX or EDAX) typically are not sensitive to species that are present below a tenth of a percent. However, elements present at even lower levels can be not only important but even dominant in determining surface electronic properties if they are properly positioned within the forbidden gap and exhibit rapid recombination kinetics. Additionally, it is possible for surfaces of different elemental compositions to assume the same geometric pattern as measured by LEED, so having the same structure should not be construed as convincing evidence of the same chemical composition and same electronic properties.

Either reconstruction or chemical reaction will introduce surface states that generally serve as traps for free carriers. These levels may be quite discrete or they may be spread over a band. Both electron and hole surface traps are known. Elements that are more electronegative than Ga and As, such as oxygen, act as electron acceptors. The more electronically deleterious states tend to lie toward the middle of the gap and are called "midgap states".

3.3.1.1 Midgap states and Fermi level pinning

The Fermi level and its position relative to the valence band maximum, VBM or E_V, and conduction band minimum, CBM or E_C, are discussed for bulk semiconductors in Section 2.2.2. The presence of midgap states at the surface or interface can have a profound effect on the Fermi level position relative to the VBM and CBM near the surface, as illustrated in FIGURE 3.3 for an n-type semiconductor.

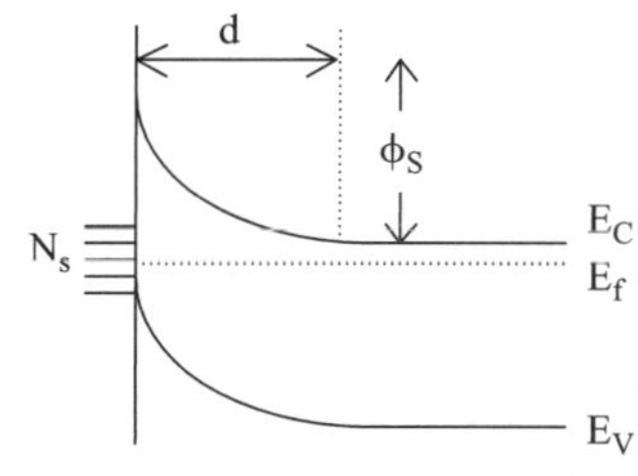

FIGURE 3.3 Fermi level pinning at energy of midgap surface states.

When there are surface states lying below the Fermi level, they will be filled by electrons from the conduction band. When there are N_s states per cm^2 below the Fermi level, the surface will acquire a negative charge equal to $-eN_s/cm^2$. This will repel electrons from the surface region and produce a depletion region nearly devoid of free electrons which will have a positive space-charge approximately equal to $+eN_dd$, where N_d is the donor atom concentration and d is the depth of the space-charge region. There is a strong electric field directed towards the surface that opposes the flow of more electrons into the unoccupied surface states. The Fermi level in the bulk is aligned with the top of the filled surface states, and the energy of the surface is raised relative to the bulk by the surface potential, ϕ_s. An exact determination of ϕ_s requires the solution of Poisson's equation with appropriate boundary conditions. For simplicity, one may assume that the space charge ρ is equal to $eN_d d$ from the surface to a depth, d, where $d = N_s/N_d$. The surface potential, ϕ_s, is then given by $eN_s^2/2\epsilon N_d$, where ϵ is the semiconductor permittivity. The energy by which the bands are raised at the surface is approximately $\Delta E = e\phi_s$. If the surface potential is sufficiently high that the valence band at the surface becomes closer to the Fermi level than the conduction band is, the near-surface region becomes effectively p-type and an inversion layer is formed. The response of these various quantities to externally applied voltages is discussed in Section 7.2.1.

When there is an extremely high density of midgap surface states, as in the case of GaAs(100), the Fermi level is effectively pinned at the energy corresponding to that of the midgap states. Changing applied voltages may merely move carriers into and out of surface states, leaving the position of the Fermi level unchanged and "pinned" at the surface-state energy. The primary goal of surface passivation is the unpinning of the Fermi level. Changing the chemical composition of the surface layer will change the energy of the dominant surface states. If the new surface states are moved close to the VBM or CBM or, even better, are moved out of the forbidden gap, the Fermi level can become unpinned. However, changing the surface composition may serve merely to pin the Fermi level at another position within the gap if there is still a high density of midgap states at the new energy position.

Fermi level pinning can occur at a wide range of interfaces. It is seen at "clean" semiconductor surfaces grown by MBE, naturally oxidised surfaces, chemically modified surfaces, semiconductor/insulator interfaces, metal/semiconductor interfaces and regrown semiconductor/semiconductor interfaces. While the remainder of this chapter will be focusing on chemical modification of exposed surfaces, most work on Fermi level pinning has been directed towards the understanding of the metal/semiconductor interface because of its importance in device performance.

A number of models have been put forth to explain the origin of Fermi level pinning and its effect on the Schottky barrier height [7] of Schottky contacts (Chapter 7). These include the unified defect (UDF) or advanced unified defect (AUD) models of Spicer et al., the metal-induced-gap state (MIGS) model of Heine and later Tersoff, the disorder-induced gap state (DIGS) model of Hasegawa and Ohno, and the effective work function (EWF) model of Woodall and Freeouf. The proposed origin of the pinning states varies with the model employed. In the UDF model, pinning is ascribed to deep levels related to stoichiometric variation, especially the As antisite. In the MIGS model, the midgap states originate in the penetration of the metal wave function into the semiconductor. In the DIGS model, the pinning states result from the loss of two-dimensional periodicity due to disorder of the bonding at the interface. In the EWF model, pinning occurs at metallic clusters formed by precipitation of As at the interface. The most appropriate model for a particular case varies with the details of the interface.

It is important to remember that band bending is caused by the total surface charge. This charge is the combination of surface-trapped charge due to surface states lying within the forbidden gap and surface-fixed charge, i.e. ionic charge on the surface that may not be related to the surface traps. With only a surface-fixed charge, the Fermi level can still move freely. In contrast, with sufficiently high densities of surface-trap states, the Fermi level is pinned and neither bulk doping nor externally applied electric fields can move it.

Band bending due to surface-fixed charge can repel minority carriers sufficiently to reduce the surface recombination velocity (Section 3.3.1.2). The change in the surface recombination velocity as a function of forward injection parameters can provide a measurement of surface-fixed charge and help distinguish contributions from fixed charge and surface states. Better results will be obtained if injection of carriers by absorption of light is avoided.

3.3.1.2 Recombination and surface recombination

The free carriers in a semiconductor participate in a dynamic process with the constant generation and recombination of individual carriers determining their concentration in a particular region of the material. The details of these processes are outside the scope of this book and are discussed quantitatively in many books on semiconductor and device physics [8]. The free carrier concentration in a particular region of the wafer or device is determined by the balance between generation and recombination. Dopant atom concentrations, temperature and illumination are the factors that dominate the generation process. The equilibrium concentrations of electrons and holes are discussed in Section 2.2.2 for the situation without illumination. Illumination will alter these concentrations by producing an equal number of electrons and holes when and where photons are being absorbed in the material. The location of creation depends on the absorption coefficient of the semiconductor; there is an exponential decrease in the number of photogenerated carriers as the light penetrates deeper into the material. The carriers generated within a region of depth dx is given by

$$\text{carriers} = \alpha I_0 \exp[-\alpha x]\, dx \tag{3.1}$$

where I_0 is the incident light intensity, x is the depth and α is the absorption coefficient. The value of α is calculated from the complex index of refraction (n + ik) as $\alpha = 4\pi k/\lambda$. The optical constants of many semiconductors for visible through UV light at room temperature are found in Aspnes and Studna [9]. When the exciting light possesses energy greatly in excess of the bandgap, the resulting carriers are initially created at sufficiently high energy that there can be carrier multiplication during their relaxation to the ground state. This may require consideration when short UV light is employed.

	Bandgaps at 300 K		
	eV	Type	$dE_g/dT \times 10^4$
AlP	2.43	i	−3.5
AlAs	2.16	i	—
AlSb	1.6	d	−4
GaP	2.25	i	−5.5
GaAs	1.43	d	−4.5
GaSb	0.69	d	−4.1
InP	1.28	d	−4.6
InAs	0.36	d	−3.3
InSb	0.17	d	−2.9
ZnS	3.8	d	−4 to −5
ZnSe	2.58	d	−7.2
ZnTe	2.28	d	−5
CdS	2.53	d	−5
CdSe	1.74	d	−4.6
CdTe	1.50	d	−4.1
Si	1.11	i	−2.3
Ge	0.67	i	−3.7
Ga_2O_3	4.3		
Al_2O_3	8–10		
SiO_2	9		

For a direct-gap semiconductor like GaAs, with an absorption coefficient in excess of 10^5/cm, a pronounced gradient in photogenerated carrier concentrations exists. For indirect semiconductors like Si and GaP, the absorption coefficients of <100/cm at energies near the bandgap lead to a relatively uniform generation of carriers throughout the material. Where light is absorbed, there is an increase in free carriers, leading to an increase in conductivity. While the light continues to be absorbed, the carrier concentrations remain at new, higher equilibrium values. When the light is removed as a carrier-generation source, the system decays back to its thermal equilibrium populations. The time dependence of the photoconductivity can be used to measure recombination lifetimes that include both radiative and non-radiative contributions. Photoluminescence measurements can also be used to study the recombination processes non-destructively in wafers that have not been fabricated into device or test structures. This will be discussed further in Section 3.3.1.3.

In general, there are three very important mechanisms whereby carriers can recombine to restore and maintain an equilibrium concentration. The first is direct recombination in which the energy released by the recombination is emitted as light. This may or may not also involve the emission or absorption of lattice phonons. The radiation may be from direct band-to-band recombination emitted at the bandgap energy or, at lower temperatures, the free carriers may form an exciton (an electron-hole pair) that emits at an energy slightly less than the bandgap energy. Radiation can be emitted from "free excitons" and from excitons bound to shallow donors or acceptors; the characteristic energy of emission can identify the type of exciton that is the source of the light. The typical radiative recombination lifetime for a direct-gap semiconductor is of the order of 10^{-9} s.

The second is trap-mediated recombination. Either holes or electrons can be "trapped", i.e. captured into a deep-level trap whose energy often lies near the middle of the forbidden energy gap. This occupied trap subsequently captures a carrier of the opposite type. When the trap energy is near the middle of the gap, recombination is non-radiative. Since crystal defects, such as As antisites, As or Ga vacancies, lattice dislocations and impurity atoms are distributed throughout the crystal, this is a characteristic mode of bulk non-radiative recombination.

The third is surface recombination. Similar to bulk non-radiative recombination, this is mediated by capture of carriers into deep levels. The primary distinction is that the traps are associated with the surface. While the radiative recombination and the bulk trap-mediated recombination are inherent properties of the wafer as grown or as modified by dopant in-diffusion or ion implantation, the surface recombination is potentially variable by the manner in which the surface is treated during processing. As will be discussed later in Chapter 5 on Dry Etching, near-surface lattice damage by energetic ions during dry etching increases surface-recombination rates and can thereby degrade device performance.

The situation is generally described for an n-type semiconductor as follows. For a rigorous mathematical treatment, we can refer the reader to most textbooks on semiconductor materials. Auger recombination processes are ignored in this discussion. The simplified presentation below is merely for illustrative purposes.

Recombination rates are proportional to the excess minority-carrier concentration since both carrier types are required for recombination to occur and there is, by definition, an ample supply of the majority carriers. For an n-type semiconductor, the net recombination rate per unit volume is given by

$$R_{tot} = (p - p_0)/\tau_p \tag{3.2}$$

where τ_p is the lifetime of the minority carriers. Direct recombination is proportional to the number of free electrons and the number of holes in the valence band

$$R_{dir} = k_{dir} np \tag{3.3}$$

In direct recombination, both the energy and momentum are conserved without the generation of any phonons to enable the conservation. While direct recombination is very important for light-emitting devices, it is not a surface-mediated process and we will not discuss it further here.

Trap-mediated recombination depends on the trap concentration and trap energy. Traps can be in the bulk or at the surface. Some general considerations apply to all trap types. To promote recombination, both types of carrier must be trapped in very close proximity at the same time. A trap can capture a carrier that passes through a certain area near it; this is the capture cross-section (σ_n and σ_p) and may differ for electrons and holes but is generally of the order of atomic dimensions, i.e. 10^{-15} cm^2. For capture to occur, the trap must be unoccupied by the particular type of carrier. Therefore, the rate of carrier capture is proportional to the concentration of traps that are unoccupied. For a trap concentration, N_t, the concentration of unoccupied traps is determined by the Fermi function, F_t, for the probability that the trap will be occupied. For electrons, the concentration of available traps is given by

$$N_{avail} = N_t(1 - F_t) \tag{3.4}$$

where

$$F_t = 1/(1 + \exp[(E_t - E_F)/kT]) \tag{3.5}$$

with E_t being the trap energy and E_F being the Fermi energy. The capture rate for electrons is given by

$$R_{capture} = v_{th}\sigma_n n N_t(1 - F_t) \tag{3.6}$$

where v_{th} is the thermal velocity of the carrier. Carriers can escape or be emitted from the traps before the opposite carrier type is captured. Although it is common to use the term "recombination centre", the more accurate term is "generation-recombination centre" since both processes can occur. At equilibrium, the rates of escape (emission) and capture are equal and the proportion of occupied traps is

$$\text{Electron-occupied traps} = N_t E_t \tag{3.7}$$

When the trap energy is closer to the conduction band, the electron has a great probability of escaping the trap before a hole is trapped and recombination can occur. This increases the efficiency of midgap states for promoting non-radiative recombination relative to states with energies nearer the conduction band.

Similar equations can be written for holes. The rate of hole capture by an electron-occupied trap is given by

$$R_{hole\ capture} = v_{th}\sigma_p p N_t F_t \tag{3.8}$$

while the rate of hole escape is given by

$$R_{hole\ escape} = v_{th}\sigma_n n_i N_t(1 - F_t)\exp[(E_i - E_t)/kT] \tag{3.9}$$

As in the case of electrons, hole escape is more likely to occur before recombination can annihilate it if the trap energy is nearer to the valence band. For both holes and electrons, trap energies in the middle of the forbidden gap are especially effective for promoting recombination. The net recombination rate at traps under low-injection conditions, assuming that $\sigma_n = \sigma_p = \sigma_0$, can be approximated as

$$R_{trap} = \frac{v_{th}\sigma_0 N_t(np - n_i^2)}{p + n + 2n_i \cosh[(E_t - E_i)/kT]} = \frac{p - p_0}{\tau} \qquad (3.10)$$

In the dark, n and p are the thermal equilibrium values, n_0 and p_0. Under illumination, these both increase to $n = (n_0 + \Delta n)$ and $p = (p_0 + \Delta p)$ where $\Delta n = \Delta p$ are the photogenerated carriers. When the light is removed, recombination proceeds to restore the thermal equilibrium concentrations.

Key points to note are the necessity for trapping both types of carriers at the same time and the dependence on the position of the Fermi level relative to the trap energy. Midgap traps are especially effective at promoting recombination.

Surface recombination is a special case of trap-mediated recombination. The loss of the third dimension leads one to discuss the effectiveness of a particular surface condition in promoting recombination in terms of a surface-recombination velocity, S (to accommodate the cm/s units), rather than a lifetime, τ, with 1/s units.

The net recombination rate at the surface at low injection, assuming equal cross-sections for holes and electrons, can be approximated by

$$R_{surf\ recomb} = \frac{v_{th}\sigma_s N_{ss}(p_s - p_0)}{1 + [2n_i/(n_0 + p_s)]\cosh[(E_{ss} - E_i)/kT]} = S(p_s - p_0) \qquad (3.11)$$

where p_s is the hole concentration at the surface and E_{ss} is the energy of the surface state. As with bulk traps, those states located in the middle of the gap are most effective in promoting recombination.

The equilibrium free carrier-concentration in the near-surface region can be markedly altered from that in the bulk through recombination mediated by surface states. This can be an especially serious problem for optoelectronic devices, since surface recombination is non-radiative in character and reduces the carrier supply for light emission through radiative recombination.

HBTs (Chapter 9) can also suffer when non-radiative recombination increases since the gain of the transistor is reduced because extra base current is required to provide the holes for recombination. While deep-level defects, such as EL2, within the bulk of the material can also promote non-radiative recombination, the mature state of GaAs technology has reduced the severity of this problem. While deep-level defects such as EL2 can promote non-radiative recombination, very little EL2 is produced during high-quality epitaxial growth. While EL2 is still present in bulk crystals, it generally does not migrate into the epitaxial layers. In present-day bulk growth, its concentration is near $10^{16}/cm^3$ and the uniformity is controlled to provide a uniform response to ion implantation. The solution to the surface state problem is not dependent on excellent growth methods because the number and character of surface states can be, and generally are, altered by each step in the fabrication sequence for a GaAs device. Midgap surface states can be especially deleterious due to their high efficiency at promoting recombination and the presence of the surface field that sweeps carriers towards the surface as long as they lie within the depletion depth (d in FIGURE 3.3). The continuing effort to find effective passivations to reduce the number of midgap states is directed at this serious problem.

The importance of a low density of midgap states is evidenced by the extreme differences in surface-recombination velocity between the SiO_2/Si and native-oxide/GaAs interfaces. Typical surface-recombination velocities for Si are less than 100 cm/s, with values even as low as 0.1 cm/s having been obtained. In contrast, typical surface-recombination velocities for GaAs range from 10^5 to 10^6 cm/s; this minimum 1000-fold difference makes the proximity of a surface a major factor in the operation of GaAs devices. GaP typically has surface-recombination velocities that are 3–10 times lower than GaAs while InP has values near 10^3 cm/s. Since surface-recombination velocity can be thought of as essentially an inverse lifetime of the carrier at the surface, the deleterious effect of such large values is obvious. A fruitful approach to reducing surface recombination velocity effects is seen in the ledge passivation of HBTs using GaAs/InGaP instead of bare GaAs and achieving a higher current gain ($\beta = I_c/I_b$).

3.3.1.3 Photoluminescence as a diagnostic of surface quality

Because of the ability of midgap surface states to shift the balance between radiative and non-radiative recombination, photoluminescence (PL) is a widely used technique to measure changes in surface-state densities that accompany effective passivation. The non-destructive nature of PL and its ease of use have made it a widely and sometimes incorrectly used tool for preliminary assessment of the quality of a particular passivation approach. The object of this section is to discuss how one can avoid an excessively simplistic use of PL that can lead to wrong conclusions about surface quality.

PL spectroscopy generally employs a relatively monochromatic light source, such as a laser, to photoexcite an excess of free carriers (equal numbers of electrons and holes); it then monitors the intensity and wavelength of light that results from the radiative recombination of the free carriers. A variation is photoluminescence excitation (PLE) spectroscopy, where the wavelength of the excitation source is swept while monitoring the luminescence output; this allows identification of specific absorption processes that lead to free carriers.

As mentioned in Section 2.5.1, PL spectroscopy can be used to identify shallow-level impurities and to qualitatively gauge the number of deep-level defects in the bulk of a wafer. Here we will focus on its use to measure surface quality and, more specifically, changes in surface quality due to passivation treatments.

Important factors for interpreting PL studies

1) Exciting wavelength.
2) AR coating effects.
3) Loss of passivant by photolysis.
4) Possible H effects on defects and dopants.

Photoluminescence intensity measurements are best used to provide a semi-quantitative rather than an absolute comparison between different treatments of samples. In general, the intensity of band-edge emission is monitored at a single wavelength corresponding to the bandgap energy with the slits on the spectrometer set with sufficient width that slight shifts in the band-edge will not affect the total intensity measurement. Intensity measurements are most informative when all the samples have been derived from a single wafer. Quantitative comparisons between different wafers can be seriously misleading due to differences in the balance between radiative and non-radiative recombination in the bulk resulting from differences in the nature and number of bulk defects in different wafers and also due to differences in dopant concentrations. This may be true even for wafers cut from the same crystal boule.

There are several very important factors to consider in analysing the meaning of PL measurements of surface quality. The most important experimental fact to consider in understanding the meaning of a particular set of PL experiments is the doping level of the material being studied. It is generally assumed that the detected light is being emitted from sample depths that are deeper than the depletion region, d, and that changes in intensity largely reflect changes in the depletion depth. Simplistically, $d = N_{ss}/N_{d,a}$, where N_d and N_a are the concentrations of donors and acceptors, respectively, and N_{ss} is the concentration of surface states. This assumption is made because carriers that are photogenerated in the depletion region are swept to the surface by the surface field and effectively consumed there by non-radiative recombination. The PL intensity, I_{PL}, is proportional to the excitation intensity at a particular depth and is given by

$$I_{PL} \propto I_0 \exp[-\alpha d] \qquad (3.12)$$

where α is the absorption coefficient at the excitation wavelength and I_0 is the incident light intensity. The normalised PL intensity ratio is given by

$$\text{PL ratio} = \exp[-\alpha(d_{\text{passivated}} - d_{\text{native oxide}})] \qquad (3.13)$$

This value will be larger for a given surface-state density when the doping level of the material is lower. There can be orders of magnitude variation in the normalised PL for slightly doped versus highly doped materials that have the same surface state density. A lightly doped or semi-insulating sample may display a 100-fold change in PL intensity while a heavily doped material with the same change in surface state density may only increase in PL intensity by a factor of 2.

The same wavelength must also be used to directly compare the results of two studies without calculating the effect of different penetration depths of the excitation light. For example, excitation at 458 nm (Ar laser) will not excite the sample to as great a depth as would excitation at 633 nm (HeNe laser) due to the larger absorption coefficient at 458 nm. Consequently, the light with the larger absorption coefficient will be more sensitive to changes in the depletion depth and will make a passivation appear more effective than it might actually be if this is not taken into account.

The intensity of the excitation light should be maintained relatively low since many species, including sulphur, can be photolytically released from the surface and higher intensities accelerate this problem. Thus, the optical measurement itself can degrade the surface, especially if the illumination is performed in the presence of oxygen. It is a wise policy to flow a curtain of nitrogen gas across the surface of the material while it is being illuminated to prevent photolysis and loss of the passivant.

An additional factor that must be considered is whether there is an alteration of the apparent PL intensity due to the presence of thin films of the appropriate thickness to serve as anti-reflection (AR) coatings. When light strikes the surface of a semiconductor, a fraction enters the material to be absorbed and another fraction is reflected from the typically specular surface of the wafer. For an uncoated GaAs wafer, typically 35% of the incident visible light is reflected. However, when a film thickness closely corresponds to a multiple of (incident wavelength/4), the reflection from the surface is reduced to nearly zero and the effective excitation intensity is increased. When passivants of some appreciable thickness are being applied or when a monolayer-type passivant is being encapsulated with, for example, a dielectric to increase its longevity, the proper interpretation of any PL intensity changes requires a knowledge of the excitation and emission wavelengths and the film thickness in case there are AR coating effects involved.

An additional effect that may lead to misleading PL intensities is the presence of atomic H during the treatment of the wafer. If the H diffuses into the material and binds to defects and/or dopants, thereby reducing bulk non-radiative recombination, the characteristic PL intensity of the bulk material can change in an uncontrolled and generally unquantifiable manner. This will make PL intensity ratios essentially uninterpretable. While this H-effect is "passivation" in the third sense as described in the introduction to this chapter, it is not the goal of the surface treatment being applied and may be a hindrance rather than a help in obtaining good device operation.

For heavily doped materials, an alternative spectroscopic approach can be used to measure changes in the depletion depth. When doping levels exceed about $3 \times 10^{17}/cm^3$, Raman scattering results from both the optical phonons (LO and TO modes) and from plasmon modes, commonly designated L_+ and L_-. These phonon modes are shifted to higher and lower frequencies relative to the LO mode depending on the doping level of the material (FIGURE 3.4). The relative intensities of the coupled plasmon modes, L_+ and

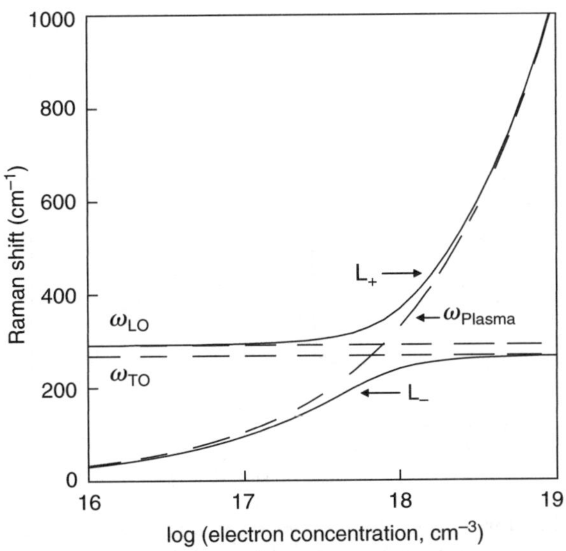

FIGURE 3.4 Doping level dependence of plasmon modes of GaAs.

L_-, change as the depletion depth changes. It is assumed that the reduced carrier concentration in the depletion region results in only the uncoupled LO phonon at 293 cm^{-1}, typical of GaAs with donor levels below about $2 \times 10^{17}/cm^2$ being present in the depletion region, and that the coupled L_+ mode is present in the bulk. Modelling the relative intensities of the two provides a measure of the depletion depth. In addition, shifts in the L_+ frequency can reveal when the passivant is also serving as a donor (FIGURE 3.5). For example, we have observed that photodeposited sulphur acts as a donor after the deposition of an ECR-plasma silicon nitride on top of the passivant and increases the near-surface doping level from 1.9×10^{18} to $2.2 \times 10^{18}/cm^3$, as measured by the L_+ line in Raman spectra excited at 514.5 nm; a more electropositive form of sulphur also appears in XPS spectra. Depletion depths were observed using the L_-/LO intensity changes at 458 and 514.5 nm to decrease from 63 nm, corresponding to a surface-state density of $1.1 \times 10^{13}/cm^2$, to 18 nm, corresponding to $3.3 \times 10^{12}/cm^2$ surface states. An increase in depletion depth to 85 nm, indicating a surface-state density increase to 1.5×10^{13}, occurred with the plasma deposition of the silicon nitride without any sulphur. This demonstrates the ability of plasma exposure to degrade the surface.

PL intensities and Raman ratios are actually indirect measurements of the band bending, so other things besides surface states that affect band bending can also alter the PL or the Raman spectrum. It is important to remember that band bending is caused by the total surface charge. This charge is the combination of

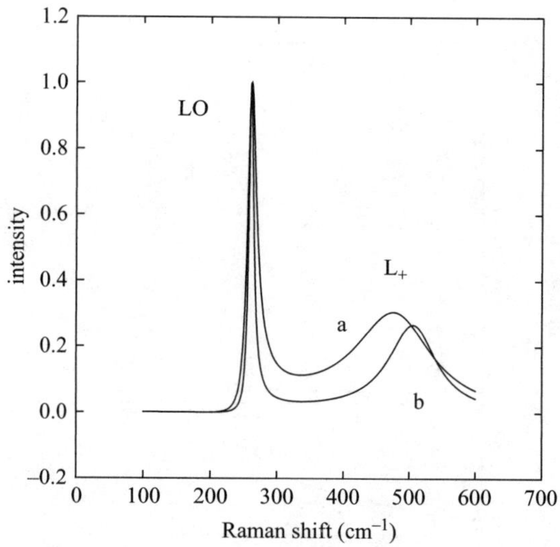

FIGURE 3.5 Normalised Raman spectra from n-GaAs doped to (a) 1.82×10^{18} cm^3 and (b) $2.0 \times 10^{18}/cm^3$. Shift in L_+ shows doping level; relative intensities of LO and L_+ show depletion depth.

surface-trapped charge due to surface states lying within the forbidden gap and surface-fixed charge, i.e. ionic charge on the surface that may not be related to the surface traps. With only a surface-fixed charge, the Fermi level can still move freely. In contrast, with sufficiently high densities of surface-trap states, the Fermi level is pinned and neither bulk doping nor externally applied electric fields can move it.

Band bending due to surface fixed charge can repel minority carriers sufficiently to reduce the surface-recombination velocity. The change in the surface-recombination velocity as a function of forward-injection parameters can provide a measurement of surface-fixed charge and help distinguish contributions from fixed charge and surface states. Better results will be obtained if injection of carriers by absorption of light is avoided when making such electrical measurements.

3.3.2 Chalcogenide passivation: S and Se

The chalcogenides, S and Se, produce advantageous changes in the surface electronic properties of GaAs. While Te is also a chalcogenide, it has not been a popular choice for improving surface performance of GaAs. Two advantages attend the sulphidation of GaAs: there is a change in the number and type of midgap surface states, which improves surface electronic properties, and the surface sulphur retards the reoxidation of a GaAs surface, which facilitates device fabrication when an oxide-free surface is required for the next processing step. Calculations for the (100) surface have predicted that the Ga-S interaction produces states outside the forbidden gap while some As-S states lie within the gap but near the conduction band where their electronic effects are less severe [10]. Consequently, even a fully sulphided ideal surface will still possess some undesirable in-gap states, but the situation should be greatly improved relative to the native oxide. Both solution-phase (wet) and gas-phase processes have been used to sulphide the surface.

3.3.2.1 Solution-phase processes

The most common solution-phase sulphur-based passivation employs ammonium sulphide. Aqueous solutions are quite basic (pH > 8) so they both dissolve the native oxide, exposing the bare GaAs surface, and provide a source of sulphur. Immersion for 15 min in aqueous $(NH_4)_2S$ (3 M or 20 wt%) at room temperature is sufficient to passivate the surface. Often some excess elemental sulphur is added (typically 4%), with the reactant then designated as $(NH_4)_2S_x$. Some heat the solution to as much as 60°C; however, this leads to etching of the surface [11], with etch rates at 60°C almost an order of magnitude higher than at 45°C. Since passivation occurs readily at room temperature, use of elevated temperatures and the associated etching can be avoided.

This is a commonly used technique because it is very easy to implement, requiring only simple glassware. However, it is very important to perform these reactions in a good fume hood. Like many sulphur compounds, these really smell horrible. It is important to avoid acidifying these solutions since they will then release hydrogen sulphide, H_2S, which is more poisonous than hydrogen cyanide, HCN. Fortunately, H_2S smells *so* bad that its presence is self-notifying at levels well below the dangerous level. However, it is also a molecule where olfactory fatigue rapidly sets in and a gradual increase in concentration to hazardous levels might not be noticed.

XPS studies of GaAs surfaces following sulphidation with ammonium sulphide solutions show two types of S: one characteristic of sulphide, S^{2-}, and the other characteristic of disulphide, S_2^{2-}. Both Ga-S and As-S bonding are observed. The surface species formed will depend on the state of the GaAs surface before introduction to the $(NH_4)_2S$ solution. If the surface was previously deoxided with NH_4OH, which effectively dissolves elemental As, only Ga-S is seen, while deoxiding with HCl or H_3PO_4, which leave elemental As, produces both Ga-S and As-S surface bonds [1].

Organic thiols (alcohols where the O has been replaced by S) have also been used as passivants. However, the organic portion of the thiol can be a source of carbon in subsequent processing. This may not be a problem, but the fabrication engineer should be aware of the possibility.

The surface can also be sulphided using an applied electric potential and a sulphide solution. While the electrochemical process is effective in applying a sulphide that appears somewhat more stable than the conventional wet dip, it requires electrical contacts to the wafer and is not likely to be generally applicable.

A thin yellowish film of excess sulphur may be observed after passivation of the GaAs surface with $(NH_4)_2S_x$. This will sublime away in vacuum, leaving only the S directly bonded to the GaAs surface. If the sample is not going to be placed under vacuum, it is customary to rinse the surface of the wafer thoroughly with DI water after removal from the sulphide solution to remove excess sulphides that can precipitate on the surface and cause problems if not rinsed away. Optimised rinsing protocols have not been reported, but use of consistent rinsing times and conditions is probably important for reproducibility.

For a GaAs sample passivated with $(NH_4)_2S_x$ and then handled without exclusion of air, surface analysis by Auger electron spectroscopy (AES) showed both oxygen and sulphur on the surface [12]. The oxygen found on the partially sulphided surface responded to temperature differently than the oxygen observed when only the native oxide is present. For a GaAs surface with native oxide only (no S), oxygen comes off in TPD (temperature-programmed desorption) over two temperature ranges: 400–500°C and 570–640°C, indicating at least two different binding sites for O. As-O is lost from 400–500°C and Ga-O is lost from 570–640°C. For the partially sulphided surface, the oxygen decreased above 400°C and disappeared by 500°C. The sulphur was unchanged to 530°C, and totally gone by 600°C.

3.3.2.2 Gas-phase processes

Unlike $(NH_4)_2S$ passivation, which also etches off the native oxide, it is necessary to remove the native oxide before the gas-phase passivations are applied. Without prior oxide removal, the greater strength of the Ga-O and As-O bonds inhibits the formation of Ga-S and As-S at the GaAs surface.

When the GaAs surface has been effectively deoxided, the initial increases in PL intensity are greater with gas-phase S than those produced with the $(NH_4)_2S_x$ treatments. When oxygen is mostly excluded, the increase in PL over the native oxide value can be about twice as great as that achieved with $(NH_4)_2S_x$ treatment.

The gas-phase processes all generate S atoms or Se atoms. This can be done thermally, photochemically or using a plasma. These have an advantage over the wet techniques of being very clean processes that do not leave any undesirable solid residue to remove before further device processing is possible.

Passivation using thermal generation of S and Se has been accomplished under UHV conditions. Under such conditions, the necessary deoxidation of the surface can be accomplished thermally, although an initial wet etch to reduce the native oxide is helpful. Chalcogen source gases are volatile disulphides or diselenides such as SnS_2 and $SnSe_2$ that decompose upon heating to release atomic S at 550°C and 340°C, respectively.

A simpler and less expensive approach employs UV photodissociation of the S_8 rings that evaporate from the surface of heated sulphur [13]. Reaction of the S atoms with GaAs occurs at room temperature under high vacuum (10^{-5} torr). This process can be used with prefabricated devices that have already been metallised since standard contact metallisations are not attacked by the NH_4OH solutions that can be used to expose the oxide-free surface. The surface can be kept mostly oxide-free by stripping the oxide under an inert atmosphere in a glove box attached to the reaction chamber. XPS studies of photosulphided GaAs show Ga-S and As-S bonding with the sulphur being intermediate between a sulphide and a disulphide in oxidation state and negatively charged, as expected.

The third approach generates S atoms by plasma-dissociation of H_2S. Like other plasma processes with H-containing source gases, this approach can inject H into the bulk of the GaAs, passivating dopants and defects. Reactivation of H-passivated dopants requires annealing at about 400°C (Section 5.7), which can result in S loss from the surface. While this is a disadvantage, the atomic H can react with the surface oxide and remove it, thereby exposing the GaAs surface to sulphur. Hydrogen-plasma removal of the native oxide is reported to produce a Ga-terminated surface depleted in As, which may be advantageous since excess As is known to cause electronic problems.

3.3.2.3 Problems with retention of passivation

The long-term retention of simple sulphur treatments in exposed regions of a device, as opposed to being confined at metal/semiconductor interfaces, presents problems. Regardless of whether the S was initially applied in solution or from the gas phase, it will be lost from the surface relatively rapidly if left in air. Our PL studies have shown virtually complete loss of S within 24 h for an unprotected surface in room light in air. The rate of loss is accelerated by light in general and especially by UV light. The light intensity employed for PL measurements is sufficient to remove enough sulphur to regain the native-oxide PL intensity after about 10 min. Photodeposited sulphur appears somewhat more stable than that deposited from an $(NH_4)_2S$ solution, but there is less than a factor of five difference. Storage in the dark slows the loss of photodeposited sulphur, but the PL intensity was still reduced by 25% after three days in the dark and had returned to the native-oxide level after three months.

Encapsulation might seem to be the solution to the rapid sulphur loss. However, there is some loss of S if one tries to encapsulate the sulphided surface with a plasma-deposited dielectric. The origin of this problem is the use of silane, SiH_4 or disilane, Si_2H_6, as the Si source gas for SiO_2 or SiN_x films. Both reactants provide an ample source of H atoms to react with the surface S and form volatile products, such as H_2S. An additional possible cause of S loss is the UV-induced dissociation of S from the surface even without reaction with H, since most plasmas emit in the UV.

Both conventional PECVD "silicon nitride" (SiN_x) and ECR-deposited "silicon nitride" or silicon oxynitride (SiO_xN_y) prevent long-term total loss of the passivation that survives the dielectric deposition process. For $7 \times 10^{16}/cm^3$ n-GaAs, for which either photosulphidation or an $(NH_4)_2S_x$ treatment produced a ten-fold increase in PL intensity, about a 2.5-fold increase was retained following dielectric deposition by either technique. No change in the encapsulated-surface PL was detected after six months, suggesting a chemically and electronically stable surface.

A sulphided surface that has been encapsulated by plasma deposition of silicon nitride or silicon oxynitride can change the near-surface doping level. Since S is a donor impurity in GaAs, it is able to increase the free-carrier concentration of the substrate. A 10% increase in electron concentrations has been observed for heavily doped GaAs using Raman spectroscopy (Section 3.3.1.3). The low-energy ion bombardment occurring during the deposition appears to be responsible. XPS studies were performed on photo-sulphided surfaces that were uncapped or capped with ECR SiN_x deposited either with line-of-sight into the plasma or shadowed from direct line-of-sight. About half of the surface S was lost with shadowed deposition while about one third was lost with line-of-sight. More significantly, while 8% of the S was transformed into the donor type with shadowed deposition, 19% underwent the transformation with line-of-sight deposition.

One promising approach to retaining an S-based passivation through the dielectric deposition process involves stabilisation of the surface-bound S by reaction with a metal ion in solution before the freshly sulphided surface is exposed to air [14]. The results displayed here were obtained with a surface that was sulphided using UV photolysis of S_8, but it should also work for surfaces sulphided using other techniques. Being careful not to expose the freshly sulphided surface to oxygen, the S-passivated sample is immersed for 30 s in a dilute (<1M) deoxygenated aqueous solution of a metal salt. The surface is then rinsed with DI water and blown dry with nitrogen. While a range of metal ions (Ga, Zn, Ni, Fe, Mn, Ca, La) produced initial PL improvements up to twice that obtained with S alone, the PL improvement gradually decreased over a period of a few days. Only Zn^{2+} produced an improvement in PL intensity that remained equal to the fresh S-only value after exposure to air for more than 700 days. Upon initial exposure to air, a 10% improvement in PL intensity is obtained and XPS studies show a (Zn, S, O) composition on the GaAs surface. With immediate encapsulation of the fresh Zn+S surface using a room-temperature ECR SiO_xN_y described more completely in Section 3.3.3.6, no significant change in PL intensity is caused by the plasma deposition and the initial high PL intensities (FIGURE 3.6) were retained indefinitely (>330 days). The slight apparent improvement in PL with the dielectric only may be largely due to its AR-coating effect, although device results in FIGURE 3.7 suggest some electronic effect as well.

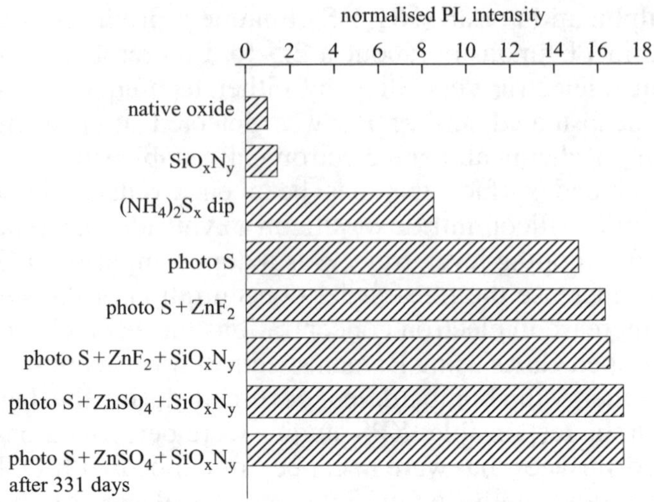

FIGURE 3.6 Normalised PL spectra from $6.9 \times 10^{16}/cm^3$ n-GaAs after different S treatments with and without dielectric deposition.

Improved device operation with an encapsulated Zn-S-GaAs surface has been obtained. Improvement in the current gain ($\beta = I_c/I_b$) for passivated $100 \times 100\,\mu m$ AlGaAs/GaAs HBTs is consistent with the improvement in PL intensity (FIGURE 3.7). The (Zn-S-O) surface that results from metal ion treatment of sulphided GaAs may be somewhat analogous to the growth of ultrathin epitaxial layers of II–VI semiconductors on the GaAs surface. When grown on an As-deficient surface, 84 nm of epitaxial ZnSe produced interface state densities in the low $10^{11}/cm^2$, which is comparable to the AlGaAs/GaAs interface. The use of standard epitaxial regrowth by MBE is not readily applicable to many partially fabricated devices, but perhaps the low interface-state density obtained with the thicker layers may be achieved by variations of the metal-ion-sulphide process. Simple multistep treatments, such as sulphidation followed by metal-ion reactions followed by dielectric encapsulation, appear to be a research area worthy of further investigation.

When the encapsulant does not have to survive elevated temperatures, an organic material may be used. Polyimide has been used successfully to cap an InGaAs/InP PIN photodetector, which displayed almost no increase in dark current a month after capping (FIGURE 3.8) [15].

Since passivation effectiveness depends intimately on the energetics of the chemical bonding at the surface, it should not be surprising that something that works very well for one material might not work as well for another. For example, while sulphur and sulphur + metal treatments improve current gain, β, for AlGaAs/GaAs HBTs, the same treatments actually appear to

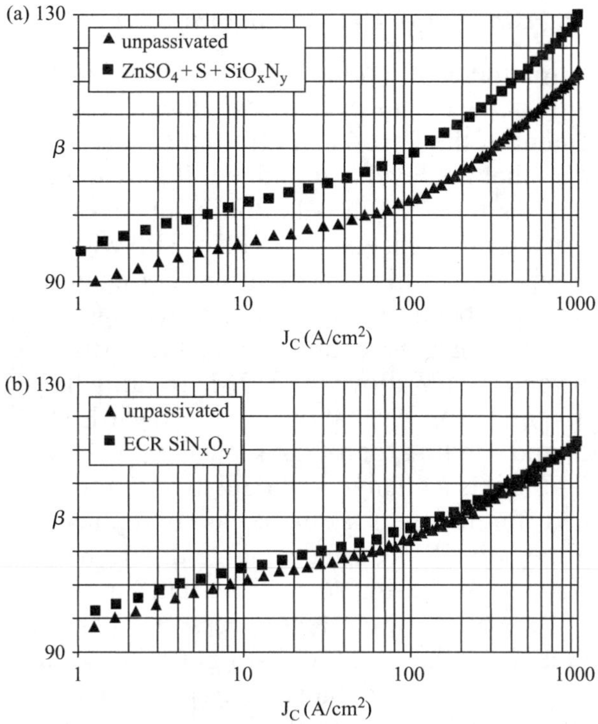

FIGURE 3.7 HBT current gain following dielectric deposition with and without metal-S passivation.

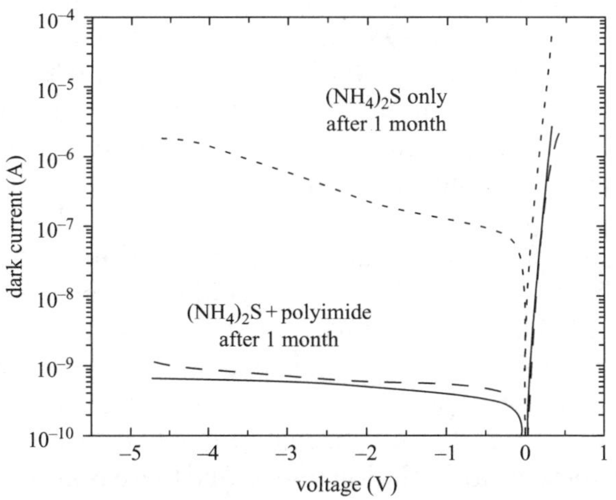

FIGURE 3.8 Retention of low dark current with polyimide encapsulation. (From Ravi et al. [15].)

introduce new gap states and lower the β for InGaAs/InP HBTs. Similarly, a passivation that works well on the (100) surface may not be as effective on an exposed (110) or (111) surface, or it might be even more effective, depending on where the chemical bonding at the surface moves the dominant surface states.

3.3.2.4 Effects on surface electronic properties

Passivation treatments are often discussed in terms of their effect on the position of the Fermi level and whether it is pinned at the surface. Movement of the Fermi level from the native-oxide position may not mean that the level is unpinned. It may merely mean that the treatment has replaced the old dominant set of midgap states with a new set at a different energy level but still within the gap. The Fermi level will pin at this new position. For example, for GaAs samples that have been cleaned using a hydrogen plasma or by thermal desorption of an As capping layer, the Fermi level is pinned at about 0.65 and 0.6 eV above the VBM. Treating the surface with S or Se shifts the Fermi level about 0.4 eV towards the CBM. This reduces the band bending for n-GaAs, but the surface is still pinned.

Whenever there is a change in the relative electronegativities of the atoms in different layers, as in the case of replacing As with the more electronegative S, a surface charge dipole will form. When charge is trapped at the interface for whatever reason, a dipole also forms. To evaluate the effectiveness of a particular passivation technique, it is important to distinguish between Fermi level pinning in the true sense and interface-trapped charge and bonding dipole effects. These effects can also be an issue for metal contacts.

The quality of an insulator/semiconductor interface can be characterised by measuring the capacitance versus applied voltage with a metal-insulator-semiconductor (MIS) or metal-oxide-semiconductor (MOS) diode structure. When generating a capacitance-voltage (C-V) curve to characterise the quality of an interface, it is always important to sweep the voltage several times in both directions in addition to performing the sweep over a range of frequencies. Without examining the response in both directions, an interface may appear much better than it actually is. Although this may make for a more impressive publication, it is misleading and of no help when making real working devices is your goal.

The ideal behaviour of an MIS diode is well described in most device textbooks [8,16]. The total capacitance is a combination of the semiconductor depletion region capacitance and the insulator capacitance in series. For n-type material, when the applied voltage is positive, there is no depletion region and there is an accumulation of electrons at the surface. This makes the total capacitance equal to that of the insulator, C_o.

As the voltage is decreased, a depletion region can begin to grow. As the voltage decreases further, an ideal interface can go from depletion into inversion; the gate voltage at which this occurs is called the threshold voltage, which is negative for an ideal MIS diode on n-type material. For p-type material, the C-V behaviour is the mirror image of n-type behaviour. The capacitance

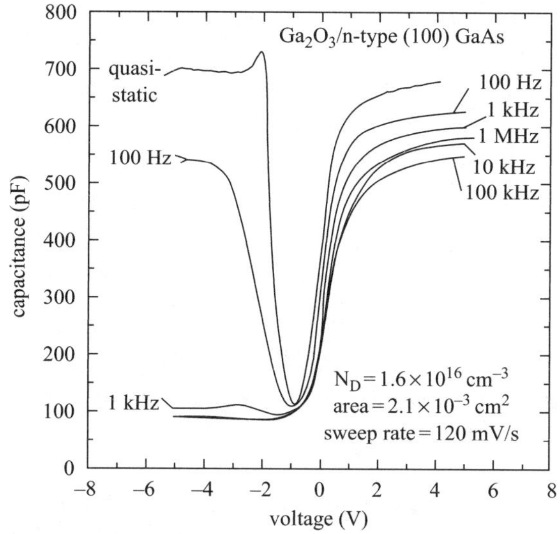

FIGURE 3.9 C-V curves for a MIS diode showing nearly ideal behaviour. (From Passlack et al. [6].)

in the inversion region depends on the measurement frequency. At high frequency, the capacitance remains low at voltages where inversion would occur. However, if the frequency is slow enough for minority-carrier generation in the depletion region to supply enough carriers to follow the gate voltage, the capacitance in inversion will be that of the insulator. This is manifest as the capacitance beginning to dip as the voltage approaches zero and then rising again to C_o of the insulator. These behaviours are displayed in FIGURE 3.9 for a UHV-deposited Gd-doped Ga_2O_3 on n-GaAs [6]. This oxide has displayed remarkably good, although not perfect, behaviour for an oxide on GaAs. It will be discussed in greater detail in Section 3.3.5.

For a non-ideal insulator, the presence of fixed charges can shift the flat-band position in a C-V curve away from zero, as illustrated in FIGURE 3.10 for 36 nm SiN_x on either $1 \times 10^{15}/cm^3$ n-GaAs or photoS-passivated $5 \times 10^{16}/cm^3$ n-GaAs. The presence of S at the interface reduced the total fixed charge in this case. The fixed charge can be thought of as a built-in voltage at the insulator/semiconductor interface. The higher value of C_{min} for the passivated interface is consistent with the higher initial doping level.

Another factor to consider, especially when dielectrics are deposited on the surface using a plasma and the density of the deposit may vary, is the possible presence of mobile charges within the dielectric. Mobile charges can lead to hysteresis in the C-V curves as shown in FIGURE 3.10. The reduction in hysteresis for the sulphided sample may be due to reduced mobile charge compared to the dielectric-only sample. The shift of the

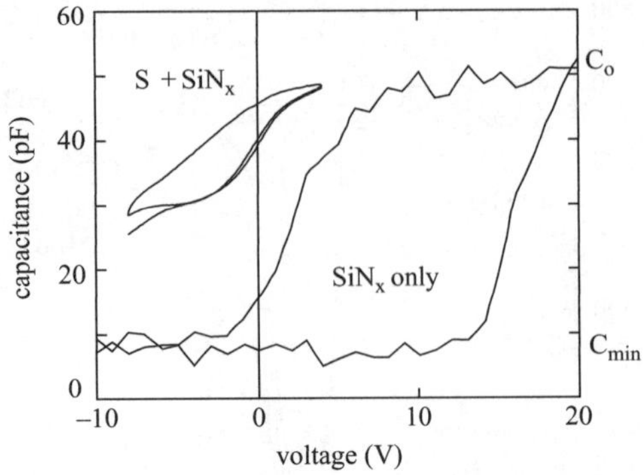

FIGURE 3.10 C-V curves with ECR SiN_x on 1×10^{15} n-GaAs and S+SiN_x on 5×10^{16} n-GaAs.

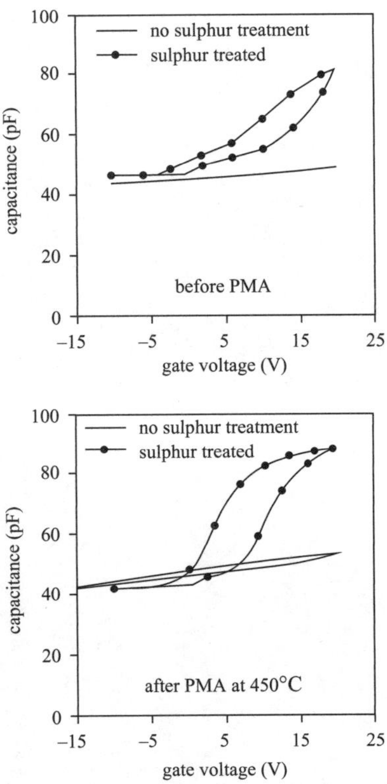

FIGURE 3.11 Effect of S on C-V curves before and after thermal anneal at 450°C. (From Remashan and Bhat [18].)

flat-band position towards zero may also be due to reduced fixed charge. However, the behaviour displayed in FIGURE 3.10 is not necessarily due to mobile charge. The same behaviour can also occur with the charging and discharging of fixed traps. For silicon nitride dielectrics, hysteresis is more likely due to charging and discharging of fixed traps within the insulator rather than to mobile charges.

To determine the surface or interface-state density from a C-V curve, a Terman analysis is performed [17]. When there is a high density of surface states, a Terman analysis generally generates a U-shaped curve of the number of surface states per unit area per unit energy (number/cm^2/eV) with the fewest number of states being located towards the middle of the gap; other curve shapes are also observed (FIGURES 3.13, 3.18 and 3.19).

The effect of PECVD SiN_x on n-GaAs with and without treatment with $(NH_4)_2S_x$ is shown in FIGURES 3.11–3.13 [18]. The nitride was grown with SiH_4, NH_3 and N_2 at the lowest RF power that would sustain the plasma. Nevertheless, there would be ion-bombardment damage and an H-rich nitride would form, as evidenced by the refractive index of 1.8 instead of the ideal of 2.0 for Si_3N_4. The nearly flat line in the C-V curve for unpassivated surfaces indicates strong pinning of the Fermi level due to their high interface-state density.

Thermal annealing did not change this behaviour for the unpassivated sample. In contrast, thermal annealing has a clear effect on the sulphided interface. Before annealing, the passivated sample does not show accumulation behaviour. After annealing at 400 and 450°C in N_2 for 5 min, well-defined depletion and accumulation regions are observed. The effects of higher-temperature anneals are shown in FIGURE 3.12. A clear improvement occurs between 450 and 500°C. However, between 500 and 550°C, results are mixed. A sharper rise is obtained in the C-V curve. The authors ascribe the improvements to H originating in the dielectric deposit. However, while annealing at 500°C did not change C_{min} from the original 35 pF, the 550°C anneal reduced it to about 20 pF. This correlates with about an order of magnitude reduction in the doping level. This may be due to H from the dielectric passivating the Si donors or to Ga diffusing into the dielectric to form acceptor-like Ga vacancies. Given the high levels of H typically found in PECVD nitrides and its facility for diffusion and passivation of dopants and defects, some role for H is almost certainly present.

The results of Terman analysis of the samples in FIGURE 3.12 are shown in FIGURE 3.13. Higher temperatures reduce the number of interface traps, D_{it}, between 0.6 and 0.8 eV above the valence band to the low 10^{11} to the high 10^{10} range with a midgap value of $8 \times 10^{11}\ cm^{-2}\ eV^{-1}$. Even though this represents a considerable improvement in the interface and its C-V behaviour, there is still a large amount of hysteresis due to slow traps at the interface. It is worth noting that the high-frequency method used to determine the relatively low D_{it} values only demonstrates that the fast interface states are reduced appreciably by the processes employed.

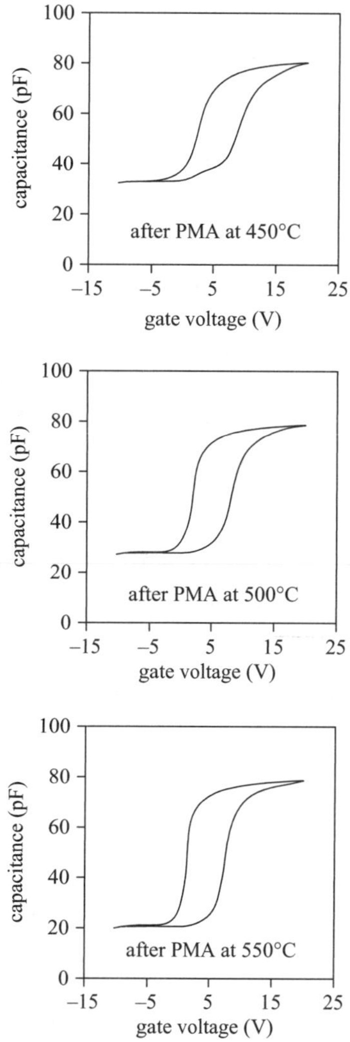

FIGURE 3.12 Effect of annealing temperature on C-V of sulphided GaAs. (From Remashan and Bhat [18].)

3.3.3 Passivation for improved semiconductor regrowth

For the growth of the highest quality crystalline semiconductors by either MOCVD or MBE, it is important to start with a surface free of oxide or other contaminants. While an oxidised surface may be rendered oxide free by heating *in vacuo* or, even better, under a flux of the group V source gas, the ability of chalcogens like S and Se to retard oxidation has been harnessed to improve surfaces for regrowth.

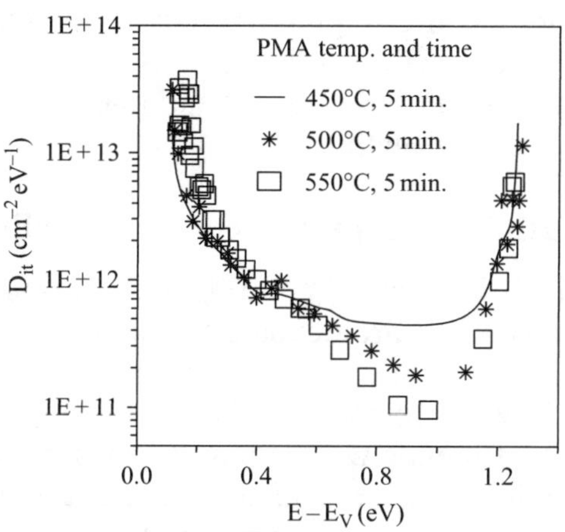

FIGURE 3.13 Density of interface states from Terman analyses of C-V curves in FIGURE 3.12. (From Remashan and Bhat [18].)

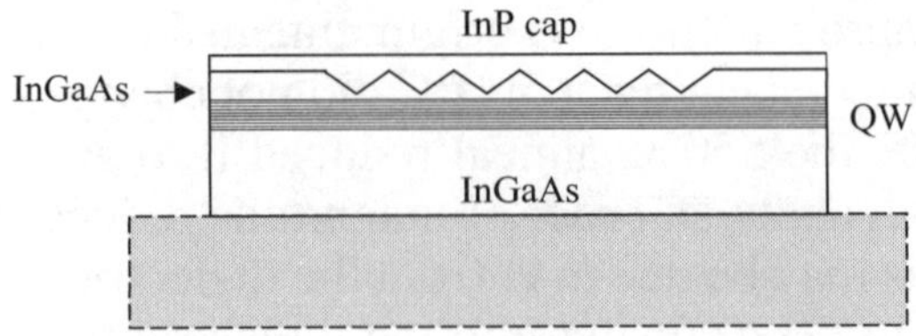

FIGURE 3.14 Regrowth for buried distributed feedback Bragg (DFB) grating.

While a new wafer fresh from its sealed container has minimal contamination, a wafer that has been etched to produce a non-planar surface morphology may have significant contamination levels prior to the final cleaning before insertion into the growth chamber for regrowth.

For example, in making edge-emitting distributed feedback (DFB) lasers, one may want to etch a Bragg grating into the surface of a wafer, placing it relatively near to the active waveguide core where the mode intensity is highest. Following etching of the grating, the wafer must be cleaned, reinserted into the growth chamber and have a cladding layer grown over the grating. A schematic of such a device is illustrated in FIGURE 3.14.

Obviously, the native oxide needs to be largely removed with either an acidic or a basic solution. Similarly, metal-ion contamination can be reduced by rinsing with NH_4OH. This has already been discussed in Section 3.2. However, unless these treatments and the loading into the growth chamber are conducted with the complete exclusion of oxygen, some surface oxide will have reformed by the time the wafer loading is completed. This reoxidation can be minimised by sulphiding the surface of the wafer to block the reactive sites where oxygen would bond; it is, nevertheless, still important to minimise oxygen exposure. Higher quality interfaces have been reported using sulphur to retard oxidation while preparing a sample for regrowth. While the sulphur does not completely prevent reoxidation, it does retard it and makes it easier to obtain a clean surface for regrowth.

3.3.4 Passivation for improved contact metallisation

A crucial part of fabricating good devices is making good contacts; Chapters 6 and 7 are devoted to this important topic. This section will discuss some of the materials issues related to metallisation and how the presence of a passivating layer may alter the interactions of metals with GaAs. The issues that are important in metallisation of unpassivated surfaces are largely the same for metallisation of passivated surfaces. The deposited metal may undergo reaction with the GaAs surface to form new chemical phases. By changing the chemical nature of the interface, the passivant may profoundly alter alloy formation, defect generation and atomic interdiffusion. There can be the removal of some surface/interface states and the introduction of new ones. The chalcogens, S and Se, are n-type dopants, so there may be a change in the near-surface doping level of the semiconductor. This may be highly desirable for making better contacts but it should be borne in mind when considering the sources and sinks of free carriers in a device design. Thermal annealing following metallisation can cause chemical reaction and defect or dopant diffusion. Extended operation under bias can induce changes. The deposition of an encapsulant can passivate dopants or defects by injection of H or, if a plasma process, it can introduce surface damage. In addition, the encapsulating process may cause loss of the passivant or change near-surface doping by generating donors from the passivant. We shall address how passivation affects several of these issues in the following section. While the standard approach at the time of writing this chapter is to metallise unpassivated surfaces, that may change with further understanding and development of suitable passivation procedures.

When a metal atom is deposited on the surface of GaAs, it often undergoes reaction with the GaAs, leading to disruption of the interface and formation of a mixed composition in the near-surface region. Sometimes extensive interdiffusion and alloying between the metal and the semiconductor occurs, changing the elemental composition of both to an appreciable distance from the nominal interface. This generally requires some heating for these reactions to proceed over more than a few atomic layers. The formation of such interfacial compounds may be an important part of forming a good contact, as will be discussed in Chapter 6 concerning ohmic contacts, but it can also be a problem, especially if it occurs in a non-uniform fashion. The common mechanisms involved in the formation of ohmic contacts are outlined in Section 6.2.3 with the example of Ge/Au/Ni, one of the most popular ohmic formulations for n-GaAs. The initial step is reaction of Ni with the native oxide to allow more uniform access of the other elements to the actual GaAs surface. It is obvious that prior replacement of oxygen with sulphur or another passivant can have a major impact here. Subsequent steps, such as reaction with Ni to disrupt the lattice with Ni-GaAs complexes, could also be modified by formation of metal sulphide interlayers. Since diffusion in semiconductors is defect-mediated, changes in defect locations and populations through passivants acting as diffusion barriers may play an important role. Diffusion of As to the contact surface can leave voids that are filled by stable intermetallics, such as AuGa and NiGe. Localised excessive reaction can produce “spiking”, the deep penetration of the conductive intermetallic in a small region, shorting the device.

To the extent to which passivation may affect any of these steps, it may be desirable or undesirable in forming ohmic contacts.

The potential importance of passivation in the formation of Schottky contacts is immediately obvious. The Schottky barrier is key to their operating characteristics (Section 7.2), and that is defined as the energy difference between the metal Fermi level and the semiconductor conduction band energy (for n-type) or valence band energy (for p-type) at the metal/semiconductor interface. Clearly, any effect a passivant may have on Fermi-level pinning through the energy and number of surface states may affect the behaviour of a Schottky contact.

A key reaction occurring in many good ohmic contacts is formation of the AuGa alloy at the interface. The formation of this alloy is affected by the presence of oxide on the surface, with alloy formation enhanced by oxide relative to that on a bare GaAs surface (deoxided in UHV before Au deposition) and As atoms segregating on the metal overlayer during Au deposition. When a thin film of Au is deposited on GaAs with its native oxide (even after nominal oxide removal with a wet etch), Au reacts with GaAs to form an AuGa alloy and releases elemental As, which segregates on the surface of the Au metal [19]. A similar reaction between the native oxide and Ni has been identified. Sulphiding the surface with $(NH_4)_2S_x$ delays both the formation of the AuGa alloy and As segregation for very thin Au layers (5 Å), but the alloy and As form as the deposition proceeds. Since the "sulphided" surface still had some oxygen bound, these results are somewhat inconclusive as to what would happen with a fully sulphided surface.

Silver, in contrast, does not react with GaAs at room temperature and a comparable AgGa alloy is not obtained with either the oxided or sulphided surface. With the less reactive Ag, there is still evidence for formation of Ag-S bonding at the interface with the S primarily being taken from As-S bonds [19]. For starting surfaces without excess As, as in the UHV annealed case, the Ag-S forms, but it is by displacement of Ga from Ga_xS_y and a small amount of Ga metal appears due to displacement by formation of Ag sulphide. With further Ag deposition, the Ag deposits as Ag metal clusters and does not react chemically with GaAs. While these differences may seem slight, the interface states depend intimately on the details of bonding at the interface and such seemingly subtle differences might actually have major consequences for the number density and energies of interface states. They can also have very different effects on the number of vacancies and other defects that can be electronically active or affect diffusion in the material.

Indium does not normally react with the GaAs-oxide so its initial deposition on a native-oxide surface proceeds by island growth. With a partially sulphided surface, layer-by-layer growth through the apparent formation of a thin InGaAs alloy was observed. On the sulphided surface, the In caused a shift in the Fermi-level position towards the Schottky limit. In contrast, for Au the position of the Fermi level did not appear to depend on whether the surface had the native oxide or was partially sulphided. Starting with an oxide-free, Ga-rich surface that has been sulphided [20], there was an initial formation of an In_xS_y interfacial layer where In appears to react with surface sulphur dimers. However, no further intermixing between In and GaAs is observed and further In addition is metallic. The In/S/GaAs structure has a higher Fermi level than In/GaAs. This change in barrier height may be in part due to a sulphur-induced interface dipole.

The results of these two sets of parallel studies illustrate how vitally important it is to be careful before assuming the results of any laboratory study may pertain directly to real device fabrication. The often "ideal" starting conditions for the lab study may be a poor model for the real device. The composition of the real device surface may be quite different in terms of Ga-to-As ratios, presence of some degree of native oxide, surface reconstruction, etc. Consequently, promising scientific results may not translate into better devices, and procedures that are not expected to work based on very careful lab studies may work quite nicely.

The specific surface treatment of an Au/GaAs Schottky diode prior to Au deposition and its thermal annealing history can strongly influence its operating characteristics. FIGURE 3.15 shows the Schottky barrier height and dark current for Au/GaAs and Au/S/GaAs with and without hydrogen plasma treatment [1]. Exposure to the plasma, presumably to remove oxide, without the thin Au protective layer damages the surface and produces poor performance relative to either acid-cleaned or S-passivated surfaces. For the Au/S/GaAs where the Au layer was only 5 nm thick, the atomic hydrogen can penetrate the thin metal film and reduce As-O and As to a volatile product that escapes from the interface. A much lower dark current results with the reduced quantity of As at the interface. Similar improvements occurred for Au/GaAs with atomic hydrogen. It is important to note that this H-exposure was performed with very thin Au films, and exposure to H with thicker Au films might not be beneficial since the As-H products would be trapped by the thicker film even though H, with its ready propensity to diffuse in virtually any material, could still be reaching the interface.

With Pd metallisation, the catalytic ability of Pd to generate atomic H from H_2 can remove the need for a plasma. The effect on dark current of thermal anneals at different temperatures is shown in FIGURE 3.16 [1]. While both acid-cleaned and S-passivated surfaces displayed improved performance with H-anneals versus

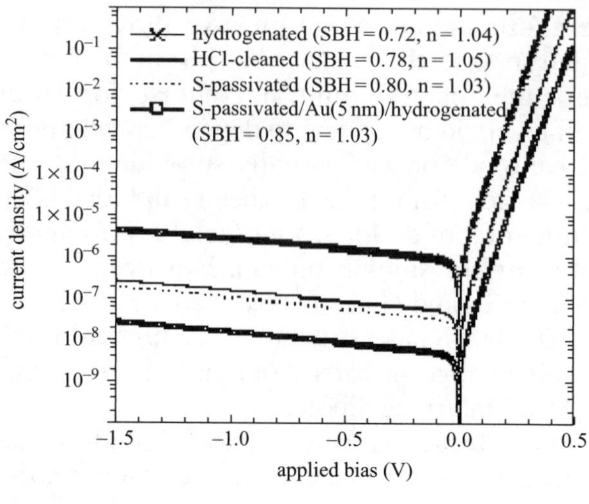

FIGURE 3.15 I-V characteristics of Au/GaAs Schottky diodes with various surface treatments. (Kang and Park [1].)

vacuum anneals, the sulphided surface retained its low dark current at appreciably higher temperatures than the acid-cleaned surface and, in general, performed better. It is, as always, important to remember that any process generating atomic H near GaAs can lead to H-atom diffusion into the semiconductor with resultant passivation of dopants and defects. This effect is increased at higher temperatures and longer exposure times, so any potentially beneficial effect of H must be weighed against possible deleterious changes it may produce.

Interest in actually preventing ill-defined interdiffusion and reaction between the metal and the semiconductor has been elevated by the possibility of integrating semiconductors and magnetic thin films to harness magnetic properties such as giant magnetoresistance (GMR). GMR permits the controlling of the electrical resistance by the application of a magnetic field and requires the precise growth of a metal superlattice. Magnetic metals such as Co and Fe are used, but these tend to intermix with GaAs. For these metals also, a monolayer of S inhibits interdiffusion of As and Ga and yields ferromagnetic behaviour at low metal coverages.

3.3.5 Special oxide passivations

The ideal solution for a semiconductor/insulator interface would be a good oxide, as with Si and SiO_2. Other approaches, as discussed above, suffer from unacceptably high interface-state densities. These occur for a variety of reasons including non-stoichiometric surfaces, structural defects, thermodynamic instability and intrinsic Fermi-level pinning due to the energy levels for interfacial bonding.

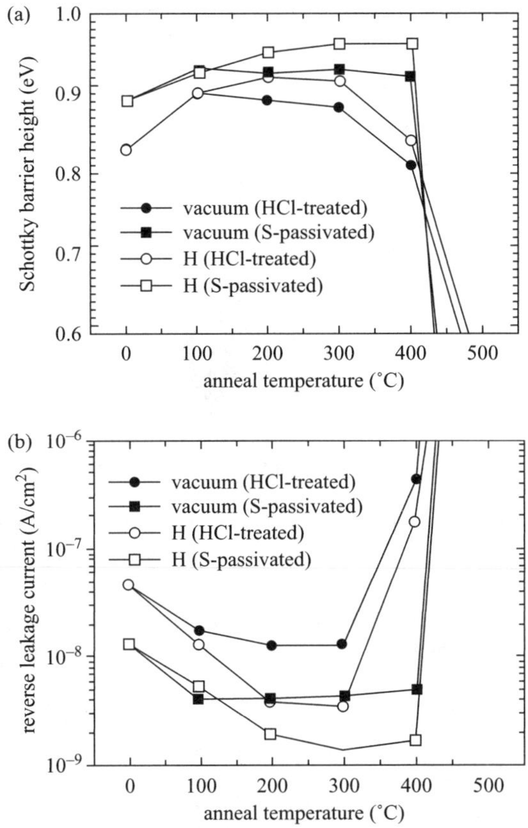

FIGURE 3.16 Schottky barriers and dark currents for Pd/GaAs Schottky diodes with various surface treatments following thermal annealing. (Kang and Park [1].)

Promising behaviour has been observed with a special oxide deposited by e-beam evaporation of single-crystal $Gd_3Ga_5O_{12}$ (a garnet) on As-stabilised (2×4) GaAs that had been transferred from a solid-source MBE III–V growth chamber to the e-beam deposition chamber at 1×10^{-10} torr [6]. Primarily Ga_2O_3 films were deposited on the clean, atomically ordered GaAs(100). These films have a non-uniform composition, with the interface being essentially Gd-free and the Gd concentration peaking at the oxide surface. This non-uniformity has consequences for the electronic properties, as discussed below.

MIS capacitors were fabricated on n-GaAs with 462 Å Ga_2O_3 and p-GaAs with 594 Å Ga_2O_3. Film resistivity was in excess of 10^{15} Ω-cm; quasistatic C-V measurements require greater than 10^{15} Ω-cm resistivity because they are very sensitive to current leakage. High-frequency (up to 1 MHz) and quasistatic C-V behaviour was measured (FIGURES 3.9 and 3.17).

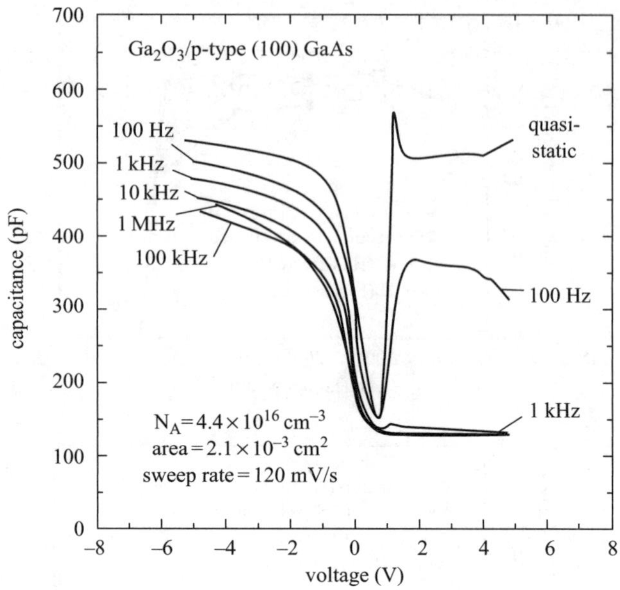

FIGURE 3.17 C-V curve for p-GaAs with e-beam-evaporated Ga_2O_3. (From Passlack et al [6].)

Inversion, depletion and accumulation are all observed in the C-V curves. Capacitance overshoot at the onset of inversion is seen for n- and p-type and may be due to filling of trapping centres in the oxide. There is a Fermi-level-independent time constant for trapping that is longer than 1 ms, which is three orders of magnitude greater than the theoretically calculated interface trap time constant. There is also a long-term drift of the capacitance with a time constant greater than 10 s. This is ascribed to oxide traps located 20 Å or so away from the interface. An approximately 350 mV hysteresis is seen in the C-V curves, indicating the presence of mobile charges.

There is frequency dispersion in accumulation that may be due to an oxide at the interface that is relatively low resistivity rather than due to a high density of interface states near the conduction band edge. Conductance-voltage measurements reveal such a region in these devices. When the insulator is inhomogeneous, containing regions of higher electrical conductivity, the observed capacitance will increase with decreasing measurement frequency (Maxwell-Wagner effect).

The non-ideal electrical behaviour of these interfaces is probably due to the non-uniform composition of the oxide as one moves away from the GaAs surface. A Terman analysis shows D_{it} in the mid 10^{10} cm^{-2} eV^{-1} midgap and trap density in the mid 10^{11}/cm^2

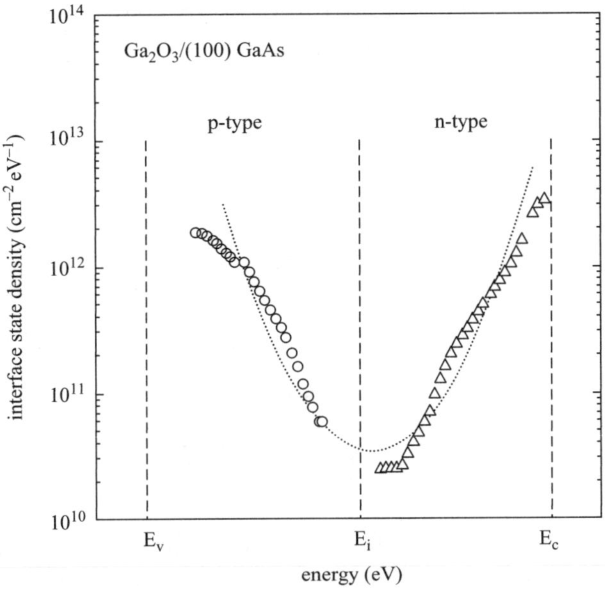

FIGURE 3.18 Density of interface states from Terman analysis C-V in FIGURES 3.9 and 3.17. (From Passlack et al. [6].)

integrated over the gap (FIGURE 3.18). While this approach produces one of the best interfaces to date, electronic problems with the oxide and the role of Gd in the oxide need to be resolved before it becomes the basis for a robust MIS technology for GaAs.

3.3.6 Dielectric passivations: PECVD and ECR SiN_x and SiO_xN_y

In this section, we will focus on the application of dielectrics as passivations without the deliberate addition of a chemically distinct monolayer passivant, such as S. Some improvements in electrical performance can be obtained with a dielectric alone and the encapsulation it provides helps to stabilise the electrical characteristics for long-term operation. An example of the improvement of an AlGaAs/GaAs HBT is shown in FIGURE 3.7.

While both silicon nitride and SiO_2 have been employed as dielectrics on GaAs devices, SiN_x has been largely the dielectric of choice due to its higher dielectric constant and its better performance as a diffusion barrier, which makes it a better environmental encapsulant. Consequently, we will focus mainly on nominally silicon nitride materials in this section.

Growth of a dielectric/GaAs structure with a highly controlled and nearly ideal interface has shown what might be inherently possible, if exceedingly difficult at present, to achieve under

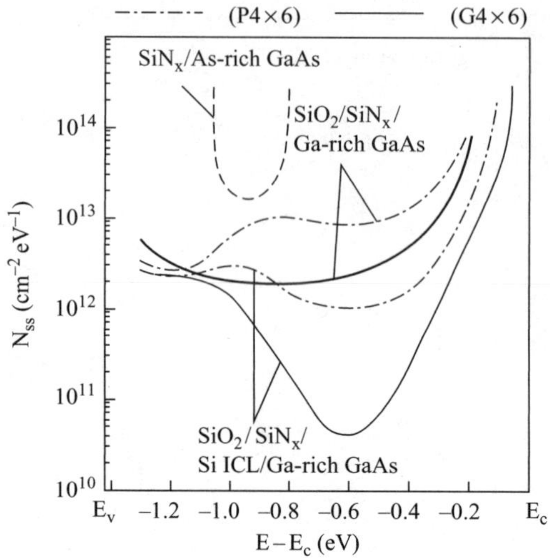

FIGURE 3.19 Surface state densities from different GaAs reconstructions at dielectric/GaAs interfaces. (From Anantathanasarn and Hasegawa [21].)

normal fabrication conditions. Using a Ga-rich (4×6) reconstruction on (100) GaAs, complete unpinning of the Fermi level over the entire bandgap with a minimum interface-state density of $4 \times 10^{10}/cm^2/eV$ has been reported [21]. In FIGURE 3.19, the results of Terman analyses of these high-quality interfaces are contrasted with nominally SiN_x-on-GaAs surfaces, which show the expected U-shaped behaviour of a poor interface. An Si interface-control layer was deposited on the clean, reconstructed surface, partially converted to SiN_x by direct nitridation, and then coated with a thick SiO_2 layer using plasma-enhanced chemical vapour deposition (PECVD). Some hysteresis and flat-band voltage shifts are observed in the C-V curves, so an ideal GaAs/dielectric combination was not totally achieved. The effectiveness of this approach has only been demonstrated on ideal (100) surfaces and may not work as well on other crystal faces, such as are exposed in mesa structures. However, it does demonstrate that a MOS-quality interface in a real device is at least theoretically and maybe actually possible.

Since these are plasma-deposited dielectrics, it is important to remember that plasmas always contain energetic ions, chemical radicals and generally UV light. Some degree of surface alteration is possible due to photodissociation or photoreactions even when the ion energy is below the atomic displacement threshold of 40 eV or so for GaAs. Damage issues that are increasingly important as energies exceed 40 eV will be discussed in several sections of Chapter 5.

There are different types of plasma hardware used to do these depositions. The most common are called plasma-enhanced chemical vapour deposition (PECVD) and electron-cyclotron resonance (ECR) or inductively coupled plasma (ICP) deposition. PECVD employs low-density plasmas, whereas ECR and ICP produce high-density plasmas (HDPs). The distinction between these plasma types is discussed in detail in Chapter 5 on dry etching. Most HDP work reported to date employed ECR plasmas, but similar results may be expected with the increasingly more popular ICP deposition systems. The most important difference for dielectric deposition is the relative degree of dissociation of the reactant gases. Low-density plasmas cause much less dissociation of the reactants into their atomic constituents than do high-density plasmas. One place where this has a very pronounced effect is on the relative amounts of H that are contained in a nominally silicon nitride deposit. For example, some PECVD "silicon nitrides" contain as much as 25% hydrogen while an ECR "silicon nitride" may contain less than 10%. An additional effect of the lower degree of dissociation is the need to heat the wafer to get a good film in PECVD. One of our PECVD reactors typically employs a wafer temperature of 250–300°C to get good films. In contrast, our ECR reactor deposits equivalent or better quality films without auxiliary heating above room temperature.

Because temperature plays such an important role in the chemistry of the less-dissociated reactants in PECVD, film properties display quite strong temperature dependence. As the wafer temperature is increased, the deposition rate goes down, the film etch rate goes down, and the refractive index goes up. These changes can result due to differences in composition, film density or a combination of the two (most likely). The hydrogen content of PECVD nitride films also goes down with increasing wafer temperature.

A second important difference between PECVD and ECR plasmas is the ion-bombardment energy. It is possible to deposit an ECR film while exposing the surface only to very low ion energies (<20 eV) since the wafer is essentially at the plasma potential. In contrast, since the wafer sits on a powered electrode in PECVD, the ion energies are typically higher; PECVD reactors may exceed the 40 eV displacement threshold and would then cause some surface disruption. Attempts are made to minimise this by controlling the relative areas of powered and grounded electrodes and selecting proper pressures and RF powers. In general, one wants the powered electrode to be large relative to the ground electrode, e.g. chamber walls, to reduce the bias voltage at the sample. Excursions to higher powers and higher DC bias voltages can occur when the plasma is first striking, and damage may result from a seemingly low-bias process during those transients.

Transient conditions while the plasmas are first striking can produce a different film composition and/or density near the GaAs interface. These are manifest in a higher refractive index for thin films. For PECVD films, the index decreases with increasing film thickness up to a thickness near 150 nm, and then remains relatively constant. Similar behaviour is seen in ECR films. It is especially important to keep this change in refractive index in mind when growing AR coatings for optical applications. Since the transients may be highly tool-dependent and can even vary significantly on the same tool, it can be helpful to grow test films to the calculated thickness, measure effective refractive index at that thickness and then fine-tune the thickness right before device fabrication to get optimum AR performance.

The ion bombardment energy can have a significant effect on the leakage current for HBTs. Surface damage can increase current conduction between the emitter-base and base-collector regions. These damage-related current channels are more important at high base-emitter voltages and high-injection levels. Ion damage will also degrade FET performance.

The films that are produced from the same nominal source gases and plasma conditions may actually be quite different when deposited in different ECR tools. Plasma tools can be constructed with different materials and with different designs. Since the plasma interacts with the chamber walls in a way that changes the actual plasma composition, it is possible to deposit a film that is NOT what you might expect based on the composition of the source gases. A good example is the deposition of SiN_x and SiO_xN_y using ECR plasmas with SiH_4 and N_2 as the source gases. In one ECR tool with minimal wall contributions, this forms partially hydrogenated SiN_x. In another tool, oxygen might be extracted from the quartz window anodized Al parts, through which the microwave radiation is launched into the plasma chamber, or from the alumina ceramic liner of the chamber. Even a little oxygen in the plasma will be incorporated at a significant level to give a partially hydrogenated oxynitride. The source gases for SiO_2 are SiH_4 and N_2O. No appreciable nitrogen incorporation is observed because of the strongly favoured formation of Si-O versus Si-N bonds (Si-O = 124.7 kcal/mol versus Si-N = 105 kcal/mol).

In virtually all cases, some amount of hydrogen is incorporated in the dielectric. We have deposited ECR films of nominally SiN_x and SiO_xN_y with as little as 7% hydrogen and PECVD "SiN_x" films that were as much as 25% hydrogen. There tends to be more H retained in PECVD films grown at lower wafer temperatures. The presence of excessive H can be a problem for post-deposition processing. For example, films with a high H content can "bubble" due to hydrogen gas release and fail when the wafer is subjected to rapid thermal annealing (RTA) to form good metal contacts.

It is possible to grow films that are Si-rich. These display higher refractive indices ($n > 2.0$) and lower resistivity. In PECVD, increasing the silane fraction increases the Si content of the film. In ECR, deposition at lower ECR powers will increase Si content for a fixed gas composition; for example, decreasing the microwave power from 400 to 160 W increases the index from 1.8 to 2.2.

Dielectric films are often characterised by measuring their refractive index and assuming that the desired composition has been attained when the "proper" refractive index is achieved. Unfortunately, this is often an erroneous assumption. Refractive index is a measurement of an optical property, not a composition. It is affected by elemental composition, lattice structure, porosity, etc. Addition of some Ar to the plasma will increase the refractive index for a fixed set of reactants and plasma conditions because it serves to densify the film. Addition of Ar also causes plasmas to run in a more stable fashion and therefore increases run-to-run reproducibility. Films with the same composition but greater porosity will have lower refractive indices. Films that contain Si-rich regions will have higher refractive indices. For example, while the refractive index of polycrystalline Si_3N_4 is 2.0, it is possible to grow ECR-deposited films with approximately the same index that are actually $Si_{2.8}N_4H_z$ films with 7% H and $SiO_xN_yH_z$ with 60% Si, 20% N, 10% H, 7% O and 3% C. One PECVD "silicon nitride" with index 2.0 was actually 35% Si, 40% N and 25% H. Although it is a poor choice as the main diagnostic for composition, it can be used very effectively as a measurement of the reproducibility of a process since it combines the effects of composition and film density. Hopefully, a specific process will produce uniform values of these parameters when it is operating reproducibly.

Lower index films tend to etch faster than higher index ones, due either to differences in composition or to the lower index films being less dense. The effect is more pronounced in wet etchants than dry, with a 5- to 20-fold increase over that for true Si_3N_4. Dry etch rates are generally less than two-fold faster for the PECVD films. Higher H-content films tend to etch faster. Having the same refractive index does not mean there will be the same wet etch rate. A factor of two spread in rate with 48% HF was obtained for several PECVD films all with a refractive index of 2.0. This all shows how important it is to test the actual etch rate of the films deposited in your own hardware before attempting to etch them during the fabrication of a valuable device.

Film stress is another important variable. ECR films tend to be lower stress than PECVD films, but variations in this can occur. Since high stress can lead to shifts in electronic properties within a chip, e.g. offset voltages, resistance of thin-film transistors, pinch-off voltages, hot electron degradation and mechanical failure, it is generally desirable to use low-stress films.

Conformality of the dielectric can be an issue especially when the dielectric is deposited on a mesa, which can be either photoresist or semiconductor. With PECVD depositions on a mesa, a profile with indentations near the mesa base (FIGURE 3.20(a)) often forms. This non-uniform film thickness can be used to advantage in some contact formation schemes. On the other hand, uniform film thickness may be desired on all surfaces. While conformal deposition is possible with PECVD, the room-temperature ECR processes tend to automatically deposit in a more uniform fashion (FIGURE 3.20(b)).

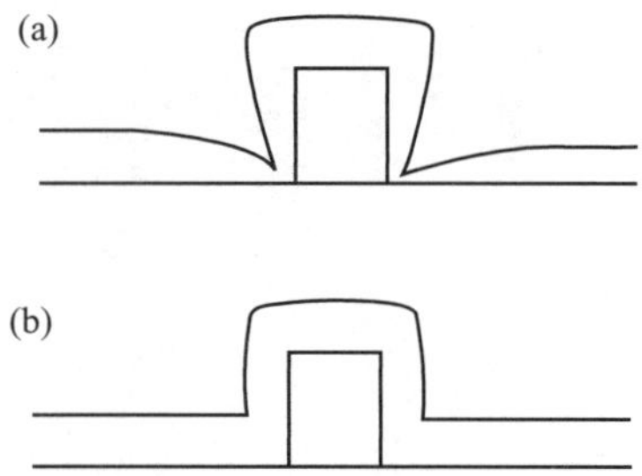

FIGURE 3.20 (a) Deposit with non-uniform thickness with PECVD. (b) Uniform thickness, conformal deposit with ECR.

Pinholes can be a serious problem in dielectric deposition. These can be caused by small particles on the wafer surface. These particles may have been on the surface of the wafer as loaded in the tool, or they may result from the deposition of particles within the reactor. It is especially important to keep the deposition system walls free of deposits that can flake and shed particles onto the wafer surface. Excessively rapid deposition of the film increases the likelihood of pinhole formation.

The use of silane as the Si source raises both safety and repeatability concerns if there is even a tiny leak in the silane gas line. Due to the pyrophoric nature of SiH_4, fires and explosions are a possibility with major leaks. Even a very small leak can lead to the deposition of "sand", i.e. tiny SiO_2 particles, in the line that can gradually clog mass flow controllers and make the plasma composition vary slowly over time, leading to irreproducibility in film composition.

3.4 CONCLUSION

The most important characteristic to remember when dealing with cleaning and passivation issues is that materials properties at the atomic level are what really matter. Composition and bonding configurations at the monolayer level can totally dominate the properties that affect device operation. Thinking "molecularly" and careful attention to process details at that level will contribute to achieving reproducible and reliable device results.

REFERENCES

[1] M.-G. Kang, H.-H. Park [*Vacuum (UK)* vol.67 (2002) p.91–100]
[2] K.R. Zavadil, C.I.H. Ashby, A.J. Howard, B.E. Hammons [*J. Vac. Sci. Technol. A (USA)* vol.12 (1994) p.1045–9]

[3] P.R. Lefebvre, L. Lai, E.A. Irene [*J. Vac. Sci. Technol. B (USA)* vol.16 (1998) p.996–1001]

[4] C.J. Sandroff, F.S. Turco-Sandroff, L.T. Florez, J.P. Harbison [*J. Appl. Phys. (USA)* vol.70 (1991) p.3632–5]

[5] S.I. Yi, P. Kruse, H.M. Hale, A.C. Kummel [*J. Chem. Phys. (USA)* vol.114 (2001) p.3215–23]

[6] M. Passlack, M. Hong, J.P. Mannaerts [*Appl. Phys. Lett. (USA)* vol.68 (1996) p.1099–101]

[7] H. Hasegawa [*Jpn. J. Appl. Phys. (Japan)* vol.38 (1999) p.1098–102]

[8] S.M. Sze [*Semiconductor Devices: Physics and Technology* (John Wiley and Sons, New York, 1985)]

[9] D.E. Aspnes, A.A. Studna [*Phys. Rev. B (USA)* vol.27 (1983) p.985–1009]

[10] T. Ohno, K. Shiraishi [*Phys. Rev. B (USA)* vol.42 (1990) p.11194–7]

[11] J.-W. Seo, T. Koker, S. Agarwala, I. Adesida [*Appl. Phys. Lett. (USA)* vol.60 (1992) p.1114–6]

[12] D. Wu, L. Liu, J. Marcano, Y. Darici, N. Paul, S. Mergui [*Mater. Sci. Eng. B (Switzerland)* vol.46 (1997) p.61–4]

[13] C.I.H. Ashby, K.R. Zavadil, A.J. Howard, B.E. Hammons [*Appl. Phys. Lett. (USA)* vol.64 (1994) p.2388–90]

[14] C.I.H. Ashby, K.R. Zavadil, A.G. Baca, P.-C. Chang, B.E. Hammons, M.J. Hafich [*Appl. Phys. Lett. (USA)* vol.76 (2000) p.327–9]

[15] M.R. Ravi, A. DasGupta, N. DasGupta [*IEEE Trans. Electron Devices (USA)* vol.50 (2003) p.532–4]

[16] S.M. Sze [*Physics of Semiconductor Devices, 2nd Ed.* (John Wiley and Sons, New York, 1981)]

[17] M. Terman [*Solid-State Electron. (UK)* vol.5 (1962) p.285–99]

[18] K. Remashan, K.N. Bhat [*Semicond. Sci. Technol. (UK)* vol.17 (2002) p.243–8]

[19] D. Shoji, T. Miura, M. Niwano, N. Miyamoto [*Appl. Surf. Sci. (Netherlands)* vol.130–132 (1998) p.441–6]

[20] S. Hohenecker, T.U. Kampen, T. Werninghaur, D.R.T. Zahn, W. Braun [*Appl. Surf. Sci. (Netherlands)* vol.142 (1999) p.28–32]

[21] S. Anantathanasarn, H. Hasegawa [*Appl. Surf. Sci. (Netherlands)* vol.216 (2003) p.275–82]

Chapter 4

Wet etching and photolithography of GaAs and related alloys

4.1 CHAPTER SCOPE

Since most III–V devices require the exposure of buried layers for metal contact formation and/or the formation of mesa structures for electrical isolation from adjacent devices on the same wafer, etching will always be a critical part of their fabrication. "Wet" (liquid-phase) etches are the simplest to implement since they require only simple benchtop glassware (or plasticware if HF is involved). There are, however, both advantages and disadvantages to using wet etching processes. The key issues to be considered in selecting a particular etch process for any application are rate, uniformity, selectivity, critical dimension control, feature profile and surface damage that can degrade electronic properties and, therefore, device performance.

The most favourable characteristic of wet etches is their ability to cause virtually no surface electronic damage. When superior surface electronic properties are most important, a wet process will clearly be preferable to a dry process, which often degrades surface electronic properties by ion bombardment.

The least desirable wet etch characteristic for many applications is the unavoidable tendency to undercut etch masks, making precise dimensional control more difficult. This is observed whether there are nearly equal vertical and lateral etch rates (isotropic etching) or etch rates that depend on the specific exposed crystal planes of the semiconductor (crystallographic etching). Precise control of critical dimensions is made more difficult by the frequently strong dependence of wet etch rates on temperature, solution agitation and crystallographic alignment of the mask. These aspects of wet etching, which will be discussed in some detail in this chapter, present both difficulties and opportunities for extra control.

A more detailed comparison between wet and dry (gas-phase) etch processes is found in Chapter 5. In general, if feature profile and tight control of dimensions are less important than surface electronic quality, a wet etch may be the process of choice. If vertical or

Advantages of wet etching
1) No surface electronic damage
2) Many chemistry choices
3) Chemical selectivity
4) Crystallographic etching
5) Inexpensive, simple equipment

Disadvantages of wet etching
1) Limited dimensional control
2) Vertical profiles hard to get
3) Solution waste disposal

controlled-angle profiles and small dimensions are needed, a dry etch that achieves the best compromise between feature profile and damage may be preferred. In this chapter, the issues of importance in selecting and using a wet etch process will be discussed.

When a pattern is to be etched, some method of protecting or masking portions of the surface from the etchants must be provided. Photolithography is the standard way of providing etch masks for pattern delineation. Many issues related to photolithography are independent of whether a wet or dry etch process is to be used. Therefore, we will address many common photoresist issues in this chapter. Some mask issues that are of importance only for dry etching will be addressed at appropriate points in Chapter 5.

4.2 MECHANISM OF WET ETCH PROCESSES

The underlying chemical mechanism of most wet etching of GaAs is the oxidation of the surface to form Ga and As oxides, followed by the dissolution of these oxides by chemical attack with acids or bases. Many GaAs etches call for the addition of hydrogen peroxide (H_2O_2) to serve as an oxidising agent to promote the formation of the surface oxides. This is especially needed for n-type material. Removal of electrons from the surface by an oxidising agent like H_2O_2 leaves an electron-deficient, "hole-rich" surface that reacts readily to form the oxide. Some other chemicals that have been used as the oxidising agent for etching GaAs are HNO_3, $K_2Cr_2O_7$, CrO_3, I_2 and Br_2. An alternative way to create a hole-rich surface in n-type material is to shine light on the surface. This can produce photochemical etching of n-type material. When ample quantities of chemical oxidising agent are present in solution, this is usually not a major effect. However, with dilute solutions, especially in bright light, photochemical etching may be an unintentional consequence if samples are left in solution for long periods of time. Under special conditions, local galvanic cells may arise that can etch GaAs without light or a specific oxidising agent.

Steps in wet etching
1) Diffusion of reactants to surface.
2) Formation of oxide on surface.
3) Acid or base attack on oxide and oxide dissolution.
4) Diffusion of products from surface.

Overall rate controlled by step 2)
Profiles controlled by step 3).

4.3 RATES AND PROFILES

The most important consideration in selecting a wet etch is the profile of the etched features. Since this is controlled primarily by whether the etching chemistry is reaction-rate limited or diffusion controlled, a brief discussion of reaction mechanisms is warranted.

A simplified representation of the chemistry typically involved in GaAs etching is given by the following equations. To initiate the etching process, an oxide is formed on the surface of the GaAs by reaction with an oxidising agent, typically H_2O_2.

$$GaAs + (\text{oxidiser}) \rightarrow GaO_x/AsO_y \tag{4.1}$$

The rate can be calculated from

$$\text{rate} = k_{ox}[\text{oxidiser}]^m\{GaAs\}\exp(-E^*_{ox}/kT) \tag{4.2}$$

Here, [oxidiser] represents the concentration of oxidiser, m is related to the number of oxidiser molecules involved in oxidising one Ga and one As, {GaAs} is the effective surface "concentration" of exposed Ga and As atoms, k_{ox} is the reaction prefactor, E^*_{ox} is the activation energy for the oxidation reaction, k is the Boltzmann factor and T is temperature in Kelvin. While this reaction may affect the overall rate, with rates being faster with higher oxidiser concentration, it is the second step, the dissolution of the oxide, that generally controls profiles.

The surface oxide is then attacked by either an acid or a base and is dissolved to give Ga^{3+} and AsO_4^{2-} ions in solution.

$$GaO_x/AsO_y + (\text{acid or base}) \rightarrow Ga^{3+} + AsO_4^{2-} \tag{4.3}$$

It is this second step in the total etching process that is usually key in determining whether the etching will be reaction-rate limited or diffusion controlled.

$$\text{rate} = k_{diss}[\text{acid or base}]^n\{\text{oxide}\}\exp(-E^*_{diss}/kT) \tag{4.4}$$

Here, [acid or base] represents the concentration of acid or base, n is related to the number of H^+ or OH^- involved up to the slowest step in the dissolution reaction, {oxide} represents the effective concentration of the surface oxide species that is slowest at dissolving, k_{diss} is the reaction prefactor and E^*_{diss} is the activation energy for the dissolution reaction.

Factors affecting diffusion versus reaction-rate control

Reactant concentration
Solution mixing
Solution viscosity
Temperature

Assuming an adequate supply of oxidiser, whether the rate of the dissolution process will be limited by the supply of acid or base (diffusion controlled) or by the activation energy of the dissolution reaction (reaction-rate limited) will be determined by the relative magnitudes of their respective terms. Any reaction has the potential to be either reaction-rate limited or diffusion controlled. It is merely a question of which term dominates under the conditions that are being employed to etch GaAs. If an activation energy for an etching process is less than 5 kcal/mol, it will probably be diffusion controlled, while a value greater than 10 kcal/mol probably means reaction-rate control.

Rates can depend on:

Doping type and level
Crystallographic facet
Heterojunctions
pn junctions
Nearby exposed metal

It is easy to see from these equations that changing concentration, solution mixing or temperature might shift the etching reaction from one dominant mode to the other. In fact, some common etching solutions perform well rather close to the switch-over point. In general, low concentrations of reactants favour reaction-rate control of etching while high concentrations favour diffusion control. Higher viscosities also favour diffusion control. Something as simple as diluting a solution to slow an etch rate can switch the dominant mode from diffusion to reaction-rate control. Consequently, caution must be used in altering "standard" recipes. Unintentional switching from one type of control to the other can produce very different results from those expected and destroy your devices.

Rate dependence on doping

$p^+ > p > n > n^+$
$n > SI$

The activation energy for the initial oxidation of the surface (E^*_{ox}) can vary appreciably depending on the crystallographic face that is exposed and on the doping type of the material. Consequently, it will always be important to determine the etch rate for a particular wafer structure rather than assuming that a literature value will apply in any specific case. The presence of heterojunctions and pn junctions in a particular device structure can profoundly affect the energetics of the oxidation and, therefore, the etch rate. Consequently, it can affect the profile, with rates generally being $p^+ > p > n > n^+$ and $n > SI$. This produces greater undercut for p-type material than n-type and a sloped profile across the depletion region near the pn junction. Doping effects are discussed in greater detail in Section 4.6 and heterojunction issues will be discussed with selective etching in Section 4.5.

Care must also be taken in aligning the mask with respect to the crystallographic axes of the wafer since there is a difference in the activation energy for oxidation of different crystal faces. This can produce totally different etch profiles for devices that are aligned differently with respect to the crystal axes. Crystallographic dependences will be discussed in Section 4.3.2.

Excessively fast rates will cause poor surface quality. Oxidation rate is generally controlled by the strength of the oxidiser so increasing the H_2O_2 fraction increases the rate; however surfaces tend to roughen with etchants that are $>15\%$ H_2O_2. Dissolution of the oxide is controlled by solution pH; surface roughness increases at pH > 7.5. At higher pHs, there is less efficient oxide removal so the thickening oxide can crack and float away in small segments. This exposes the underlying surface for further oxidation. This cracking and flaking away is inherently a non-uniform process and roughening is the result.

4.3.1 Diffusion control

The rate of diffusion-controlled reactions can be limited by the diffusion of reactants to the surface or of products away from the surface. For relatively narrow features, this rate limitation by diffusion can lead to decreases in the reaction rate as features are etched deeper and diffusion times into and out of the deepening trench increase. The exponential term ($\exp[-E^*/k_BT]$) is of less importance so these rates are relatively insensitive to temperature. However, they are highly sensitive to the degree of agitation so controlled stirring or rotation is important to achieve consistent and predictable results.

Slow rotation (<100 rpm) permits very uniform thinning of a wafer. Fast rotation in a centrifuge can decrease lateral undercutting. In practical terms, "rotation" often consists of manually swirling the sample holder in solution, which may not be very reproducible. Stirring at a reproducible rate with a magnetic stirrer or an over-head propeller design can produce more predictable results. Automated equipment is available for higher volume applications.

The classic profile of a diffusion-controlled reaction is shown in FIGURE 4.1. Lateral undercutting proceeds at a rate comparable to the vertical etch rate in the field region away from the mask. However, a shallow trench or downward bulge forms near the edge of the mask and slopes back away from the edge. This profile is well understood and has been quantitatively modelled [1] assuming no closely adjacent additional mask edges. The degree of trenching can be quite dependent on the areal ratio of exposed and masked areas. The classic example of a diffusion-controlled etchant for GaAs is the $HCl/H_2O_2/H_2O$ system. Under "ideal" diffusion conditions, i.e. with no convection, the ratio of the bulge depth to the lateral undercut beneath the mask is about 1.33 and the profile would remain constant with time (FIGURE 4.1(a)). However, some convection is always present during etching due to temperature and/or concentration gradients generated near the surface by the etching reaction itself. For short etch times, the bulge-to-undercut ratio is essentially 1.33, but as the etch proceeds in time, the mask-edge region appears more and more like the field region more remote from the mask edge and the downward bulge becomes less prominent with time (FIGURE 4.1(b)).

The potential problem of trenching should be considered before using these etches. For example, a trench-prone etch would be totally unsuitable for gate-recess etching. For acid/peroxide ratios of 1 HCl : y H_2O_2 : 1 H_2O, negligible trenching is obtained for $y \leq 10$ and up to 10–20% deeper etching near mask edges is obtained than at the centre between two masked areas for $y = 40$

(a)

(b)

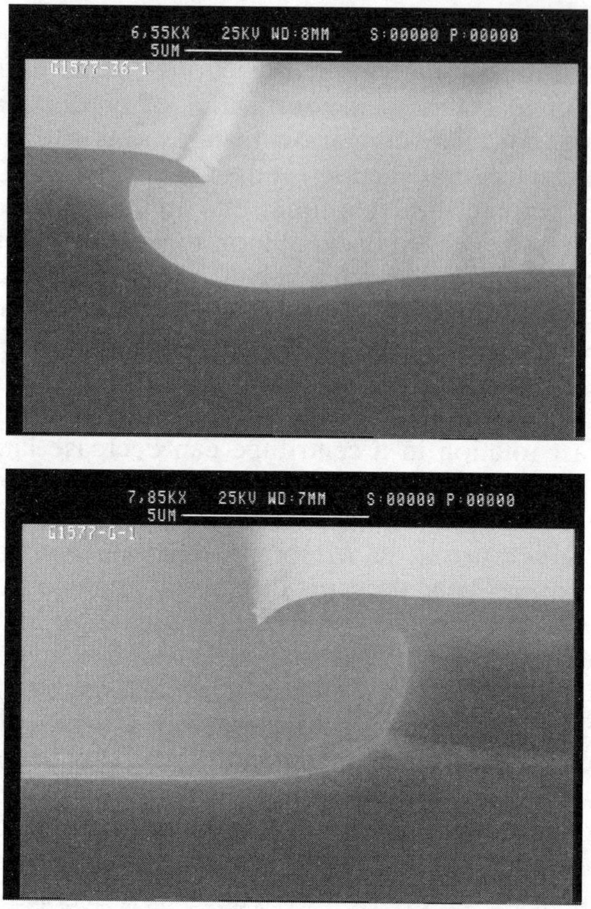

FIGURE 4.1 Diffusion-limited etch profiles: (a) classic profile; (b) profile with flatter bulge.

when the mask pattern consists of 100-μm squares with 100-μm spacing.

Trenching is also affected by the age of the etching solution. When etching is performed within a few minutes after mixing, little trenching is observed. In contrast, it can become particularly prominent with increasing age of the solution as the Cl_2 concentration increases in the solution with age. For dilute solutions ($y = 40$) used more than 20 min after mixing, trench depths can be more than 50% deeper near the mask edges. These solutions age rapidly. They start with a relatively deep yellow colour but become pale within 20 min. The temperature of the solution is also difficult to stabilise during this time. This time-dependent behaviour limits the usefulness of these etchants.

While classic profiles are generally obtained with diffusion-controlled etches, local galvanic effects can produce crystallographic effects using normally diffusion-controlled etchants.

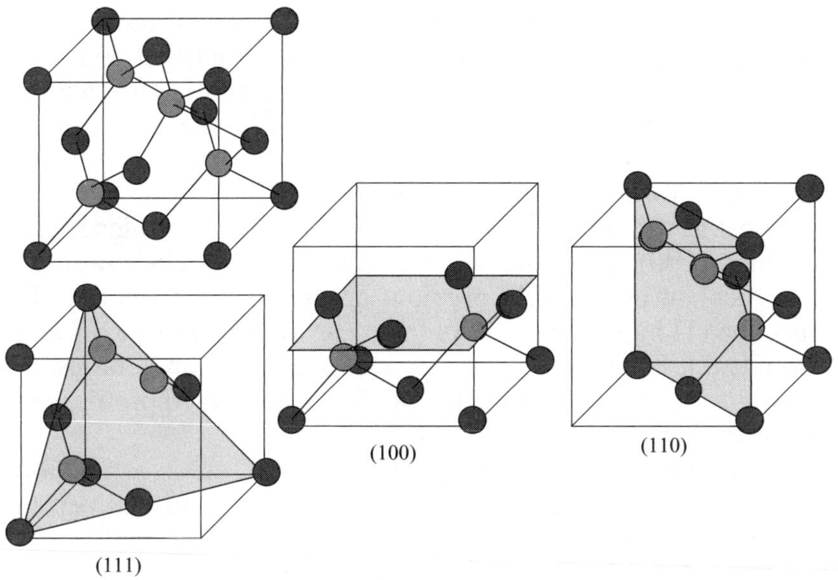

FIGURE 4.2 Crystallographic planes of GaAs.

When such effects occur, changing the concentration of reactants can sometimes move one back into the diffusion-controlled regime.

4.3.2 Reaction-rate control

For reaction-rate-limited etching, the exponential term of EQN 4.4 is of dominant importance. Consequently, these reactions display a pronounced dependence on temperature and are relatively insensitive to agitation. They also display differences in the etch rates of different crystal faces, which have different activation energies for their oxidation. This leads to the alternative name of crystallographic etchants for these solutions. This anisotropic etching leads to a variety of profile possibilities depending on how the etch mask is aligned relative to the wafer axes.

GaAs has the cubic zinc sulphide or zinc blende structure, with two interpenetrating face-centred cubic (fcc) lattices. Within the cube, there are four "molecules" of GaAs. Each Ga is bonded to four As in a tetrahedral configuration. This leads to chemically inequivalent (111) planes with the (111)A plane being composed on the top layer of Ga atoms and the (111)B plane being As atoms. In general, the (111)Ga face, also called the (111)A face, etches a factor of 2–5 slower than the following faces: (100), (110) or (111)As, also called (111)B. The orientations of the planes are shown in FIGURE 4.2. A wide range of profiles are possible; a number of possibilities are illustrated in FIGURE 4.3. Undesirable etch pits are more likely to form on (111) faces.

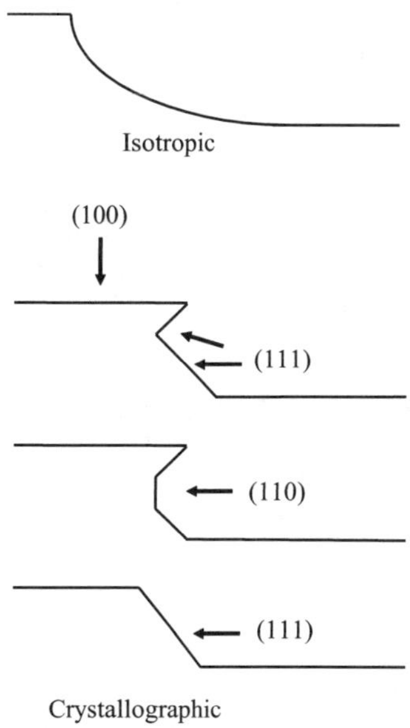

FIGURE 4.3 Crystallographic etch profiles in GaAs for (100) wafer.

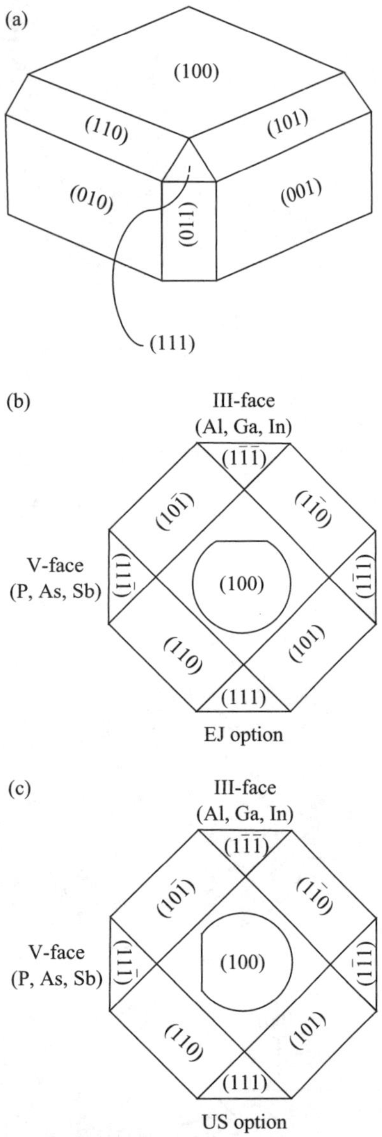

FIGURE 4.4 Crystal plane conventions.

Some comments about crystallographic notation are in order before discussing the crystallographic dependences of some wet etchants. While the flat edge of a GaAs wafer is a {110} cleavage plane, there are two alternative conventions for orienting the flat relative to the (111) faces (FIGURE 4.4). In the European-Japanese (EJ) option for a boule shape as shown in FIGURE 4.4, the flat is closest to the vertex of an upward-facing (111)Ga plane when the shiny (100) surface is pointed upward. In the US option, the (111)As plane assumes the previous orientation relative to the flat. Since the (111)Ga and (111)As faces can have markedly different etch rates, it is important to know which convention pertains to a specific wafer when aligning the mask or unexpected profiles may ruin your devices.

The profile of the etched device depends on its orientation relative to the (110) crystal planes (the cleavage planes perpendicular to a (100) wafer surface [2]). When etching with 1 : 1 : 8 $NH_4OH : H_2O_2 : H_2O$, placing the mask edges perpendicular to the $\langle 011 \rangle$ direction yields sloped walls due to a slower etch rate for (111)Ga planes. The (111) surface is at a 55° angle to the (100) surface along the direction of the natural cleavage planes. Mask edges placed perpendicular to $\langle \bar{1}10 \rangle$ produce vertical profiles due to the fast (111)As rate. Alignment of a linear pattern at 45 degrees to the natural (110) cleavage planes permits vertical (100) sidewalls with (100) wafers.

These crystallographic properties can be used to form surface gratings for distributed feedback (DFB) lasers, usually in combination with a subsequent regrowth, or to etch vertical facets for edge-emitting lasers.

4.3.3 Aging of etching solution

Etching solutions can change composition with the passage of time, with a resulting change in their associated etch rates. In general, rates decrease with increasing time even though the solution may visually appear the same. The rate at which this occurs depends on the details of the etching solution; if precise control of rates is required, it is a good policy to remeasure the actual rate of a less-than-fresh etchant immediately before using it.

Sometimes what seems like etch-rate variation is mainly due to temperature variations when the etchant is mixed. A good example is the $H_3PO_4/H_2O_2/H_2O$ system. The etch rate varies if the solution is used too soon after it is mixed. The temperature of the solution can remain above room temperature for about an hour due to the heat of mixing. After thermal equilibration, no change in etch rate is observed for the next week. After one week, it is probably a good idea to discard the etching solution and start afresh due to the possible instability of H_2O_2. If an etching solution is to be stored, it is very important to store it so water evaporation will be minimised to minimise changes in reactant concentrations.

Quite different behaviour is observed with HCl/H_2O_2 solutions. For these, the rate first increases, then decreases. The proposed origin for this behaviour is the initial formation of Cl atoms and Cl_2, followed by escape of the gas from the solution. The aging of HCl/H_2O_2-based solutions can be observed as visible colour changes from bright yellow to pale yellow over only a few minutes. The rate remains diffusion controlled, but trenching near the mask edges can increase as the solution ages.

4.4 PRACTICAL WET ETCHING

There are many different etchants that have been employed for GaAs and related compounds. A relatively comprehensive tabulation of wet etch rates for GaAs through 1995 is available in Chapter 18 of "Properties of Gallium Arsenide" [3]. Regardless of the specific etchant and the details of how it works, certain practical considerations must be kept in mind. Faster etchants are more difficult to control precisely than are slower ones due to etch-time uncertainties. In general, we prefer etchants with rates on the order of 0.2–0.5 μm/min over those with rates in excess of 1 μm/min. Patience is a virtue in etching as in most of life, and a few more minutes spent etching are preferable to overetched and ruined devices that require starting over. Final etch depth depends on the time that the wafer is in contact with the etching solution. When a wafer is first removed from the etching container, etching solution will remain on parts of the surface. It is very important to remove that solution quickly or etching may proceed for longer times than intended where the etchant remains. This can produce localised overetching and poor uniformity across the wafer. The faster the etching rate, the more significant this effect can be. There are two easy ways to solve this variable-etch-time problem. One can quickly stop the etching by rinsing the etchant solution from the surface with running DI water, taking care to ensure that the entire surface is very quickly rinsed and continuing to play the running water across the entire surface for a short period of time ($\leq$1 min). Alternatively, if a source of running DI water is not conveniently available or if it is crucially important to stop etching everywhere "simultaneously", the wafer can be completely immersed in an ample container of DI water, swished around and then immersed in a second container of clean DI water. This can be further repeated. The rapid dilution of the etchant solution immediately suppresses the rate of etching to a negligible level.

In many cases, the etch rate of a semiconductor depends strongly on pH and one is instructed to mix up the etchant and adjust the pH to a certain value. This is usually done by adding NH_4OH if the pH was too low or HCl if it was too high. Phosphoric acid might be a better choice when H_2O_2 is present. Note that in adjusting the pH, the composition of the solution is also changed by the added NH_4^+, PO_4^{3-} or Cl^- ions. This can be especially significant when the added ions are capable of forming complexes with metal ions, such as $M(NH_3)_x^{y+}$. These additional soluble products can increase the rate of etching. When a ternary or quaternary material is being etched, this effect can produce unexpected changes in etch selectivity as well as rate.

When the operation of the final device will not be adversely affected, it is sometimes possible to grow into the structure a thin etch-stop layer of another semiconductor composition that etches very slowly in the etching solution. This can yield very precise control of etch depth with considerable variation in etch time, but one must always remember that the semiconductor material is etching laterally while it is etching vertically, so final device dimensions must always be kept in mind even when an etch-stop layer is used. Issues related to such compositional selectivity are discussed in more detail in Section 4.5.

Another important consideration is the chemical character of the wafer surface before one begins to etch. All III–V semiconductors form native oxides whose thickness and exact composition can vary depending on what atmosphere they have experienced and how long they have been exposed to it. In addition to oxidation, residues of volatile organic compounds can also be present. These issues have been discussed extensively in Chapter 3. The native oxide may react quite differently from the oxide formed in the solution oxidation reaction. This can produce unexpected variations in the initial etch rate, depending on the sample history. The time before the etch rate assumes its "normal" value is sometimes referred to as the incubation time and can vary appreciably depending on the specific wafer history. It can also vary across the surface of a single wafer, leading to etch non-uniformity. This effect tends to be more important for compositions containing Al or especially Sb, which has several oxide forms with some being highly resistant to some etchants. Faster etches tend to leave a thicker residue of oxidised semiconductor on the surface than slower etches do. If multistep etching is to be performed, especially for thin layers where etch initiation may be a major fraction of total etch time, it is important to remove this residue and allow a native oxide to reform uniformly across the surface. To obtain the most uniform results, precleaning of the surface, as discussed briefly in Section 4.4.2, is generally desirable.

Although some etches reputedly don't etch a particular semiconductor, they may actually be forming an insoluble oxide at the semiconductor surface. This will usually appear as a "stain" compared to a pristine wafer surface. Subsequent immersion in a solution that can etch that oxide, e.g. phosphoric or hydrochloric acids, may etch that oxide away. Even if the stain remains unaltered through subsequent processing, this is a sign that there has been degradation of your wafer and there may be deleterious electronic consequences for the devices.

When there is a high concentration of peroxide in an etchant, there will sometimes be a problem with bubbles forming on the surface of the semiconductor during the etching process. When this occurs, the bubbles will locally mask the semiconductor until they release from the surface and this can lead to degraded surface quality. Solution agitation can help prevent this problem.

Many etch rates are reported as values at "room temperature". However, "room temperature" can typically be anywhere between 18 and 25°C, so reproducible etch results require knowledge and control of the actual temperature, especially for reaction-rate-limited etchants. A rule-of-thumb is that a 10 degree rise in temperature will double the rate of a reaction with an activation energy of 15 kcal/mol, so attention to temperature can be very important when etch depths must be carefully controlled. A non-obvious source of temperature variation is the evaporative cooling of a solution standing in a fume hood with substantial air flow. This can alter the temperature of a standing solution by as much as three degrees while standing for only a few minutes, so it is important to measure solution temperatures immediately before etching when precise depth control is critical. It is also important to maintain a constant airflow environment to keep the evaporative cooling constant.

4.4.1 Photoresist issues

Patterned etching requires a mask to protect regions of a wafer where no etching is desired. A brief discussion of photoresists is, therefore, appropriate before considering specific wet etch chemistries. The different formulations of photoresist can be divided into two general types: positive tone and negative tone. In either case, a thin film of photoresist that has been spun onto the wafer surface is exposed to light. This light causes a chemical change in the exposed regions that alters the solubility of the photoresist film in a basic solution commonly referred to as the developer. When the exposed regions are dissolved by the

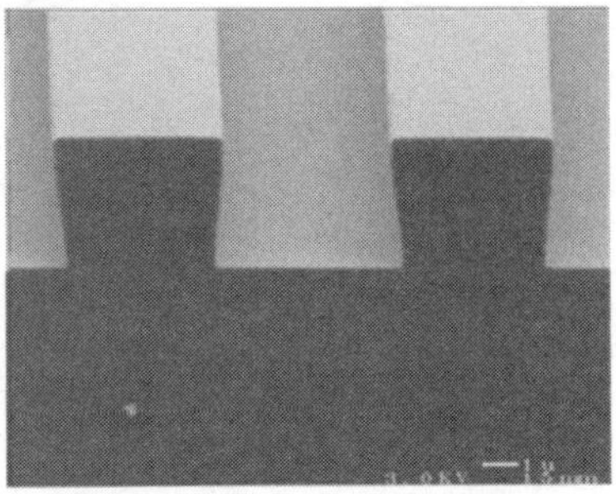

FIGURE 4.5 Re-entrant profile of a negative-tone photoresist.

Mercury lamp lines for photoresist exposure

G-line = 436 nm
H-line = 405 nm
I-line = 365 nm

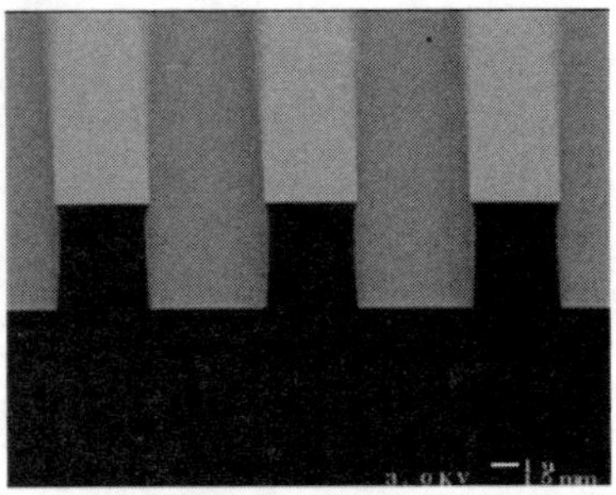

FIGURE 4.6 Typical profile of a positive-tone photoresist suitable for wet or dry etching.

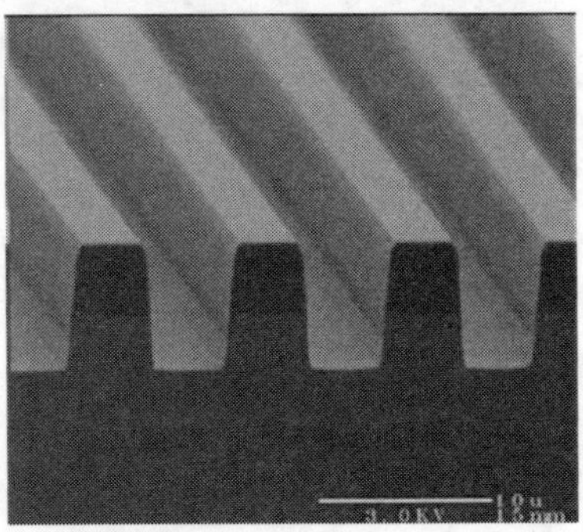

FIGURE 4.7 Sloped-wall profile of positive-tone resist suitable for wet etching.

developer, i.e. the exposed regions have become more soluble, the resist is designated a positive resist. Conversely, when the exposed regions have become less soluble, the resist is called a negative resist. The earliest photoresists, based on photocross-linking of a rubber by an additive containing two azide groups, were negative-tone resists. Photoresists based on cross-linking of poly(vinyl cinnamate) are also negative-tone resists. Negative-tone photoresists tend to have a re-entrant profile (FIGURE 4.5), where the top of the resist is wider than the bottom at the wafer surface. This type of profile is very useful for metal lift off, as will be discussed in Section 6.3.1.

More commonly, positive tone photoresists based on Novolac polymers are used today. Such materials can achieve sub-micron resolution. Chemically, they consist of phenol-formaldehyde polymers with a small amount of diazonaphthaquinone dissolved in them. When irradiated, photoreactions produce a chemical rearrangement that eventually produces a carboxylic acid when exposed to the basic (OH^- containing) developer solution. While these photoaltered regions are highly soluble in the basic developer, the unirradiated material has a very low solubility in basic developer.

Photoresist exposure parameters are often described in terms of particular lines from a mercury lamp. The most common lines so referenced are G-line at 436 nm, H-line at 405 nm and I-line at 365 nm. Since the absorption coefficient for a particular resist may vary significantly at the different exposure wavelengths, different exposure times may be required depending on the wavelength being employed.

Ideally, a uniform amount of absorption and chemical reaction would occur through the entire thickness of the photoresist. This produces a relatively straight-walled profile, as shown in FIGURE 4.6. Resists are developed for optimum use at a particular wavelength of light where modest absorption coefficients make the uniform-absorption assumption approximately true and straight walls should be possible. However, light absorption is an exponential process, so variations in exposure conditions and subtle differences in resist composition can produce varied profiles from nominally similar resists. FIGURE 4.7 shows a sloped-wall profile obtained with a similar Novolac-type positive-tone resist to that in FIGURE 4.6. The angled walls would be equally suitable for wet etching, but the angle in the resist profile would be copied into the semiconductor during dry etching. The deliberate introduction of significant wall slope to dry-etch sloped-wall mesas will be discussed in Chapter 5.

Light absorption is an exponential process, so higher absorption coefficients at a shorter wavelength than recommended can lead to heavy absorption near the top surface, with extensive reaction in this region, and less light reaching the depths of the photoresist, resulting in much less reaction near the wafer surface. This can produce quite different solubilities in the developer for the surface region and the lower region of the resist. Taken to extreme, this produces resist failures. For a negative resist, the mask may even be released from the wafer surface. For a positive resist, the areas one wishes to expose to etchant may remain masked.

Another optical phenomenon is the formation of standing waves due to the reflection of light from the wafer surface (they are moderately good mirrors at typical exposure wavelengths). This produces alternate regions of higher and lower light intensity throughout the photoresist thickness. Depending on the absorption coefficient and the exposure time, this may lead to differences in solubility in the developer, as described above, and affect resist profile. Antireflection treatments can be applied prior to application of the photoresist if this seems to be a problem. This is not a significant problem with multiwavelength light sources, whose superposition of standing waves washes out the effect.

A typical process sequence for preparing a photoresist mask consists of the following steps. First, an adhesion promoter, such as HMDS (hexamethyldisilane), is applied to the surface. Although HMDS was initially developed as an adhesion promoter for Si and SiO_2 surfaces and it is not clear that it assists adhesion on III–V materials, it is nevertheless used by many. Then the photoresist is spin-coated onto the surface. The thickness of the film is determined by the photoresist viscosity and the spinning speed. The thickness is generally proportional to (spin speed in rpm)$^{-1/2}$. A pre-bake (or “soft bake”) is performed before exposure to remove excess solvent and densify the photoresist. Although an oven can be used, it is generally better to use a semiconductor hotplate to get greater uniformity in photoresist behaviour. Pre-bake conditions depend on the particulars of the resist, but typical conditions might be 1 min at 90–110°C. Resist can pile up at the wafer edges, causing an “edge bead” problem that makes proper exposure during contact lithography difficult. One can pre-expose the beaded-up area and remove it before proceeding with the patterning lithography. After exposure to light for a carefully controlled time to insure sufficiently complete reaction in the exposed areas, the wafer is ready for development.

Typical photoresist sequence

1) Adhesion promoter (optional)
2) Spin on resist
3) Pre-bake
4) Edge-bead removal
5) Patterned exposure
6) Develop
7) Inspect with microscope
8) Post-bake
9) Residual organics removal
10) Inspect with microscope

When the photoactive compound in the resist is exposed to UV light, N_2 gas is a reaction product. If this N_2 is not able to escape before the bulk of the resist becomes heavily cross-linked, pockets of gas will be trapped. These “blisters” can pop and hurl resist debris onto the wafer surface, producing masking in unintended locations. Using the right intensity of light to produce a rate of reaction that allows sufficient gas escape will avoid this.

The development time must be selected to provide adequate clearing of the exposed areas without degradation of the resist. In general, one can see the pattern appear as the development proceeds. Gentle agitation during this process helps with uniformity. Depending on the density of the pattern, the proper development time may vary somewhat. The wafer should remain briefly in the developer after the pattern appears to be clearly developed. Then it should be thoroughly rinsed in DI water. Rinsing with flowing DI water until the resistivity of the water that has been in contact with the wafer is above 6 MΩ insures complete removal of any residual developer. The wafer should be dried with N_2 before proceeding with the etching process or the concentration of etchant in contact with the wafer surface may be diluted from that which was intended.

The wafer should always be inspected with a microscope at this point. If there is a problem with the pattern development at this point, it is still possible to strip the resist with acetone and try again. For example, if a feature appears to be larger or smaller than the desired size, either overdevelopment or underdevelopment has occurred. These can be caused by errors in exposure time, pre-bake temperature, and development time or temperature. Troubleshooting involves systematically varying these one at a time until the source of the problem is identified and the proper conditions are identified.

Post-development baking of the patterned wafer can slightly toughen the photoresist and enhance its resistance to the subsequent etch process. This is generally done between 90 and 110°C for a few (5 or so) minutes. Baking at too high a temperature (near or above the glass transition temperature) may cause the resist to "reflow" and change the sidewall profile (Section 5.10.3), so careful attention to temperature is important. Post-development baking is especially beneficial for dry etching with plasmas but is less important for wet processing. Photoresists are always etched somewhat during plasma processing. The photoresist issues related to their behaviour in plasma environments are deferred to Chapter 5, where we discuss dry etching processes in detail.

If a large quantity of solvent is too suddenly evaporated from the resist, the top portion of a resist pattern may shrink excessively; this is called pullback. In general, too rapid temperature changes are not good for resist integrity and should be avoided whenever possible.

Following the post-bake, a final clean-up of residual organics from the exposed GaAs surface is important to avoid unwanted surface roughening as the residues act as local micromasks. Atomic oxygen is the method of choice for this. An ozone clean can be used, or a very-low-energy oxygen plasma etch, where the energy of the ions striking the surface is less than 10 eV, can be used. The plasma etching approach is discussed in greater detail in Section 5.5.

Some etchants, such as those involving HF, can attack photoresist rather aggressively. Other etchants, such as citric acid, produce no mask attack. When employing an etchant for the first time, it is important to check the behaviour of the photoresist to insure that it won't fail during a time that is at least twice as long as you intend to etch the GaAs wafer. Chemical incompatibilities should be identified before starting lithography with real device wafers.

After etching is completed, the wafer should be soaked in acetone. This swells the photoresist and removes it from the wafer. If the resist does not come off easily with simple soaking, ultrasonic agitation for a few minutes will usually effect its removal. A few monolayers of photoresist are virtually always left on the surface after acetone dissolution. These can be completely removed by a low-energy oxygen plasma "descum" process (Section 5.6).

Modifications in the processing of positive-tone resists can also produce re-entrant profiles suitable for metal liftoff applications. The simplest method can be overexposure of the resist. The same resist that produces a sloped-wall profile in FIGURE 4.7 is shown with a re-entrant profile after overexposure in FIGURE 4.8. The discontinuous metal film that results from deposition on re-entrant photoresist is also seen in the figure.

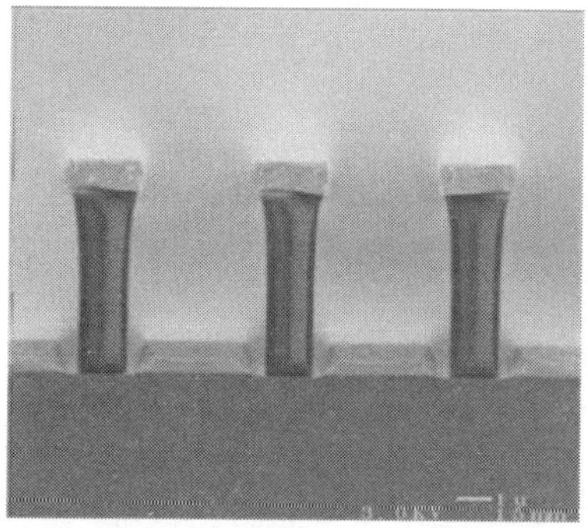

FIGURE 4.8 Re-entrant profile from overexposure of positive-tone resist used in FIGURE 4.7.

A technique that is generally useful to obtain a re-entrant profile with any positive-tone resist is image reversal. For image reversal, the resist is first exposed to create an internal pattern. It is then baked at high temperature (110°C for 90–180 s). Before development, the resist is then flood-exposed over the entire surface. Following development, the resist is dissolved except where the first exposure and baking has produced some crosslinking. An image-reversed profile is shown in FIGURE 4.9. Virtually any Novolac-type resist can be made to image-reverse if treated with ammonia vapour before the first exposure.

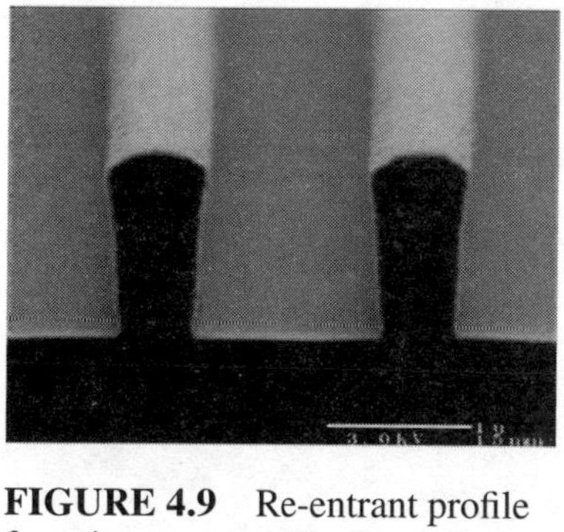

FIGURE 4.9 Re-entrant profile from image-reversal of a positive-tone resist.

While resist failure in dry etching is generally due to direct etching or sputtering of the resist, failure in wet etching is commonly by loss of adhesion to the GaAs surface. This potential problem becomes more likely with increasing undercut as the etch proceeds deeper into the surface. The depth at which this may become a problem will depend on the size of the feature with larger ratios of undercut-area-to-unetched-area being more susceptible. Solution agitation can help rip off the resist when undercutting is extensive, so diffusion-limited etchants, where agitation is important for uniform etching, are more susceptible to this problem.

4.4.2 Basic wet etches

Basic wet etches
1) Etch oxide.
2) Degrease surface.
3) Complex and dissolve metallic impurities if NH_4OH-based.
4) Generally don't attack contact metals.

Basic hydroxide-based etchants can do more than merely etch the semiconductor. They provide some degreasing of the surface by saponification of greasy organic residues that may be present. Saponification is the process of hydrolysing a fat (a fatty acid ester) with OH^- to form glycerol and a fatty acid salt (soap). NH_4OH-based etches are preferred over ones based on KOH or NaOH because they can also remove metallic impurities, like Cu, Cr and Al, from the surface by forming chemical complexes with the metal ions. The NH_3 in NH_4OH etchants forms soluble complexes with many other metal ions as well. (NH_4OH is formed by dissolving NH_3 in water: $NH_3 + H_2O \leftrightarrow NH_4OH$.) Some typical basic etchants are listed in TABLE 4.1.

Etching with $NH_4OH/H_2O_2/H_2O$ can yield either isotropic or crystallographic profiles. For example, a ratio of 3 : 1 : 1 produces

TABLE 4.1 Basic wet etches.

Etchant	Etch rate (μm/min)	T (°C)	(111)Ga/ (111)As	Ref.	Comments
NH_4OH/H_2O_2	0.1	25	–	[A]	pH = 7.04 ± 0.02, agitated
NH_4OH/H_2O_2	0.017	25	–	[A]	pH = 7.04 ± 0.02, stagnant
NH_4OH/H_2O_2	4.1	25	–	[B]	pH = 8.4, 32 : 1 vs. $Al_{0.16}Ga_{0.84}As$
0.3N NH_4OH/	0.18–0.2	r.t.	1/6–1/4	[C]	n-type
0.1N H_2O_2	0.12–0.14	r.t.	1/6–1/4	[C]	p-type
3N NH_4OH/ 1N H_2O_2	1.4	r.t.	<1	[C]	little mask attack
NH_4OH/ H_2O_2/H_2O	0.5–5	25	f(ratios)	[D]	very smooth
0.08M NH_4OH/ 0.2M H_2O_2	0.3	25	1/4	[E]	pH 10–12
1NH_4OH/ 1H_2O_2/8H_2O	1.5	r.t.	–	[F]	
250 H_2O_2/ 1 NH_4OH	0.1	20	–	[G]	GaAs/$In_{0.1}Ga_{0.9}$As >50 : 1

[A] R.A. Logan, F.K. Reinhart [*J. Appl. Phys. (USA)* vol.44 (1973) p.4172–6]
[B] K. Kenefick [*Electron. Lett. (UK)* vol.21 (1985) p.558–9]
[C] J.J. Gannon, C.J. Nuese [*J. Electrochem. Soc. (USA)* vol.121 (1974) p.1215]
[D] L.P. Molodyakova, V.M. Andreev [*Inorg. Mater. (USA)* vol.12 (1977) p.956]
[E] J.J. Kelly, A.C. Reynders [*Appl. Surf. Sci. (Netherlands)* vol.29 (1987) p.149]
[F] S.H. Jones, D.K. Walker [*J. Electrochem. Soc. (USA)* vol.137 (1990) p.1653]
[G] D.G. Hill, K.L. Lear, J.S. Harris Jr. [*J. Electrochem. Soc. (USA)* vol.137 (1990) p.2912]

crystallographic profiles while a ratio of 2 : 5 : 3 produces isotropic ones at "room temperature".

In the absence of H_2O_2, dilute NH_4OH solutions are often used to gently remove native oxide and any greasy residues from GaAs as a surface preparation step for both wet and dry etching processes (Section 3.2). This precleaning produces a much more uniform surface for the etchant to attack. This can be very important for etch uniformity when short etching times are involved and variation in initial etch rates at different places on the wafer might produce differences in total etch depth. Typical dilutions for this application are 1 : 10 to 1 : 20 $NH_4OH : H_2O$. Whether acidic or basic etchants are to be used, this precleaning step can improve etch reproducibility and uniformity. It is especially important when relatively small thicknesses are to be removed. Under such conditions, any non-uniformity in the initial etch rate due to variations in either native-oxide thickness or virtual masking by localised organic residues can produce unacceptable lateral variation in the total etch depth.

Since a basic solution will etch glass at a finite etch rate, the container used for wet etching is important. With basic etchants, use of a suitable plastic container is advisable.

4.4.3 Acidic wet etches

Acidic wet etches
1) Etch oxide
2) Good compositional selectivity with organic acids.

GaAs is not soluble in non-oxidising acids from pH 1 to 7, so an oxidiser such as H_2O_2 is combined with the acid to produce an effective etchant. Examples of non-oxidising acids include HCl, H_2SO_4, H_3PO_4 and organic acids such as citric and succinic acids. Nitric acid can itself serve as an oxidising agent but additional oxidising agent in the form of H_2O_2 is still usually added. TABLE 4.2 lists some acidic etchants. The listed ratios are volume ratios and refer to volumes of concentrated acids and bases as purchased (i.e. undiluted), 30% H_2O_2, 50% by weight citric acid in water or 20% by weight succinic acid in water. When specific pHs are listed, NH_4OH is generally used to adjust the pH. Note that adjusting the pH also introduces NH_4^+ ions into solution, which may play a role in the process.

The phosphoric acid family of etchants are especially useful because they produce very smooth surfaces and can provide control of etch depths at the level of tens of angstroms. Some of these, including our personal favourites, are listed in TABLE 4.2. The most commonly used compositions etch anisotropically. The 1 $H_3PO_4 : 4\ H_2O_2 : 45\ H_2O$ mixture is non-selective for As-based materials and is useful for both the emitter mesa and base mesa etching of AlGaAs/GaAs HBTs, as discussed in Section 9.3, and for mesa etching of FETs. They etch P-containing materials much

TABLE 4.2 Acidic wet etches.

Etchant	Etch rate (μm/min)	T (°C)	(111)Ga/(111)As	Ref.	Comments
GaAs etchants					
$H_3PO_4/H_2O_2/H_2O$	0.01–4	30	f(ratios)	[a]	reproducible to 10s of Å for below 0.1 μm/min
$H_3PO_4/4H_2O_2/45H_2O$	0.3	19		#	very smooth
$H_3PO_4/4H_2O_2/495H_2O$	0.03	19		#	reproducible to 10s of Å
*Citric acid/H_2O_2	0.6	24	1/2	[b]	no mask attack
*10 citric acid/$1H_2O_2$	2	18	–	[c]	95 : 1 vs. $Al_{0.3}Ga_{0.7}As$
*Citric acid/H_2O_2	0.15–0.4	27–100	–	[d]	select vs. $Al_{0.28}Ga_{0.72}As$: >80 : 1 to 0.7 : 1 as f(H_2O_2)
*4 citric acid/$1H_2O_2$	0.36	20	–	[e]	select vs. $Al_xGa_{1-x}As$: x = 0.3, 0.45, 1 : 155, 260, 1450
*Citric acid/H_2O_2:	0.006–0.040	r.t.	–	[f]	116 : 1 to 1 : 1 vs. thick $Al_{0.3}Ga_{0.7}As$ as f(citric : H_2O_2)
0.5 : 1 to 50 : 1		–	–		
(5 : 1 max rate)	(0.31)				
*4 citric/$1H_2O_2/1H_2O$	0.35	300–273	–	[d]	1 : 1 vs. $Al_{0.28}Ga_{0.72}As$:
**15 succinic/$1H_2O_2$	0.15	r.t.	–	[g,h]	select vs. $Al_xGa_{1-x}As$ as f(succinic : H_2O_2, x, pH, T)
$HCl/4H_2O_2/40H_2O$	0.22	r.t.	below 1	[i]	
$40HCl/4H_2O_2/H_2O$	>5.0	300	1	[i]	isotropic
$1HCl/20CH_3OOH/3H_2O_2$	0.27	293	–	[j]	select. vs. GaInP as f(H_2O_2)
HF	0.1	353	–	[k]	1 :>10 vs. Al fraction ≥0.4
$HF/H_2O_2/H_2O$	10–0.1	298	f(H_2O_2)	[l]	high ($H_2O_2 : H_2O$) = xtallog. low ($H_2O_2 : H_2O$) = isotropic
$2HNO_3/3H_2O_2$	20.5	–	4/5	[m]	(111)A/(111)B≈1 for 1 : 9 mix
$H_2SO_4/H_2O_2/H_2O$	3.5	273	up to 1	[n]	orientation dependence f($H_2SO_4 : H_2O_2$)
$1H_2SO_4/8H_2O_2/1H_2O$	14	r.t.	1/4	[i]	
Other III–V etchants					
AlGaAs					
$1HF/6H_2O$	3.8	r.t.		#	lateral undercut rate for 70% Al
$1HF/3$ isopropanol/$6H_2O$	2.9–3.1	r.t.		#	lateral undercut rate for 90% Al
InGaP					
$7H_3PO_4/1HCl$	0.12	r.t.		#	for InGaP vs. GaAs or InGaAs
InGaAsP					
$3H_2SO_4/1H_2O_2/1H_2O$	2.0	r.t.		#	for $In_{0.58}Ga_{0.42}As_{0.9}P_{0.1}$ vs. InP
$3H_2SO_4/1H_2O_2/10H_2O$	0.25	35		#	for $In_{0.74}Ga_{0.26}As_{0.57}P_{0.43}$ vs. InP

Precleaning in 10 : 1–20 : 1 H_2O/NH_4OH improves etching reproducibility
Ratios are volume ratios using concentrated acid sources unless otherwise indicated
* Citric = 1 g citric acid/1 g H_2O (50 wt% solution); H_2O_2 = 30 wt% solution; ** Succinic = 1 g succinic acid/5 g H_2O; H_2O_2 = 30 wt% solution; NH_4OH adjusts pH; r.t. = assumed "room temperature", i.e. from 19–25°C
# Authors' data

[a] Y. Mori, N. Watanabe [*J. Electrochem. Soc. (USA)* vol.125 (1978) p.1510]
[b] M. Otsubo, T. Oda, H. Kumabe, H. Miki [*J. Electrochem. Soc. (USA)* vol.123 (1976) p.676]
[c] C. Juang, K.J. Kuhn, R.B. Darling [*J. Vac. Sci. Technol. B (USA)* vol.8 (1990) p.1122]
[d] B.-Y. Mao, J.A. Nielsen, R.A. Friedman, G.Y. Lee [*J. Electrochem. Soc. (USA)* vol.141 (1994) p.1082]
[e] M. Tong, D.G. Ballegeer, A. Ketterson, E.J. Roan, K.Y. Cheng, I. Adesida [*J. Electron. Mater. (USA)* vol.21 (1992) p.9]
[f] G.C. DeSalvo, W.F. Tseng, J. Comas [*J. Electrochem. Soc. (USA)* vol.139 (1992) p.831]
[g] A.J. Tang, K. Sadra, B.G. Streetman [*J. Electrochem. Soc. (USA)* vol.140 (1993) p.L82]
[h] S.A. Merritt, M. Dagenais [*J. Electrochem. Soc. (USA)* vol.140 (1993) p.L138]
[i] D.W. Shaw [*J. Electrochem. Soc. (USA)* vol.128 (1981) p.874]
[j] J.R. Flemish, K.A. Jones [*J. Electrochem. Soc. (USA)* vol.140 (1993) p.844]
[k] X.S. Wu, L.A. Coldren, J.L. Merz [*Electron. Lett. (UK)* vol.21 (1985) p.558–9]
[l] T. Takebe, T. Yamamoto, M. Fijii, K. Kobayashi [*J. Electrochem. Soc. (USA)* vol.140 (1993) p.1169]
[m] M.M. Berdichenko, L.N. Vozmilova, V.G. Malyarova [*Inorg. Mater. (USA)* vol.24 (1989) p.1028–9]
[n] S. Iida, K. Ito [*J. Electrochem. Soc. (USA)* vol.118 (1971) p.768]

more slowly and a thin InGaP layer can be used as an etch-stop layer.

The hydrochloric acid family of etchants generally produce isotropic profiles. Because these etching solutions tend to change composition, and therefore change etching behaviour, over a relatively short period of time, these would generally not be the etchants of choice for precisely controlling etch depths. Consistent agitation and conscious consideration of the age of the solution are essential if these etchants are used. Hydrochloric acid alone, without the addition of oxidiser, can etch P-based materials selectively versus As-based ones, as discussed in Section 4.5.

The sulphuric acid family of etchants generally produce crystallographic profiles. Consequently, they are not generally the best etchants to use if one wants crystallographic independence. However, this feature may be advantageously employed if one wants to etch vertical facets, as discussed in Section 4.3.2. They have proven useful for etching smooth, vertical laser facets when properly aligned on the wafer.

Etchants based on the organic acids, citric and succinic, can produce very smooth surfaces and do not attack photoresist masks. These etchants have the added advantage that they can provide selective etching for GaAs versus AlGaAs, as will be discussed in greater detail in Section 4.5.

The hydrofluoric acid family of etchants require the addition of an oxidiser to etch GaAs. However, HF without added oxidiser can selectively etch AlGaAs with an Al mole fraction >0.45. With added H_2O_2, the etching of GaAs will be crystallographic if the H_2O_2/H_2O ratio is high but isotropic if the ratio is low. HF is an especially aggressive etchant and attacks photoresist readily. CAUTION!! It is also the most DANGEROUS of the acids to use, even when quite dilute. When in contact with human skin, it is an anaesthetic so a person often does not realise they have been exposed until major damage has occurred. Attack on tissue will continue until the F ions are tied up by reaction with Ca, such as at the bones, producing deep acid burns that are horribly painful. Immediate treatment with Ca-containing medications such as Ca-gluconate cream is vital to minimise damage and a physician should be seen immediately.

4.4.4 Metal etches

There are occasions when an error in metallisation makes its removal necessary, hopefully without damaging the semiconductor excessively. Many metals can be simply dissolved by dilute or concentrated acid without the addition of an oxidiser such as H_2O_2. Unless otherwise specified, a "dilute" acid is often defined as having a 6N concentration. For these metals, removal of metal with little damage to the GaAs layers is possible. If InP is exposed, there is a greater problem since it will etch in some acids without addition of oxidiser.

Unfortunately, many of the best metallisations for GaAs-based devices employ noble metals such as Au, Pt or Pd, as discussed in Chapter 7. These pose more of a problem for removal without damage. Gold and Pt can be etched in aqua regia (3 : 1 HCl/HNO_3) at rates in excess of 20 μm/min, but display poor selectivity and significant GaAs sidewall attack will occur. Pd may be etched in 1 : 10 : 10 HCl/HNO_3 : acetic acid at a rate of 0.1 μm/min.

Gold and Pd can also be etched with 4 : 1 : 40 KI : I_2 : H_2O at a rate of 0.5–1 μm/min. This etchant is compatible with photoresist, but the etch will also attack GaAs. A deionised water rinse after the etch should be followed by a methanol rinse or soak to ensure adequate removal of residual iodine. Although the metal may be successfully removed, it is unlikely that the device will not be adversely affected by the process.

4.5 COMPOSITIONAL SELECTIVITY

The vast majority of GaAs devices are heterostructure devices, which makes the issue of how a particular etchant etches ternary and quaternary compounds of varying mole fraction a matter of great importance when choosing an etchant for a particular application. Sometimes one wants equirate etching of all the semiconductor layers in a device; alternatively, virtually total selectivity for one composition over another is sometimes the goal. The important issues for compositional selectivity are the topic of this section.

Since wet etching is based on holes being generated in the semiconductor surface region by an oxidising agent in solution followed by reaction with water, a redox chemistry capable of injecting holes into the valence band of the semiconductor is required. Because every semiconductor composition has its valence band at a different energy relative to the vacuum energy level, it is possible to select a redox solution (etchant solution) that will inject holes into one semiconductor composition while not being able to inject into another one. For example, while a half-reaction potential for oxidation greater than +0.90 V versus the standard hydrogen electrode is needed to etch p-GaAs, a potential of only +0.54 V is required to etch p-InP. (Note that, by definition, the standard hydrogen electrode has an oxidation potential of $E^0 = 0.00$ for a pressure (fugacity) of 1 atmosphere and $[H^+]$ of 1 M, but its actual potential relative to the vacuum level is -4.5 ± 0.2 V.)

The potential of the standard hydrogen electrode is -4.5 ± 0.2 V below the vacuum level.

The Gibbs free energy, ΔG, for a half-reaction is given by

$$\Delta G = -n\mathcal{F}E \qquad (4.5)$$

where n is the number of equivalents reacting, and $\mathcal{F}$ is Faraday's constant (96,485 C/mol). The half-potential (electromotive force or emf), E, is affected by both the pH and the concentration of the oxidant. For example, E^0 for 1 M H_2O_2 in water is +1.776 V if acidic and +0.87 V if alkaline (basic). Spontaneous reactions have positive emfs, $E^0 > 0$, and negative changes in free energy, $\Delta G^0 < 0$. The dependence of E on concentration is given by the Nernst equation (EQN 4.6) for reversible potentials. It is used to account for the effects of the ionic species activity when the reactants and products are not in their standard states (1.0 M, 1.0 atmosphere):

$$E = E^0 - (RT/n\mathcal{F})\ln[\Pi a(\text{red})^x/\Pi a(\text{ox})^y] \qquad (4.6)$$

for oxidant + n electrons = reduced species, where E^0 is the standard emf for the half-reaction, n is the number of equivalents of charge reacting, T is the temperature in Kelvin, R is the gas constant, $\mathcal{F}$ is Faraday's constant, a is the activity (concentration) of either oxidant or reductant (activity = concentration at lower concentration values) and the exponents x and y correspond to the stoichiomentry of the oxidants and reductants in the reaction. The Π symbols indicate that the product of the reductant concentration terms are divided by the product of the oxidant concentration terms. Solids are treated as having a "concentration" of 1. At 25°C, $RT/\mathcal{F} = 0.059$. The standard emf, E^0, is the oxidation potential for the half-reaction at 1 unit of activity (typically activity ≈ molar concentration) and f = 1 (fugacity ≈ pressure in atmospheres). These values are available in standard emf tables for a wide variety of half-reactions. Be very careful to make sure to use the proper sign for the emf for the direction in which your reaction is proceeding. If the reaction you are considering is written in the opposite direction of that in the table, the sign of the emf must be reversed as well.

It is important to note that the emf for a reaction depends on pH, so be sure to use basic emf tables for basic solution calculations and acidic tables for acidic solution calculations. Additional details are available in many textbooks [4]; tables of emfs versus pH are also available [5,6].

4.5.1 GaAs versus AlGaAs selectivity

We will discuss selectivity between GaAs and AlGaAs in some detail because many of the same considerations pertain to the selectivity for InGaAs, InGaP, InGaAsP and InP. The primary determining factors for selectivity include the relative amounts of oxidant and acid, the degree of dilution at a fixed acid/oxidant ratio and the pH. Often the rates for both materials may increase as a process parameter is adjusted, but one will increase much faster. Such is the case with pH, where the rate for GaAs increases faster than that for AlGaAs as pH is increased. Dilution of a particular citric/H_2O_2 combination by addition of water will change the pH, and attempts to slow the etch rate by reducing the total citric and H_2O_2 concentrations may have unintended effects on selectivity.

In many inorganic acid/H_2O_2 solutions where the anion doesn't preferentially form complexes with Ga or Al, rates for AlGaAs are somewhat faster than for GaAs, with the rates increasing roughly linearly with Al mole fraction. For example, at pH 1.1 GaAs may etch at 0.3 μm/min while $Al_{0.5}Ga_{0.5}As$ etches at 0.65 μm/min. This difference may be enough to give isotropic profiles for AlGaAs under conditions that produce crystallographic etching of GaAs.

Selective etching of $Al_xGa_{1-x}As$ with $x > 0.4$ is possible with HF; selectivities greater than 10 : 1 are possible. Adding H_2O_2 to HF allows GaAs to etch as well, and the selectivity will depend on the acid/oxidant ratio. When one wants to selectively etch GaAs versus AlGaAs, a less dangerous etchant is to be preferred.

Organic acids such as citric, succinic, and malonic display a selectivity that is the opposite of that obtained with HF. They can provide a high degree of selectivity for GaAs versus AlGaAs. In addition, InP is virtually unetched by these etchants. Selectivities in excess of 100 : 1 for GaAs versus $Al_{0.3}Ga_{0.7}As$ have been achieved, while equirate etching is also possible merely by changing the citric acid/peroxide ratio. The maximum rate of 0.3 μm/min and a selectivity of 100 : 1 for GaAs versus $Al_{0.3}Ga_{0.7}As$ is obtained with a 5 : 1 ratio [7]. At ratios above 8 : 1, the etch rate of $Al_{0.3}Ga_{0.3}As$ increases rapidly; at ratios greater than 20 : 1, equirate etching is obtained.

The citric/peroxide etchants start with a stock citric acid solution that is composed of 1 g of citric acid in 1 ml DI water. This should be mixed at least one day early to insure all the citric acid dissolves since the dissolution is an endothermic reaction and the solution cools noticeably as the citric acid dissolves. The desired volume ratio of this solution is mixed with 30% H_2O_2 at least 15 min before etching to allow the solution to return to room temperature before using (this combination heats upon mixing). For citric/peroxide ratios greater than 1, the rate is reaction-rate limited and crystallographic etching occurs. At lower citric/peroxide ratios, bubble formation on the surface can be a problem.

Succinic acid can produce nearly equirate etching up to x $\approx$ 0.3 for an air-exposed surface, and up to nearly x $\approx$ 0.4 when the surface oxide is freshly removed. The higher Al-content surfaces allow formation of a thicker native oxide that inhibits initial etching. This vividly demonstrates how important surface preconditioning is for some etchants. There is a hundredfold decrease in rate between x = 0.4 and 0.5 and rates drop even more as Al fraction increases further. Typical succinic acid solutions combine 1 g succinic acid in 5 ml water. Proper etchant preparation is similar to the citric case described above. The pH is typically adjusted with NH_4OH.

If one plans to use compositional selectivity for etch-stop purposes, the greater retardation that can be obtained by relatively small changes in Al content can increase the process window in terms of the overetch time for etching the entire wafer to a uniform depth. While overetching in solution, it is always important to remember that lateral etching is proceeding at a rate comparable to vertical etching, so significant differences in device dimension can occur for different overetch times. The use of this characteristic for selective sidewall recessing for channel isolation from a contact metal will be discussed in the next section. It is also important to remember that a very thin etch-stop layer may not actually stop etching. It is important to remember that compositional selectivity is often determined by comparing the etch rates of relatively thick layers. If an etch-stop layer is thin enough for carriers to diffuse easily through the layer or if pn junctions are affecting current flow, the supposed "etch stop" may have a negligible effect and may be etched away at a rate greatly in excess of that expected from the corresponding thick-layer etch rate.

Selectivity can also be achieved using H_2O_2 with the pH adjusted with either NH_4OH or H_3PO_4, although higher selectivity is achieved with the organic acids. For $6 < pH < 7.1$, GaAs can be etched selectively versus AlGaAs using NH_4OH/H_2O_2, with the etch rate for GaAs increasing more rapidly with increasing pH than does the rate for AlGaAs. The H_3PO_4/H_2O_2 solutions listed in TABLE 4.2 are non-selective.

4.5.2 Selectivity versus other materials

A very important combination is InGaP/GaAs for HBTs. The different group V element simplifies the selective etching of InGaP versus GaAs since P-containing materials can be etched with HCl-based etchants without the addition of any peroxide or other oxidising agent. A particular favourite of ours employs a ratio of 7 : 1 H_3PO_4/HCl for selective etching of InGaP.

Another important family of devices are those based on lattice-matched InGaAs/InAlAs/InP or InGaAs/InAlAs grown metamorphically on GaAs. For these materials, a suitable pre-etch clean is a dip in 1 : 2 HCl/H_2O. A useful non-selective etch for InGaAs/InAlAs is 1 : 1 : 30 H_3PO_4/H_2O_2/H_2O.

The organic acid/peroxide etchants that provide selectivity between GaAs and AlGaAs can also provide selectivity between InGaAs and InAlAs. The selectivity obtained will depend on the ternary mole fractions, the pH, the degree of dilution and the acid/peroxide ratio. For example, a dilute citric solution is only weakly selective. A 5 : 1 : 30 mixture of citric/H_2O_2/H_2O at pH 5.1 etches $In_{0.53}Ga_{0.47}As$ at 0.034 μm/min and $In_{0.52}Al_{0.48}As$ at 0.002 μm/min. A 5 : 1 : 0 ratio produces rates of 0.143 and 0.004 μm/min, respectively [7]. At low citric/peroxide ratios (0.5–2.0), $In_{0.2}Ga_{0.8}As$ etches faster than GaAs. From 3.0 to 7.0, their rates are about the same but GaAs etches much faster than AlGaAs. Above 8.0, all three materials etch at comparable rates. As with GaAs/AlGaAs, higher citric fractions decrease selectivity. At a 15 : 1 ratio, the difference between InGaAs and InAlAs rates is only 3%.

Use of an isotropic etchant doesn't guarantee smooth sidewalls when etching heterostructures. If an etchant exhibits compositional selectivity between the layers, a corrogated sidewall can result. This is illustrated in FIGURE 4.10 for the etching of InGaAs/InAlAs mirror stacks. With a variety of compositions of HCl/H_2O_2/H_2O ranging from 10 : 4 : 1 to 40 : 4 : 1, classic diffusion-controlled profiles result (FIGURE 4.10(a)), indicating equirate etching of the $In_{0.33}Ga_{0.67}As$ and $In_{0.32}Al_{0.68}As$ mirror stack. Rates are constant for InGaAs with In mole fractions between 0.38 and 0.15 while rates for In fractions >0.15 were more than 10 times slower than for GaAs. Similar selectivities are obtained with H_2O_2/NH_4OH for In fractions >0.10.

A distinctly different profile is obtained with another isotropic etchant, HCl/HNO_3/H_2O_2, in ratios of 1 : 1 : 1 and 4 : 1 : 1. These etchants produce corrugated profiles in the InGaAs/InAsAl mirror-stack region, with the Al-containing layers etching 7–12% faster (FIGURE 4.10(b)). This clearly indicates that seemingly small differences in etch rates can produce striking effects on wall profile and may make a particular etchant quite unsuitable for a particular application.

Selective etching in the InGaAs/InAlAs system has been used to reduce or eliminate mesa-sidewall leakage when gates cross the mesa edge. The following approach will work for other material combinations, too. Succinic-based etchants (6 : 1 succininc/H_2O_2 adjusted to pH 5.5 with NH_4OH) have been used to selectively recess the InGaAs channel to avoid contact with the gate metals

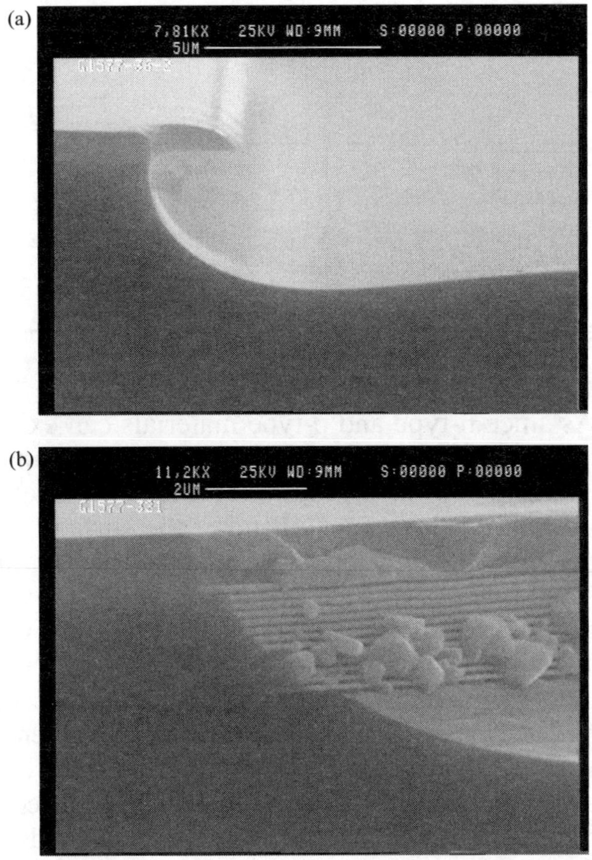

FIGURE 4.10 Compositional selectivity: (a) non-selective diffusion-controlled etch of an InGaAs/InAlAs heterostructure, (b) mostly diffusion control with slight compositional selectivity between InGaAs and InAlAs.

that cross the mesa edge in HFETs, as illustrated in FIGURE 4.11. As will be discussed more extensively in Chapter 10, the reaction rate in a narrow layer is lowered relative to the rate at an open surface due to limitation of reactant access in very thin layers. Consequently, the depth of the recess is less than predicted by free surface rates. Because these are reaction-rate-limited processes, the sidewall leakage, I, depends on the orientation of the device, with the leakage decreasing as $I[0\bar{1}1] > I[001] > I[011]$. Selective channel recessing for mesa isolation has also been used for InGaAs/InAlAs/InP MODFETs using citric/peroxide [8] and NH_4OH/H_2O_2 [9] etchants.

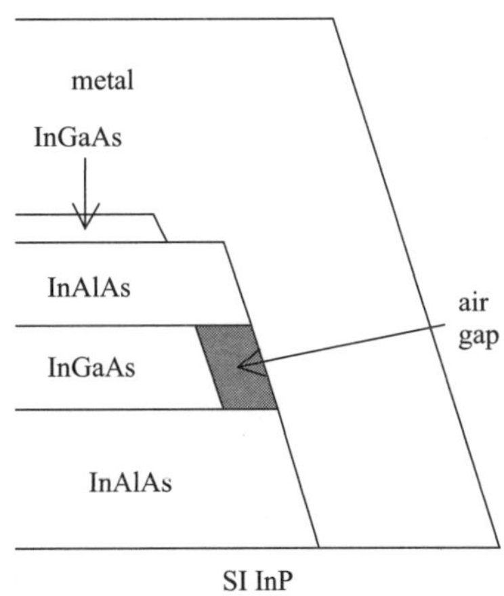

FIGURE 4.11 Selective channel recess for HFETs.

4.6 EFFECTS OF DOPING TYPE

Because wet etching relies on the oxidation of the surface, the rate at which the reaction proceeds can depend on the doping type

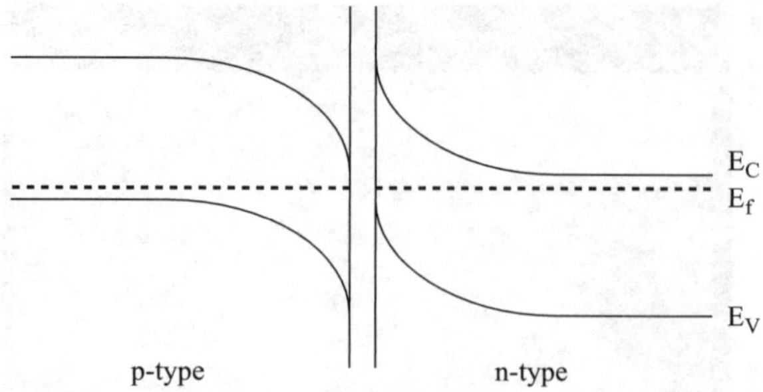

FIGURE 4.12 VBM for n-type and p-type with same Fermi energy.

of the GaAs since n-type and p-type materials can exhibit different electrochemical potentials in contact with the same solution (FIGURE 4.12). For a fixed position of the Fermi level, it is clear that the potential of the valence band at the surface is different. For example, the effect of dopant type can be seen in the etching of n-GaAs and p-GaAs by 0.3N NH_4OH/0.1N H_2O_2 where room temperature rates for n-type are about 0.19 μm/min while p-type etches at about 0.13 μm/min.

Exposure to bright light while in a wet etchant can produce photogenerated electron-hole pairs. The holes in the valence band of n-GaAs move to the surface due to the surface field and can enhance the etch rate (photochemical etching.) The photons effectively play the same role as the chemical oxidant, such as H_2O_2. Illumination of p-type GaAs has a minimal effect since its surface band bending is in the opposite direction of that in n-type GaAs.

Additional doping-dependent differences can occur when a metal is in close proximity and exposed to the etching solution. Local electrochemical effects can lead to unexpected etching profiles with either enhanced or suppressed etching due to the completion of a local galvanic cell; this is discussed further in Section 4.7.

4.7 ELECTROLYTIC EFFECTS IN WET ETCHING

Electrochemical etching is often considered in terms of the driving of a chemical reaction by the deliberate application of an applied bias from an external voltage source. In the fabrication of GaAs devices, this type of deliberate electrochemical etching does not play a major role and will not be discussed here. Excellent books dealing with this topic exist [10], and the reader is encouraged to consult them should this topic be of interest.

However, unintentional electrochemical etching can occur when a galvanic circuit is completed by the unintentional exposure of

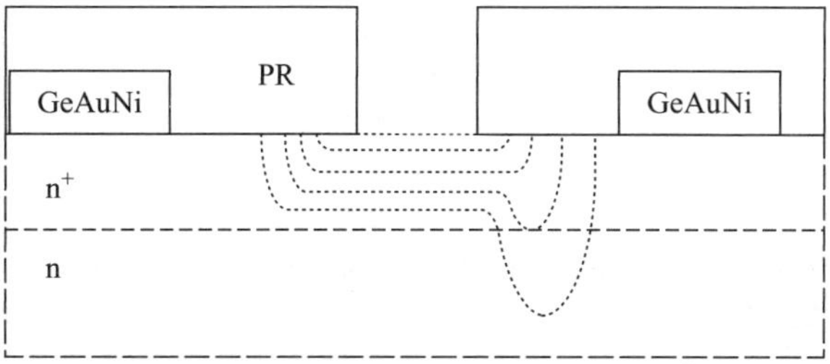

FIGURE 4.13 Effect of etching too close to exposed contact metal in FET.

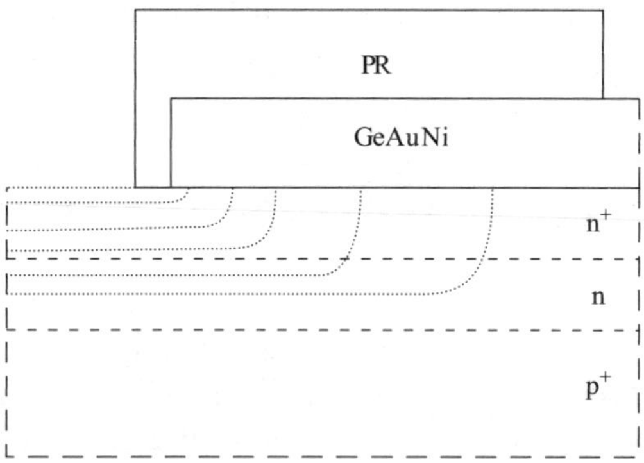

FIGURE 4.14 Rapid undercutting when approaching pn juction in HBT if metal contact exposed to etching solution.

contact metal to the etching solution. Three examples where this might occur are discussed here.

The first two cases may occur when a portion of the ohmic metal is not isolated from the etching solution with photoresist. The dotted lines in FIGURE 4.13 represent the profile of the etching surface as the etch proceeds. When the surface profile reaches the point where separation between the etching solution and the metal contact is of the order of 0.25 μm, a galvanic circuit through the solution, semiconductor and metal is completed and the etch rate rapidly and uncontrollably accelerates. This leads to rapid trenching into the substrate.

The second example where exposed ohmic metal can cause accelerated etching is shown in FIGURE 4.14. As the etch front approaches the pn junction, the lateral etch rate under the GeAuNi metal, which is normally equal to the vertical etch rate into the substrate, undergoes rapid acceleration. This leads to uncontrollable undercutting. This problem can be avoided by using a different metallisation, such as WSi, instead of GeAuNi.

The third example is more insidious since the sample doesn't have to be in a solution that one would expect to be an etchant; it occurs in DI water. A local galvanic reaction can occur where

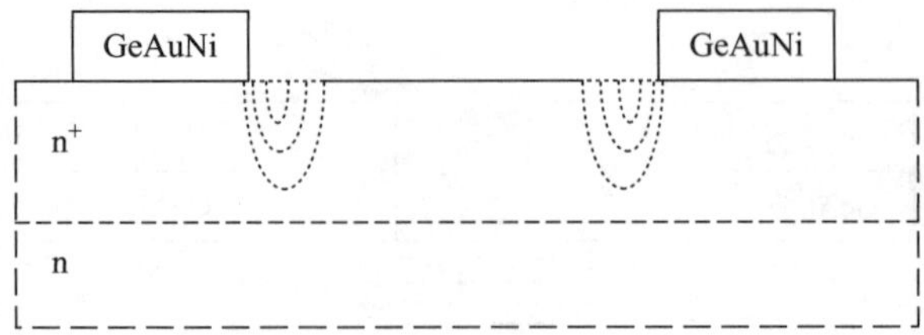

FIGURE 4.15 Trenching in DI water with exposed metal contacts.

GeAuNi contacts the semiconductor, leading to trenching, as shown in FIGURE 4.15. The depth of the trenching increases with time, so it is very important to minimise rinse time in DI water. Trenching can also be reduced by deoxygenating the water by bubbling N_2 through the water (gas sparging) before rinsing.

4.8 EFFECTS OF DEFECTS AND DAMAGE

Disruption of the GaAs lattice stucture by any means will, in general, increase the ease with which the damaged area etches. This can be due to grown-in defects, mechanical damage, e.g. scratches, or lattice dislocations and defects, including those generated by ion implantation. Grown-in crystal defects tend to manifest as etch pits with characteristic shapes. Patterned implantation can be used to selectively enhance etching in the damaged regions. This is something to keep in mind when subsequently etching wafers employing implant isolation.

4.9 CONCLUSION

While wet etching has been replaced by dry etching when vertical profiles are essential, there remain many cases where wet etching remains useful. The electronic quality of the surface is superior to that of dry-etched surfaces, and the higher degree of compositional selectivity than generally available with dry processes will continue to make wet etching the process of choice for many applications.

REFERENCES

[1] H.K. Kuiken, J.J. Kelly, P.H.L. Notten [*J. Electrochem. Soc. (USA)* vol.133 (1986) p.1217–26]

[2] T. Takabe, T. Yamamoto, M. Fujii, K. Kobayashi [*J. Electrochem. Soc. (USA)* vol.140 (1993) p.1169–80]

[3] C.I.H. Ashby [in *Properties of GaAs*, Eds. M.R. Brozel, G.E. Stillman (Institution of Electrical Engineers, London, 1995) ch.18, p.707–16]

[4] C.H. Hamann, A. Hamnett, W. Vielstich [*Electrochemistry* (Wiley-VCH, New York, 1998)]
[5] M. Pourbaix [*Atlas of Electrochemical Equilibria in Aqueous Solutions* Translator J. Franklin (National Association of Corrosion Engineers, 1975)]
[6] [*Standard Guide for Development and Use of a Galvanic Series for Predicting Galvanic Corrosion Performance*, ASTM Designation: G 82–83, *Annual Book of ASTM Standards*]
[7] G.C. DeSalvo, W.F. Tseng, J. Comas [*J. Electrochem. Soc.(USA)* vol.139 (1992) p.831–5]
[8] M. Tong, K. Nummila, A. Ketterson, I. Adesida, C. Caneau, R. Bhat [*IEEE Electron Device Lett. (USA)* vol.13 (1992) p.525–7]
[9] K. Kenefick [*Electron. Lett. (UK)* vol.21 (1985) p.558–9]
[10] J. McHardy, F. Ludwig [*Electrochemistry of Semiconductors and Electronics: Processes and Device* (Noyes Publications, New Jersey, 1992)]

Chapter 5

Dry etching of GaAs and related alloys

5.1 CHAPTER SCOPE

There are five important characteristics to consider in selecting any dry etch process. These are rate, feature profiles, compositional selectivity, damage and uniformity across the wafer. In general, the balance between chemical and physical contributions to the etching for each type of process will largely determine these characteristics. Chemical contributions generally produce rapid rates, compositional selectivity and lower damage. However, they also can decrease anisotropy (ratio of vertical to horizontal etch rates) and can even produce either isotropic etching or etch rates that depend on the relative reactivity of different crystallographic facets, as observed in wet etching. Physical sputtering contributions generally increase anisotropy, but they also damage the surface, degrading device performance that depends on near-surface crystal quality. Consequently, the simultaneous goals of an etch process that is both highly anisotropic and damage free are frequently in conflict and an acceptable compromise between the two will drive the process selection for GaAs device fabrication. In this chapter, the many practical issues that must be considered to successfully dry etch GaAs will be discussed.

5.2 COMPARISON OF WET AND DRY ETCHING

It is possible to etch device structures using either liquid-phase "wet" etch processes or gas-phase "dry" processes. Both approaches have their own advantages and disadvantages and the best choice depends on the specifics of the application. (For advantages and disadvantages of dry etching, see side note in page 149.)

Wet etch processes, in general, produce better surface electronic properties than many dry etch processes because they don't damage the underlying crystal lattice of the GaAs. In contrast, the most commonly used dry processes bombard the surface with ions having enough energy to locally disrupt the lattice, and some of this damage may remain after the etch is completed. This problem

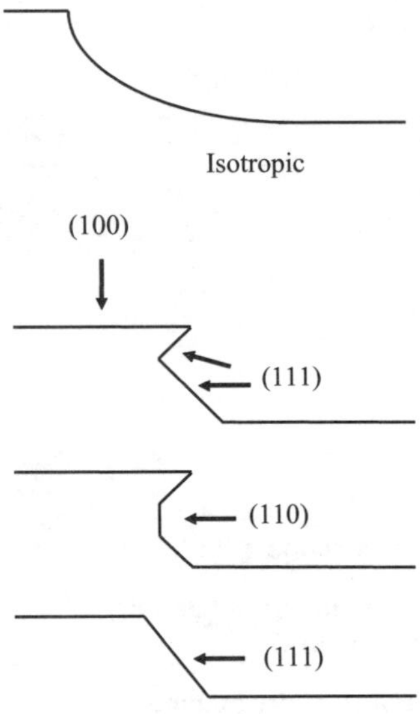

FIGURE 5.1 Isotropic profile and crystallographic profiles of wet etch.

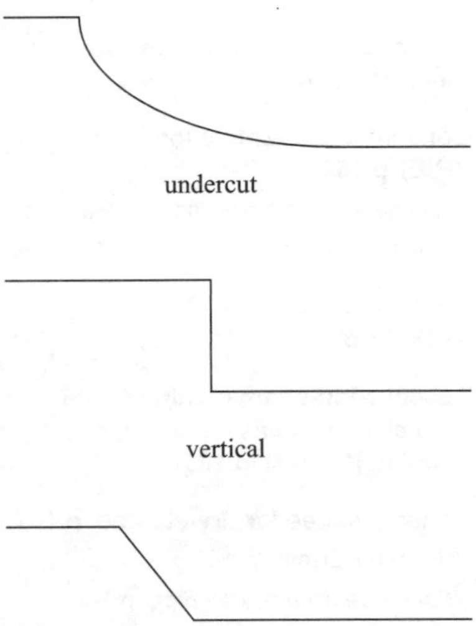

FIGURE 5.2 Range of profiles from undercut (isotropic) to vertical to angled for RIE etching.

can be reduced by post-etch thermal annealing, but some damage may remain.

Another advantage of wet etching is the ease with which it can be used. While dry processes require gas-handling and vacuum systems and usually complex electronic hardware to establish a low-temperature plasma discharge, wet etching can be performed in simple glassware on a benchtop. Chemical reagents for wet etching are also inexpensive and it is very easy to vary the chemistry. The chemistry can be based on one of many different schemes that oxidise the surface to yield a soluble product.

However, there are some limitations encountered with wet etch processes. It is much harder to control the profile of a feature that has been wet-etched. The etch undercuts the mask to a depth comparable to the vertical etch depth. This limits the smallest critical dimensions that can be achieved. The profile may be diffusion controlled (isotropic) or it may depend on the different etch rates of different crystallographic faces (FIGURE 5.1). This can introduce a strong dependence of the final features on how the pattern is aligned with the GaAs wafer's crystal planes.

Wet etch rates often vary significantly with changing temperature and with how recently the solution has been mixed. This can make run-to-run reproducibility difficult to achieve.

Dry etch processes can provide excellent profile control. The energetic ion bombardment increases the etch rate on the exposed surface relative to those regions protected by the mask, so vertical sidewalls with negligible undercutting are readily achieved. Depending on the balance between chemical and physical contributions to the dry etch, it is possible to vary the profile from isotropic/crystallographic to vertical to angled (FIGURE 5.2). With vertical profiles, smaller critical dimensions are achievable. The ion enhancement also removes the dependence on the pattern alignment relative to the wafer crystal planes.

The primary disadvantage of dry etch processing is the degradation of electronic properties that results from both the mechanical displacement damage and the charge-induced damage to the lattice that are produced by ion bombardment. An additional disadvantage is the limited number of possible chemistries that are useful for etching GaAs and related compounds. In general, one is restricted to halogen and hydrocarbon etch chemistries since gas-phase reactants and volatile products are required for successful dry etching. Another disadvantage of an ion-bombardment-dependent etch is the erosion and heating of the resist by sputtering and sometimes chemical etching. This increases the probability of ruining a device because the resist fails before the etch process is complete.

In general, if feature profile and tight control of dimensions are less important than surface electronic quality, a wet etch may be the process of choice. If vertical or controlled-angle profiles and small dimensions are needed, a dry etch that achieves the best compromise between feature profile and damage may be preferred.

Advantages of dry etching
1) Good dimensional control
2) Excellent profile control: vertical to controlled angle.

Disadvantages of dry etching
1) Ion-induced surface damage
2) Limited number of chemistries
3) Resist erosion can limit depth
4) Hazardous reactant gases
5) Expensive hardware

5.3 OVERVIEW OF DRY ETCHING PROCESSES

There are several types of dry etch process. Each will be discussed separately below, but there are many features common to several and a thorough knowledge of the important characteristics of each will help one select the best process for a particular application.

The simplest type is ion-beam or sputter etching. No reactive gas is used and physical sputtering of the GaAs by the incident high-energy ions is the only mechanism involved.

In chemical dry etching, no plasma is employed or a plasma is remotely located without line-of-sight to the sample so no ion bombardment is involved. A component of this will be present in every plasma process and must be considered as a possible effect before striking the plasma or after the plasma is quenched.

Plasma etching is used here to designate the situation where the wafer sits within the plasma but not on a powered electrode so the incident ion energies will usually be less than 20 eV, and typically of the order of 10–15 eV. Highly reactive radicals, formed by dissociating the source gas, react faster with the surface than what one sees in simple chemical dry etching, but ion bombardment effects are negligible.

Reactive ion etching (RIE) has long been the standard dry process where reactive radicals and ions are formed and the ions are accelerated into the wafer at energies >50 eV to produce a mechanism that combines both chemical etching and physical sputtering. The same radio frequency (RF) power supply is used to generate the plasma and accelerate the ions into the wafer.

Types of dry etching
1) Ion-beam or sputtering
2) Chemical dry etching
3) Plasma etching
4) Reactive ion etching
5) High-density plasma etching (ECR or ICP)
6) Reactive-ion-beam etching
7) Chemically-assisted ion-beam etching

High-density-plasma etching (HDPE) uses two separate power supplies with one generating the plasma and the second one providing the RF power that controls the ion bombardment energy. The source gases are more efficiently dissociated than in regular RIE (by up to two orders of magnitude) using microwaves and magnets to establish electron cyclotron resonance (ECR) or using inductively coupled RF power to generate the plasma. The latter is called an inductively coupled plasma (ICP) system. HDPE provides semi-independent control of the plasma chemical species, both ions and neutrals, and the energy with which the ions hit the wafer surface.

Reactive-ion-beam etching (RIBE) and chemically assisted ion-beam etching (CAIBE) do not use an RF supply to provide the incident ion energy. Ions are accelerated from a "remote" plasma (not in direct contact with the wafer) using ion-acceleration techniques similar to those in simple ion-beam etching. In RIBE, a neutral gas is predissociated to provide a beam of highly reactive radicals, such as Cl from Cl_2, that strike the surface in concert with an ion beam. In CAIBE, a beam of ions, usually Ar^+ or Xe^+, is directed at the wafer while it is immersed in an ambient of neutral gas, such as Cl_2. The smallest critical dimensions are achieved with these techniques.

In the following sections, each of these dry etch techniques will be discussed in greater detail. For a comprehensive tabulation of etch rates for all these processes through 1995, see "Properties of Gallium Arsenide" [1]. A useful compilation of reviews of various aspects of plasma processing has been recently published [2]. Since ion bombardment effects are so important in RIE, HDPE, RIBE and CAIBE, we will begin with a discussion of ion beam etching.

5.4 ION-BEAM ETCHING (IBE) AND ION EFFECTS IN OTHER PLASMA PROCESSES

Ion-beam etching (IBE) is essentially a momentum-transfer sputtering process where the kinetic energy of the incoming ions is transferred to the surface atoms, ejecting them from the surface. Although non-reactive IBE is not generally used to etch GaAs devices, the sputtering process which can be studied in isolation in IBE is a key feature of most other dry processes and requires consideration whenever the ion bombardment energies exceed 40 eV, as they usually do in RIE, HDPE, RIBE, and CAIBE.

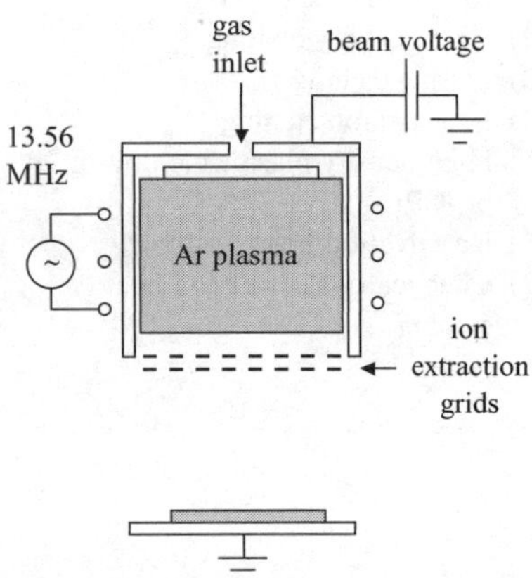

FIGURE 5.3 Ion beam etching (sputtering).

No reactive gases are used in simple IBE. A heavy inert gas such as Ar is typically employed (FIGURE 5.3). Lack of a chemical component to the etch means that all materials etch at similar rates, depending only on their relative bond strength and atomic mass. Sputtering is more efficient when the mass of the incident ion is closely matched to the mass of the sputtered atoms. Typical rates for a wide variety of materials range between 0.5 and 2.5 atoms/ion. For 100 and 500 eV Ar^+, this corresponds to a removal rate of the order of 0.01 and 0.1 $\mu m/min/mA/cm^2$, respectively. There is a non-linear dependence of rate on ion energy but generally a linear dependence on ion current density.

There is a pronounced dependence of sputtering rate on the ion incidence angle with a maximum etch rate near an incidence angle of 60 degrees, where surface atoms acquire a large momentum component directed away from the surface. This leads to a characteristic angled profile. When angled profiles occur in plasma-based etches, it is generally an indication that a sputtering component dominates the etch.

For a given incident ion energy, IBE is the most damaging of processes. There is appreciable disruption of the near-surface lattice by the impacting ions if they exceed the atomic displacement energy threshold for GaAs (~40 eV). Since this damaged region is not partially removed by chemical reaction, more damage remains after an IBE etch than after a plasma etch at the same ion energy. In general, the higher the ion energy in any etch process, the more significant the surface damage will be in degrading device performance. An Ar ion energy of only 40 eV can increase a diode ideality factor (Section 7.2.1) to 1.25 while energies of 80 and 200 eV can produce increases to 1.4–1.5 and 2.3–2.7, respectively. In plasma processes where some of the damage is removed by chemical etching, lesser device degradation will occur, but even in these, ion energies over 100 eV generally produce some degradation and even 2 min with 60 eV ions can produce threshold voltage shifts and transconductance decreases in FETs.

An additional effect to keep in mind is the heating of the surface by the depositing ion energy. When it is important to avoid heating a wafer too much, high ion energies during etching should be avoided and, if possible, the wafer should be actively cooled during the etch.

5.5 CHEMICAL DRY ETCHING

While device fabrication relies on plasma-based dry etching to produce useful etch rates and the desired etch profiles for devices, it is important to remember that GaAs can be etched by gas-phase reactants through ordinary chemical reactions in the absence of any plasma. Such reactions are isotropic or crystallographically rate-dependent, similar to wet etching reactions. Since these reactions are thermally activated reactions of molecular species rather than relatively temperature-independent reactions of highly reactive atoms or other radicals, they tend to be quite slow at room temperature. However, they can become significant at higher temperatures. Since wafers tend to heat up under ion bombardment, temperatures over 100°C may occur rather routinely after long etches and simple thermal etching may be appreciable. A rate of 0.01 μm/min has been observed at 100°C with tenths of a mTorr of Cl_2 and rates up to 0.2 μm/min are possible at 300°C. Consequently, letting a sample sit for extended periods without plasma but still exposed to Cl_2-containing gas mixtures may produce unintended etching whose isotropic character could degrade device profiles. Leaving the wafer in the

Cl-containing gas for very long after etching, when the wafer has already been heated by ion bombardment, should generally be avoided. However, a slight amount of chemical etching may serve to remove some of the ion-induced near-surface damage so a brief opportunity for thermal reaction may improve surface electronic properties.

5.6 PLASMA ETCHING AT VERY LOW ION ENERGIES

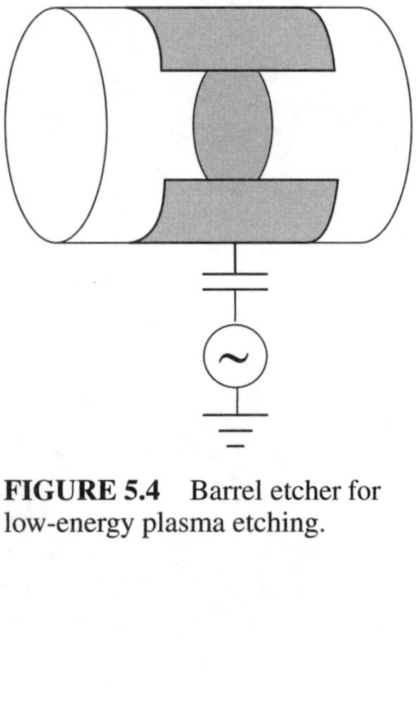

FIGURE 5.4 Barrel etcher for low-energy plasma etching.

Although most dry etch processes employ a plasma to generate highly reactive atoms and radicals, the term plasma etching (PE) is generally applied to processes where there is negligible bombardment with high-energy ions. In PE, the wafer sits within the plasma but it is not on a powered electrode and few if any ions with energies over 10 eV strike the GaAs surface (FIGURE 5.4). The atomic-displacement threshold energy for GaAs is of the order of 40 V, so processes at energies below that value do not produce lattice damage that can degrade device operation.

Rates can be fast when samples are heated but profiles are isotropic. Rates up to 5 μm/min can result at 5 mtorr and 60°C. However, surfaces are virtually damage free. PE is not generally used for regular GaAs etching because it lacks the primary advantages of profile control provided by energetic ion bombardment in other dry etching techniques while involving much more expensive and complicated hardware than wet etching.

Because PE is a low-damage process, it can sometimes be used to remove surface damage produced by more commonly employed dry processes. A PE process can be implemented in a high-density plasma etcher while the reactant-generating ICP or ECR power is on by reducing the power of the RF-powered electrode to produce very low bias voltages (<10 V) at the end of the etch. This can remove the thin surface layer that has been damaged by ion bombardment during the etch. Etch rates are low due to the low pressures (<10 mTorr) used in these processes, so profile degradation is avoided.

While PE is not very useful for GaAs patterning, it is very useful for removal of residual organics from PR processes when an oxygen plasma is used. The hardware for this process is sometimes referred to as an asher. The low ion energy avoids degradation of the surface electronic properties of the GaAs while the oxygen atoms remove trace organic contamination from the surface where photoresist has been removed during PR pattern development before etching. This permits better pattern definition during the etch process. It is equally valuable after solvent removal of most of the photoresist following an etch. Without this PE cleaning of the surface, it would be difficult to obtain highly reproducible metal-semiconductor contacts.

5.7 CONVENTIONAL REACTIVE ION ETCHING (RIE)

Reactive ion etching (RIE) is the workhorse of dry etching techniques for GaAs. The equipment involved is widely available in most clean rooms and uses the most reasonably priced basic hardware of any of the dry processes. Two geometries have been commonly employed: planar and hexode (FIGURE 5.5). In the planar geometry, which is by far the most common, the radio frequency (RF) plasma is formed between two parallel plates with one electrode powered and the other grounded. In the hexode geometry, wafers are mounted on a hexagonal column located within a grounded chamber. Most reactors today are of the planar variety. In both cases, RF power is applied to the electrode upon which the wafers sit.

Because of the RF character of the applied power, the actual voltages present at the powered electrode vary with time during the RF cycle, making the environment experienced by the wafer more complex than is implied when describing the plasma system in terms of input power or even applied bias voltage (FIGURE 5.6). Typical commercial RF plasma hardware operates at a frequency of 13.56 MHz. The rapidly varying potential allows the electrons, which have very high mobilities, to respond to the varying RF potential. However, the ions, being much heavier and possessing much lower mobilities, respond only to the time-averaged potential, V_{dc}.

The time-averaged potentials at different locations between the powered and grounded electrodes are shown in FIGURE 5.7. Ions are accelerated in the field and strike the wafer at energies between $e(|V_{dc}| + V_p)$ and zero. In a typical reactor, where the area of the grounded surfaces is large compared to the area of the powered electrode, V_p will be relatively low (20–50 V) and the maximum energy of ions incident on the wafer is reasonably well described, within about 10%, by V_{dc}. Because it is the energy of the incident ions that is so important in determining the etching characteristics, it is important to measure the DC potential of the cathode (powered electrode) whenever possible rather than using the RF power alone to describe an etching process.

The same RF power supply both dissociates the source gases into more reactive radicals and ions and provides the ion acceleration voltage that determines the ion bombardment energy during the etch cycle. This means that there is no independent or even semi-independent control of the degree of dissociation in the plasma and

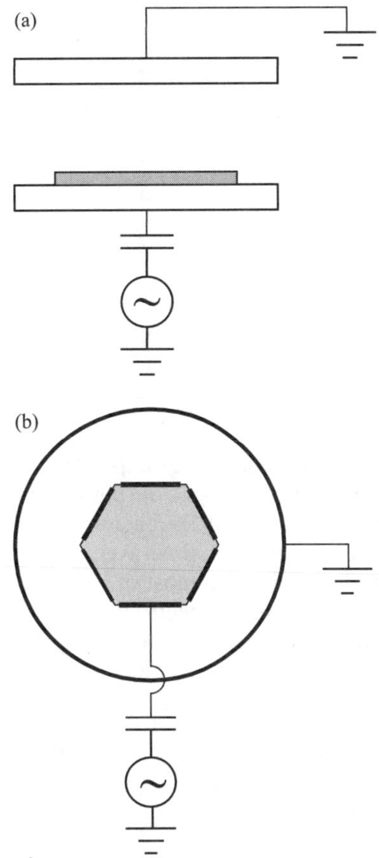

FIGURE 5.5 Typical RIE etchers. Wafer sits on powered electrode: (a) Planar (b) Hexode.

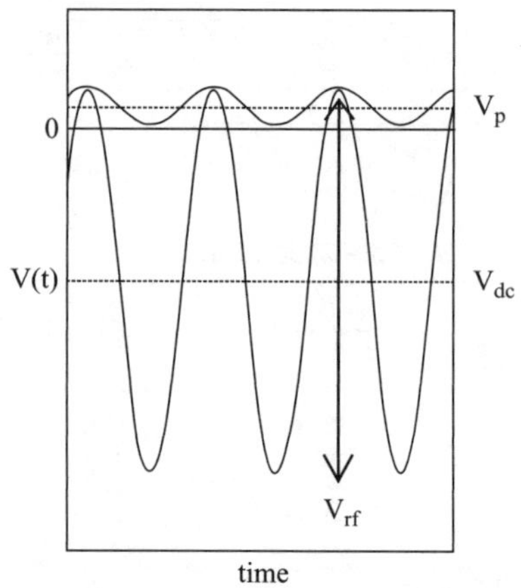

FIGURE 5.6 Time-dependent variation of voltages during several RF cycles in an RIE reactor. V_p = plasma potential: time-average of potential through most of plasma and at grounded electrode; V_{dc} = self-bias potential: time-average of potential at powered electrode.

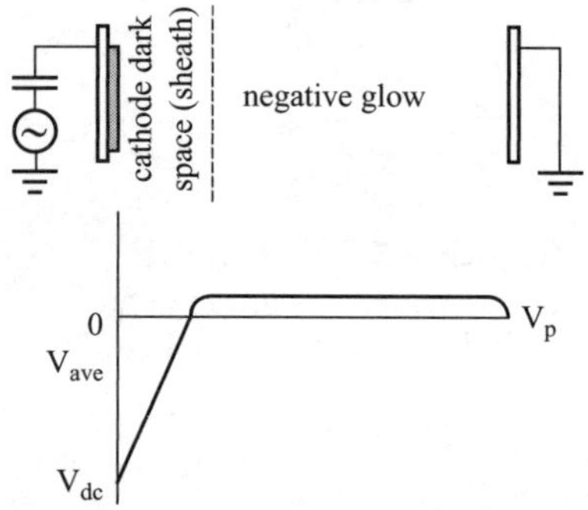

FIGURE 5.7 Time-averaged potential in a planar RIE etcher.

the energies of the ions striking the wafer. The wafer is etched by neutrals and by ions whose energies are greater than the positive space-charge potential of the plasma.

It is important to remember that a plasma is actually a complex mixture of ions, electrons and neutral species. In etching plasmas, the neutrals greatly outnumber the ions, and a complete description of the plasma would require knowledge of the number and identity of the neutral species in the plasma as well as the range of ion energies and ion densities. In addition, there are energetic neutrals which form when electrons are transferred between energetic ions and neutrals. Depending on where in the sheath the neutralisation occurs, the incident neutral energy will vary widely. The degree of ionisation (ion/neutral ratio) is usually in the range of 10^{-4}–10^{-6}, so it is clear why etch contributions due to neutrals are so important and why it is so difficult to fully characterise what species are actually reacting at the wafer surface.

The energies of individual ions striking the wafer will depend on their history. As an ion passes through the sheath as it accelerates towards the cathode, it may undergo collisions that alter its energy and charge state. Some ions may be neutralised by collisions with neutral species. This can produce an energetic neutral that moves towards the wafer with an energy determined by how much it had been accelerated prior to the neutralisation. The new ion resulting from the collision will accelerate towards the surface and strike with a fraction of the maximum possible energy. The result is an energy spread of up to several hundred eV for the chemical species striking the surface. Consequently, reaction at the surface will involve a complex mixture of low-energy neutrals and ions and higher energy neutrals and ions.

Since charge neutrality in the plasma must be maintained, most ions are positively charged. In a typical RF plasma there may be 10^9–10^{11} ions/cm^3. In contrast, the neutral density at 10 mTorr (a typical RIE pressure) is 10^{14}–10^{15} molecules/cm^3. Clearly, neutral species are predominant and will play a vital part in any RIE process while ions largely affect relative product desorption rates. Ions accelerate product desorption where they strike, so rates on bombarded surfaces become faster than rates on sidewalls, leading to anisotropic profiles. In general, chemical etching by plasma-generated radicals with its associated high rates, low anisotropy and low damage is dominant when pressures are high and sheath voltages are low. As pressure increases, reactant concentrations increase, but at the same time ion energy decreases due to the decreasing self-bias voltage. Since rates tend to decrease with decreasing ion energy, increasing pressure first increases etch rates by increasing reactant supply, then decreases them as the rates shift from reactant-supply dominated to ion-energy dominated.

In contrast, sputtering is dominant at low pressures and high sheath voltages. This increases both the anisotropy and the damage while reducing the material selectivity of an etch process. Higher ion fluxes at a fixed energy and higher ion energies at a fixed flux increase the sputtering rates. Higher fluxes and energies also deposit more energy into the wafer and can produce substantial heating if the wafer is not actively cooled. This will accelerate the chemical reaction rates, including those with neutral molecular species. The profile evolution from the mostly chemical (low ion energy) to mostly sputtering (high ion energy) is illustrated in FIGURE 5.8.

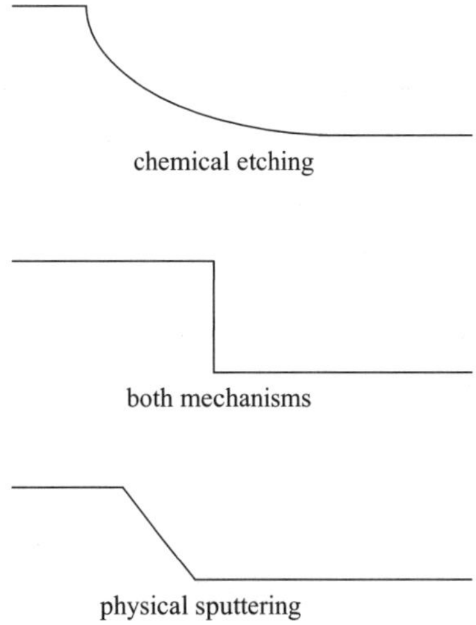

FIGURE 5.8 Profile evolution depending on ion and neutral contributions.

It is customary to describe process conditions in terms of pressure, gas flow rate, RF power and applied bias, but this is grossly inadequate to define the chemical characteristics of a plasma. The great importance of the neutral species, which are not directly measured by any of the diagnostics on standard reactors, and the variation in ion densities at different energies are the probable origin of the markedly different etching rates obtained with nominally identical RF power even in nominally identical reactors from the same manufacturer. Rates reported in the literature for a specific set of conditions should always be taken only as guidelines and starting points when transferring a process into a different reactor. Each reactor must be individually characterised with a process for reliable etching results to be obtained. Some RIE rates and their associated reaction conditions are listed in TABLE 5.1.

5.7.1 Halogen-based plasmas for RIE

A variety of chemistries have been employed in RIE of GaAs, but the most commonly used ones are halogen-based (F, Cl, Br, I). The most common ones employ Cl_2 and BCl_3, separately or in combination. $SiCl_4$ is another good Cl source gas. Rates can vary by two orders of magnitude. Processes with intermediate rates tend to produce good surfaces. Higher etch rates, such as those produced when Cl_2 is the sole source of Cl, tend to produce rougher surfaces and decreased anisotropy.

Typical chemistries employ Cl-based source chemicals, such as Cl_2 and BCl_3, that are dissociatively excited in the plasma to produce Cl atoms that react readily with the GaAs surface. The removal of the products $GaCl_3$ and $AsCl_3$ is facilitated by the simultaneous ion bombardment that sputters away the products from the surface, producing much faster etch rates where the ions strike than in regions without ion bombardment. Sputtering can remove even partially chlorinated $GaCl_x$ and $AsCl_x$ ($x = 1$–3),

TABLE 5.1 Reactive ion etching rates for GaAs.

Etchant	Rate (μm/min)	Pressure (mTorr)	Flow (sccm)	RF power density (W/cm^2)	Self bias (−V)	Ref.	Comments
Cl_2	4.5	150	3–23	0.44		[a]	via drilling
	3.0	85		0.8	400	[b]	
	1.4	5			270	[c]	vertical walls
	0.15	2	40	0.1	150–160	[d]	low damage, semi-isotropic
$5Cl_2 : 3Ar$	0.09	5	20	0.14	300	[e]	equirate versus AlGaAs and vertical profile
$4Cl_2 : 1Ar$	0.80	5			200	[c]	severe undercut
$3Cl_2 : 1Ar$	3.4	15	12	0.75	330	[f]	isotropic
$1Cl_2 : 5BCl_3$	0.6	60		0.42	270	[g]	high reproducibility
$1Cl_2 : 1BCl_3 : 3Ar$	0.9	15	100	–	300	[h]	very smooth vias
BCl_3	0.017	11	20	0.20		[i]	very smooth
$1BCl_3 : 9Ar$	0.28	15	0.27		300	[j]	equirate versus AlGaAs profiles, resolution, rates f(W,P gases)
$SiCl_4$	0.129	10	10	0.81	600	[k]	no polymer film
	0.067	10	10	0.27	275	[k]	no polymer film, max. at 10 mtorr
$1SiCl_4/4SiF_4$	0.31	60	10	–	60	[l]	490 : 1 versus $Al_{0.35}Ga_{0.65}As$
$1CH_4 : 5H_2$	0.016	10	24	0.75	500	[m]	donor passivation
$1C_2H_6 : 3H_2$	0.036	10	43	1.1	500	[n,o]	
$1C_3H_8 : 5H_2$	0.038	10	41	1.1	500	[n,o]	2 : 1GaAs/AlGaAs

[a] K.P. Hilton, J. Woodward [*Electron. Lett. (UK)* vol.21 (1985) p.962]
[b] C. Gagne [*Solid State Technol. (USA)* (May 1988) p.187]
[c] E.L. Hu, R.E. Howard [*J. Vac. Sci. Technol. B (USA)* vol.2 (1984) p.85]
[d] B.S. Lee, H. Baratte [*J. Electrochem. Soc. (USA)* vol.137 (1990) p.980]
[e] S.S. Cooperman, H.K. Choi, H.H. Sawin, D.F. Kolesar [*J. Vac. Sci. Technol. B (USA)* vol.7 (1989) p.41]
[f] H. Yamada, H. Ito, H. Inaba [*J. Vac. Sci. Technol. B (USA)* vol.3 (1985) p.884]
[g] H. Tamura, H. Kurihara [*Jpn. J. Appl. Phys. (Japan)* vol.23 (1984) p.L731]
[h] K.J. Nordheden, D.W. Ferguson, P.M. Smith [*J. Vac. Sci. Technol. B (USA)* vol.11 (1993) p.1879]
[i] G.J. Sonek, J.M. Ballantyne [*J. Vac. Sci. Technol. B (USA)* vol.2 (1984) p.653]
[j] A. Scherer, H.G. Craighead, E.D. Beebe [*J. Vac. Sci. Technol. B (USA)* vol.5 (1987) p.1599]
[k] M.B. Stern, P.F. Liao [*J. Vac. Sci. Technol. B (USA)* vol.1 (1983) p.1053]
[l] W.H. Guggina, A.A. Ketterson, A. Andideh, J. Hughes, I. Adesida, S. Caracci, J. Kolodzey [*J. Vac. Sci. Technol. B (USA)* vol.8 (1990) p.1956]
[m] J. Werking, J. Schramm, C. Nguyen, E.L. Hu, H. Kroemer [*Appl. Phys. Lett. (USA)* vol.58, no.18 (1991) p.2003–5]
[n] V.J. Law, G.A.C. Jones, M. Tewordt [*Semicond. Sci. Technol. (UK)* vol.5 (1990) p.1001–3]
[o] V.J. Law, M. Tewordt, S.G. Ingram, G.A.C. Jones [*J. Vac. Sci. Technol. B. (USA)* vol.9, no.3 (1991) p.1449–55]

accelerating the total GaAs etch rate. This difference in etch rate between surfaces with and without ion bombardment is the origin of the profile control available with dry etching. The undercutting of the mask that results in wet etching from nearly equal vertical and horizontal etch rates is avoided by increasing the vertical etch rate through sputtering away the intermediate and final reaction products rather than relying solely on the natural volatility of the final products. Depending on the degree of sputtering in a given process, profiles can be varied from isotropic undercut, as in purely chemical processes, through vertical to angled, as in IBE without any chemical contributions.

The highest rates are achieved with Cl_2 as the Cl atom source, but Cl_2 can produce surface roughness and may exhibit low reproducibility in etch rates from run to run. The very high rates are due to a strong chemical component, and reduced anisotropy and surface roughness can be a problem. The performance of Cl_2 is improved by the addition of other gases. Addition of Ar or BCl_3 increases anisotropy by increasing the sputtering contribution while addition of some H_2 to a Cl_2/Ar mixture increases both surface smoothness and anisotropy while decreasing the etch rate somewhat. Addition of CH_4 to Cl_2/Ar yields smoothness, anisotropy and a rapid etch rate. However, addition of H-containing sources to the gas mix produces H atoms that can passivate dopants and may require post-etch annealing to reactivate them.

BCl_3 or BCl_3/Cl_2 mixtures tend to produce smoother surfaces than Cl_2-only plasmas. The smooth morphology, good anisotropy and high degree of reproducibility make the BCl_3-based etches some of the best. BCl_3 can also act as a getter for oxygen atoms that are formed in the plasma from residual O_2 or H_2O. The product of this reaction is the volatile trimer $B_3Cl_3O_3$. $SiCl_4$ may also serve this gettering purpose. Therefore, if one must work with a non-loadlocked reactor, use of a BCl_3-based chemistry is especially advantageous since even trace oxygen can retard the etching of AlGaAs.

When etching heterostructures of GaAs and AlGaAs, either highly selective etching of one versus the other or equirate etching of the two materials may be desired. Either may be achieved with the selection of proper reactants and RF plasma conditions. Selective removal of GaAs is needed for the gate recess of HEMTs where the ohmic contact layer must be removed to expose the donor layer for gate oxide deposition. Similarly in the emitter mesa process for HBTs, the GaAs contact layer is removed from the wide-bandgap emitter first before etching the emitter and exposing the base for contact formation.

One of the best ways to achieve selective etching of GaAs versus AlGaAs is through the addition of an F-containing reactant to the gas mixture. The formation of relatively non-volatile AlF_3 on the surface of the AlGaAs greatly reduces the etch rate. In contrast, GaF_3 is readily removed from the surface so etching of GaAs proceeds relatively unimpeded. The origin of this pronounced difference is found in the structures of GaF_3 and AlF_3. While GaF_3 forms the dimer, $(GaF_3)_2$, which is quite volatile, AlF_3 forms an extended octahedral network that must be broken up and mechanically ejected by physical sputtering (FIGURE 5.9). Consequently, at the lower bias voltages that one generally employs for selective etching, the octahedral AlF_3 network can serve as a barrier to further etching. The formation of a similar octahedral network

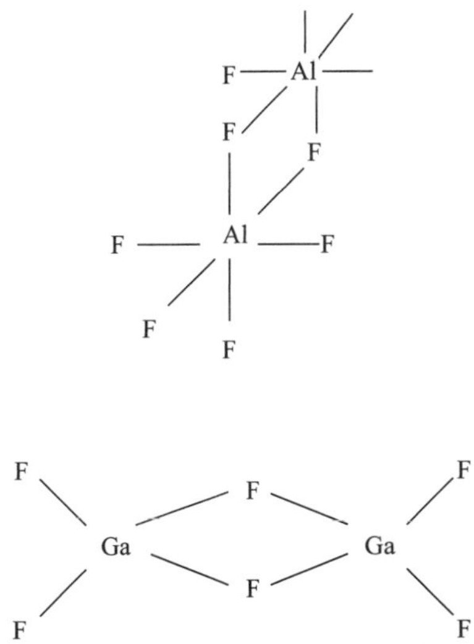

FIGURE 5.9 GaF_3 dimer and AlF_3 octahedral matrix.

by $InCl_3$ accounts for the slower etch rates of InAs, InGaAs and InGaP versus GaAs. Selectivity will be degraded at high ion energies where physical sputtering can dominate over the formation of non-volatile surface chemical products.

A good source of F atoms is SF_6. With $SiCl_4/SF_6$ mixtures of varying mole ratios, GaAs : AlGaAs etch ratios from 1 : 1 to 500 : 1 have been demonstrated. This large difference in etch rate permits the use of AlGaAs as an RIE etch-stop layer, with the required thickness of the AlGaAs layer being lower for higher Al mole fractions. However, the ion-induced damage below the etch stop may be problematic for some applications.

Etches employing $SiCl_4$ generally produce better surface morphologies, greater anisotropy and less change in surface stoichiometry than the analogous Cl_2-based etches. $SiCl_4$ produces little As loss, while Cl_2-etched surfaces are quite As deficient. Since the surface stoichiometry can strongly affect the surface electronic properties through the presence of a different distribution of surface states, the choice of etchant can be important to final device performance. There also appears to be less Cl-containing residue on an $SiCl_4$-etched surface. However, Si can be incorporated into the near-surface region by ion implantation at biases above 250 V and can be present at the $10^{19}/cm^3$ level at the surface, although the level drops rapidly to $<10^{15}/cm^3$ at 30 nm with a 250-V etch bias. While this may be advantageous for forming good ohmic contacts on n-type material, it could be a problem for p-type if the top surface were not etched away before contact formation.

A frequently unintentional way to obtain selective GaAs etching is to have appreciable amounts of water or oxygen adsorb on the reactor walls during sample loading without the aid of a loadlock. A loadlock is an extremely valuable equipment feature when any readily oxidised material, such as AlGaAs, is to be etched. It consists of an antechamber to the actual etching chamber where the sample can be inserted and the moist, oxygen-containing atmosphere can be pumped away before transferring the wafer into the plasma chamber. This avoids adsorption of significant quantities of water or oxygen on the plasma chamber walls, which subsequently serve as sources of oxygen when the plasma discharge is on. The great ease of oxidation of Al leads to the formation of a thin aluminium oxide surface layer that retards etching of the underlying AlGaAs even at the low oxygen concentrations available from wall desorption. While GaAs will etch well in non-loadlocked systems, more reproducible results are favoured by using a loadlock system for these materials as well. With a loadlock, equirate etching of AlGaAs and GaAs is possible when desired.

When equirate etching is desired, the most important requirement is a load-locked reactor to avoid formation of aluminium oxide on the surface. It is also important not to have residual F present in the reactor. If F-based chemistries have been used in a reactor prior to an equirate etching process, it will be important to run a reactor-chamber cleaning process before subjecting valuable substrates to RIE processing in the reactor.

Other halogen-based plasmas that generate Br or I atoms will etch GaAs, but these do not, in general, have major advantages relative to Cl-based chemistries. A major practical problem is the greater rate of corrosion of the process tool with Br or I.

There may be an advantage to performing a brief water rinse after Cl-based etching to remove any residual Cl compounds from the surface that might later contribute to corrosion problems.

5.7.2 Alkane-based plasmas for RIE

A useful alternative to Cl-based plasmas are plasmas employing alkanes (C_xH_y) as the source of reactive radicals. The simplest alkyl radical formed in these plasmas is the methyl radical, CH_3. Organometallic compounds, such as $Ga(CH_3)_3$, are formed by reaction with Al, Ga or In. These are the same compounds used as sources of the Column III atoms in metalorganic chemical vapour deposition (MOCVD). While In does not etch well with Cl-based chemistries due to the formation of the non-volatile octahedral $(InCl_3)_x$ structure (similar in structure to AlF_3 in FIGURE 5.9), it does etch well with the alkyl radicals formed in alkane plasmas. Consequently, these chemistries are important despite some practical disadvantages.

The primary problem with alkane plasmas is the ready formation of polymers with the general composition $(CH_2)_x$. Care must be taken to minimise polymer deposition on the GaAs wafer or etching will stop as a thick polymer layer blocks access of the GaAs to reactive alkyl radicals. This problem can be avoided or minimised by the addition of H_2 to the gas mixture. The hydrogen helps remove polymer by reacting with the polymer to form volatile alkanes that readily desorb from the surface. The addition of heavy ions, such as Ar, delays the onset of polymerisation and improves surface morphology, but the chemical effects of hydrogen predominate in controlling polymer buildup. Rates initially increase with increasing hydrogen as polymerisation is reduced, but further increases in hydrogen lead to lower rates due to dilution effects. In general, alkane/H_2 ratios between 0.2 and 0.4 work well, with the optimum ratio following the order propane $<$ ethane $<$ methane.

While hydrogen reduces the polymer buildup, hydrogen passivation of dopants requires post-etch thermal anneals at temperatures near 400°C to restore electrical activity. The activation energy for recovery of donor activity is about 1.75 eV so a 360°C anneal for 5 min gives almost complete reactivation of Si dopants. The depth of Si dopant passivation by H increases with increasing substrate temperature during etching and with increasing etch time. Higher ion energies make reactivation more difficult. With C_2H_6/H_2 etching at a 410 V self-bias, almost 500°C is required for restoration of initial carrier concentrations. This is the combined effect of H dopant passivation and the damage caused by deep penetration of the light H ion into the substrate by ion implantation.

While thermal annealing above 400°C can reactivate dopants by dissociating the bound H, the released H can diffuse deeper into the device so trying to minimise H diffusion during etching is advisable. In addition, annealing can cause vacancies to diffuse deeper in the wafer, resulting in electrically active defects. Defect reactions are discussed in some detail in Section 10.2.2. Annealing techniques are discussed in Section 7.2.1.

An important practical problem with alkane etches is the build-up of polymer on chamber walls where ion bombardment does not occur. Even under optimised etching conditions, frequent and regular oxygen plasma cleaning is essential.

5.8 HIGH-DENSITY PLASMA ETCHING (HDPE)

A single power supply is used in conventional RIE both to generate the plasma and to provide the bias potential for accelerating the ions into the wafer surface. Consequently, one does not have independent control of the energy of the incident ions and of the degree of dissociation of the source gases into highly reactive neutrals and ions. To achieve a high degree of dissociation, high RF powers are needed but high powers also produce high bias voltages and more damage. Since it is highly desirable in most cases to minimise damage by minimising ion energy, one would like to be able to produce a high degree of dissociation to get rapid etching while keeping bias voltages below 200 V.

This problem has been solved in high-density-plasma etching where the reactors employ separate power supplies to generate the reactive plasma species and the ion acceleration voltage. The use of two separate supplies permits semi-independent control of the degree of dissociation and the incident ion energy so high etch rates can be achieved with minimal damage. The ion acceleration voltage is still provided by a capacitively coupled RF power supply operating at 13.56 Mz and the wafer sits on the powered electrode. However, the reactant source gases are dissociated into reactive neutrals and ions using either electron cyclotron resonance (ECR) excitation with microwaves, typically at 2.45 GHz (FIGURE 5.10), or using an inductively coupled plasma (ICP) typically operating at 2 MHz (FIGURE 5.11).

Although many of the same source gases are used in RIE and HDPE, the actual plasma composition is radically different. HDPE plasmas are much more highly dissociated into ions, free atoms and other radicals, producing much faster etching at lower pressures than with RIE. Ion densities in an HDPE reactor can exceed those in conventional RIE by two to four orders of magnitude (up to 10^{13} ions/cm^3). Typical pressures in HDPE reactors are between 1 and 10 mTorr while RIE employs pressures of the order of 5–100 mTorr. Some typical rates and their associated reaction conditions are listed in TABLES 5.2 and 5.3.

The greatest advantage of HDPE versus RIE is the much lower ion bombardment energies that are used. Whereas is it difficult to achieve good etch rates and feature profiles at 100-V bias with RIE, a 100-V bias produces both good rates and vertical profiles with HDPE.

In general, a reactive gas mixture and pressure that performs well in ECR will also perform well in ICP. Comparable etch rates can be obtained by proper selection of microwave and ICP powers to permit translation from one platform to the other. A significant practical difference between ICP and ECR etchers is the use of magnets to confine the plasma and collimate the ions in the ECR. This gives some extra degrees of control that can be effectively used to tune the composition of an ECR-plasma deposition, such as changing the O : N ratio in SiO_xN_y films. For etching, however, the need to tune the magnets in ECR is more bothersome than helpful, and ICP hardware is easier to use.

From a practical perspective, the tuning of the magnets that may be required to minimise reflected microwave power when the plasma first ignites can lead to some variation in performance for very shallow etches where the plasma on-time must be short. For very shallow etches, an appreciable fraction of total plasma time may be involved in the tuning, leading to reproducibility problems. For deeper etches requiring longer etch times, this is not a significant factor. ICP systems, in contrast, do not require manual tuning and tend to stabilise faster, but in any plasma system there are voltage excursions when the plasma first strikes. Since these are not identical from run to run, some variation will occur for very shallow etching in ICP and RIE etches as well.

Selective etching of GaAs versus AlGaAs or InGaP can be achieved in HDPE with the same chemistries that work in RIE, but some selectivity is lost with the higher density plasmas. For example, using 20% SF_6 in BCl_3 at 5 mTorr and only 50 W RF power, selectivities of GaAs versus AlGaAs or InGaP that exceed

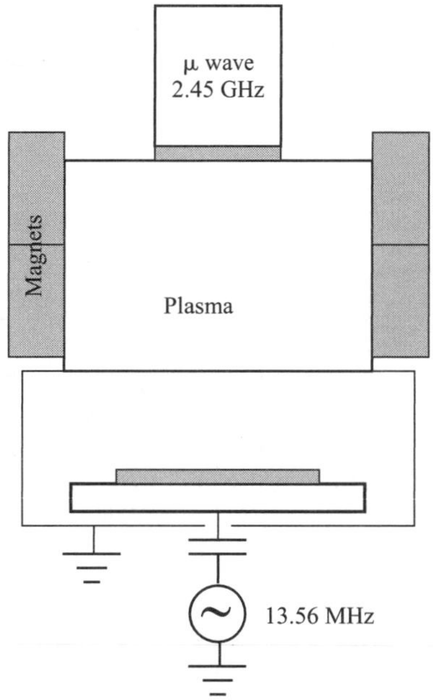

FIGURE 5.10 ECR-HDPE reactor.

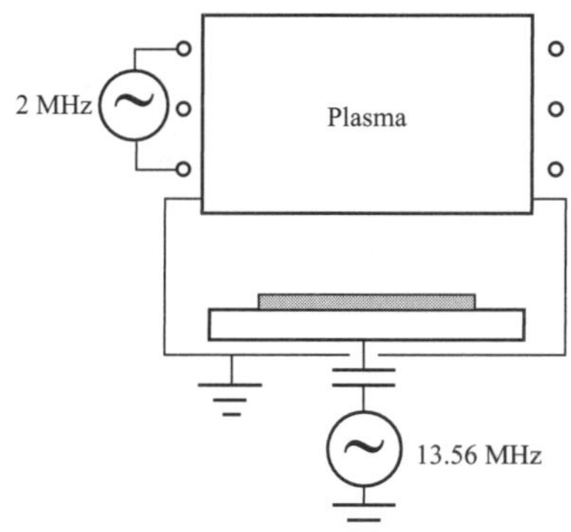

FIGURE 5.11 ICP-HDPE reactor.

TABLE 5.2 ECR-high-density-plasma etching (ECR-HDPE) rates for GaAs.

Etchant	Rate (μm/min)	T* (°C)	Pressure (mTorr)	Flow (sccm)	μ-wave power (W)	RF power (W)	Self bias (−V)	Ref.	Comments
Cl_2	0.23	r.t.	2	10		50	0	[a]	damage removal
$4Cl_2 : 1Ar$	0.9	r.t.	2	10		250	10	[b]	
$Cl_2 : 9Ar$	0.037		0.5		25	70	80	[b]	smooth, vertical
$4.3BCl_3 : 1Cl_2$	0.8–1.1	–	10–20	80	300	–	150	[c]	30 μm vias
BCl_3	0.080	–	1	30	200	–	150	[d]	GaAs ≅ AlGaAs
$2BCl_3 : 3Ar$	0.5	20	1	25	500	–	50	[e]	GaAs ≅ AlGaAs
$BCl_3 : N_2$	0.4	25	1	30	850	150	250	[f]	
$3BCl_3 : 1N_2$	1.3	25	1	30	850	150	250	[f]	
$2BCl_3 : 1SF_6$	0.071	–	1	30	200	–	150	[d]	GaAs/AlGaAs >600/1
$CH_4 : 3H_2/Ar$	0.008		1	27	250	0–60	50	[g]	GaAs≫ AlGaAs InGaAs/GaAs 2 : 1
$1CH_4 : 5H_2$	0.01	40–50	0.75	36	400	0	<5–10	[h]	
$1C_2H_6 : 10H_2$	0.015	40–50	0.75	33	600	0	<5–10	[h]	

* Nominal platen temperature; wafer surface probably hotter

[a] K.K. Ko, S.W. Pang [*J. Electrochem. Soc. (USA)* vol.141 (1994) p.255]

[b] S.W. Pang, K.K. Ko [*J. Vac. Sci. Technol. B (USA)* vol.10 (1992) p.2703]

[c] S.J. Pearton, F. Ren, A. Katz, J.R. Lothian, T.R. Fullowan, B. Tseng [*J. Vac. Sci. Technol. B (USA)* vol.11 (1993) p.152]

[d] S.J. Pearton, W.S. Hobson, C.R. Abernathy, F. Ren, T.R. Fullowan, B. Tseng [*Plasma Chem. Plasma Process. (USA)* vol.13 (1993) p.311]

[e] S.J. Pearton, C.R. Abernathy, R.F. Kopf, F. Ren [*J. Electrochem. Soc. (USA)* vol.141 (1994) p.2250]

[f] R.J. Shul, G.B. McClellan, R.D. Briggs, D.J. Rieger, S.J. Pearton, C.R. Abernathy, J.W. Lee, C. Constantine, C. Barratt [*J. Vac. Sci. Technol. A (USA)* vol.15 (1997) p.633]

[g] C. Constantine, D. Johnson, S.J. Pearton, U.K. Chakrabarti, A.B. Emerson, W.S. Hobson, A.P. Kinsella [*J. Vac. Sci. Technol. B (USA)* vol.8 (1990) p.596]

[h] V.J. Law, M. Tewordt, S.G. Ingram, G.A.C. Jones [*J. Vac. Sci. Technol. B (USA)* vol.9 (1991) p.1449]

TABLE 5.3 ICP-high-density-plasma etching (ICP-HDPE) rates for GaAs.

Etchant	Rate (μm/ min)	T* (°C)	Pressure (mTorr)	Flow (sccm)	ICP power (W)	RF power (W)	Self bias (−V)	Ref.	Comments
$Cl_2 : Ar$	1.2	25	1	30	500	150	250	[a]	
$3Cl_2 : 1Ar$	1.6	25	1	30	500	150	250	[a]	
$Cl_2 : N_2$	0.7	25	1	30	500	150	250	[a]	
$3Cl_2 : 1N_2$	1.1	25	1	30	500	150	250	[a]	
BCl_3	0.30	10	2.5	40	500	100	250	#	smooth
BCl_3	0.10	10	2.5	40	160	130	150	#	
$BCl_3 : N_2$	0.4	25	1	30	500	150	250	[a]	
$3BCl_3 : 1N_2$	0.95	25	1	30	500	150	250	[a]	

* Nominal platen temperature; wafer surface probably hotter

Author's data

[a] R.J. Shul, G.B. McClellan, R.D. Briggs, D.J. Rieger, S.J. Pearton, C.R. Abernathy, J.W. Lee, C. Constantine, C. Barratt [*J. Vac. Sci. Technol. A (USA)* vol.15 (1997) p.633]

80 are attained in the conventional RIE mode. With the addition of 200 W ICP power to switch the reactor mode to HDPE, the selectivity of GaAs versus AlGaAs is reduced to 45 and versus

InGaP to about 8. Increasing ion flux by increasing ICP source power decreases selectivity while increasing ion energy by increasing RF power also decreases selectivity. Both effects are due to the better sputtering of AlF_3 or InF_3 with higher ion energies and fluxes.

Hydrogen atoms in HDPE produce the same desirable and deleterious effects observed in RIE. However, the use of much lower incident ion energies makes hydrogen diffusion more important than H implantation effects in HDPE. Consequently, any H-related problems can be more easily rectified with HDPE etching.

Photoresist erosion rates tend to be higher in HDPE than RIE due to the higher fraction of radicals in the HDPE plasmas that increase the chemical contribution to resist removal. This loss of selectivity for GaAs versus resist must be considered when translating a process from RIE to HDPE. Selectivity versus photoresist in HDPE is generally less than 10 and often of the order of 2–4. In contrast, selectivity versus resist greater than 10 is generally possible with RIE.

5.9 REACTIVE-ION-BEAM ETCHING (RIBE) AND CHEMICALLY ASSISTED ION-BEAM ETCHING (CAIBE)

Reactive-ion-beam etching (RIBE) and chemically assisted ion-beam etching (CAIBE) combine a collimated ion beam with a chemically reactive gas to provide highly anisotropic etching. Unlike RIE and HDPE, the ions in RIBE and CAIBE are generated in an ion source rather than in a plasma that is in direct contact with the sample (FIGURE 5.12). Use of a remote ion source provides

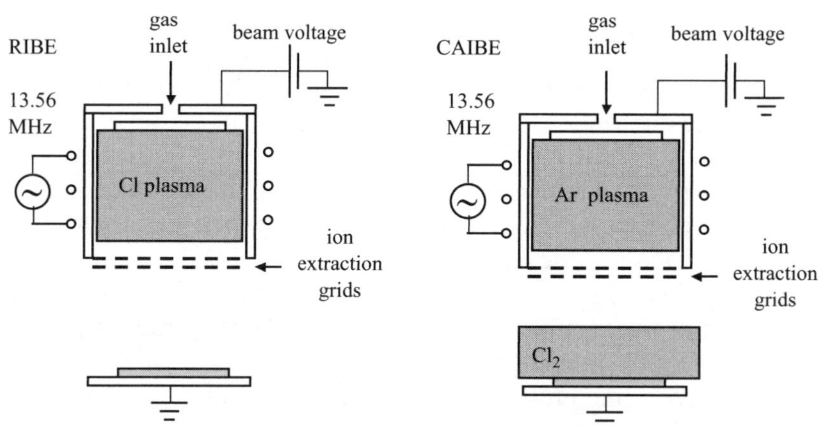

FIGURE 5.12 RIBE and CAIBE reactors.

TABLE 5.4 Reactive-ion-beam etching (RIBE) rates for GaAs.

Etchant	Normalised rate (micron/min/ (1 mA/cm^2))	Pressure (mTorr)	Ion energy (eV)	Current density (mA/cm^2)	Beam angle off normal (°)	Ref.	Comments
Cl_2	0.6	0.25	550	0.5	0	[a]	
	8.9	1	200	0.09	0	[b]	473 K
	2.2	1	200	0.09	0	[b]	373 K
	0.067	0.2	200	0.05	30	[c]	no loss of As
	0.405	0.3	500	0.21	30	[c]	no loss of As
CCl_4	0.06	0.23	500	0.4	0	[d]	
BCl_3	0.12*	1.5	600	–	0	[e]	p damaged more than n
2% O_2/Ar	0.02	0.14	100	0.15	0	[f]	

* Rate not current-normalised

[a] K. Asakawa, S. Sugasa [*Jpn. J. Appl. Phys. 2 (Japan)* vol.22 (1983) p.L653–5]

[b] T. Tadokoro, F. Koyama, K. Iga [*J. Vac. Sci. Technol. B (USA)* vol.7 (1989) p.1111–4]

[c] R.A. Barker, T.M. Mayer, R.H. Burton [*Appl. Phys. Lett. (USA)* vol.40 (1982) p.583–6]

[d] W.-X. Chen, L.M. Walpita, C.C. Sun, W.S.C. Chang [*J. Vac. Sci. Technol. B (USA)* vol.4 (1986) p.701–5]

[e] K. Nagata, O. Nakajima, T. Ishibashi [*Jpn. J. Appl. Phys. (Japan)* vol.25 (1986) p.L510–12]

[f] H. Kinoshita, T. Ishida, K. Kaminishi [*Appl. Phys. Lett. (USA)* vol.49 (1986) p.204–6]

independent control of ion energy and of the flux of ions and neutrals at the wafer. It also reduces the spread in ion energies to a few eV compared with spreads of up to several hundred eV in RIE.

RIBE involves the generation of reactive chemical species, such as Cl^+ and Cl_2^+, in an ion source using gases such as Cl_2 and BCl_3. A beam of these ions is used to etch GaAs. CAIBE, also called ion-beam-assisted etching (IBAE), uses a beam of inert ions, such as Ar^+, Xe^+ or Ga^+, that is directed at the wafer, which is immersed in a neutral gas, such as Cl_2. A variation called radical-beam-ion-beam etching (RBIBE) predissociates the Cl_2 to provide a beam of highly reactive Cl atoms that strikes the wafer in concert with an ion beam. Focused-ion-beam (FIB) etching is another variant in which the collimated ion beam from a broad-area ion gun is replaced by a scanned, tightly focused beam that can be used for direct-write etching.

Etch rates (TABLES 5.4–5.6) with these processes are intermediate between those of simple physical sputtering and plasma etching in RIE systems. Higher rates occur when parameters are selected to increase the importance of chemical etching versus ion sputtering. Damage is also reduced when chemical contributions increase, but undercutting may become a problem. Increasing the fraction of total ion energy that is deposited very near the surface increases rates, and etching with Ar^+ produces etch rates an order of magnitude faster than with He^+ at the same incident energy. While it is customary to normalise etch rates to fluxes of 1 mA/cm^2,

TABLE 5.5 Chemically-assisted-ion-beam etching (CAIBE) rates for GaAs.

Etchant	Rate (μm/min/ (1 mA/cm^2))	Pressure (mTorr)	Ion energy (eV)	Current density (mA/cm^2)	Angle off normal (°)	Ref.	Comments
Broad area							
$Cl_2 + Ar^+$	5	10	1000	1	0	[a]	±5% uniformity
	0.085	10	500	0.02	0	[a]	
	5	0.08	500	0.25	0	[b]	heating effects
	3	0.08	500	0.02	0	[b]	
	1.32	0.25	200	0.025	–	[c]	
	0.50	0.25	200	0.100	–	[c]	
	1.36	0.25	400	0.025	–	[c]	
	0.65	0.25	400	0.100	–	[c]	
Focused ion beam (FIB)			(keV)				
$Cl_2 + Ga^+$	1.4	20	35	80	0	[d]	
	3.6	20	35	80	70	[d]	

[a] G.A. Lincoln, M.W. Geis, S. Pang, N.N. Efremow [*J. Vac. Sci. Technol. B (USA)* vol.1 (1983) p.1043–6]
[b] M.W. Geis, G.A. Lincoln, N. Efremow, W.J. Piacentini [*J. Vac. Sci. Technol. (USA)* vol.19 (1981) p.1390–3]
[c] J.A. Skidmore, D.L. Green, D. Young, J.A. Olsen, E.L. Hu, L.A. Coldren, P.M. Petroff [*J. Vac. Sci. Technol. B (USA)* vol.9 (1991) p.3516–20]
[d] Y. Ochiai, K. Gamo, S. Namba [*J. Vac. Sci. Technol. B (USA)* vol.3 (1985) p.67–70]

TABLE 5.6 Radical-beam-ion-beam etching (RBIBE) rates for GaAs.

Etchant	Rate (μm/min/ (1 mA/cm^2))	Flow (sccm)	Pressure (mTorr)	Ion energy (eV)	Current density (mA/cm^2)	Angle off normal (°)	Ref.	Comments
$Cl_2 + Ar^+$	2.8	9	0.8	200	0.2	10	[a]	straight, smooth facets
	6.4	5	0.25	200	0.025	–	[b]	low damage, 30°C
	1.9	5	0.25	200	0.100	–	[b]	equirate versus $Al_{0.6}Ga_{0.4}As$
	8.0	5	0.25	400	0.025	–	[b]	
	1.6	5	0.25	400	0.100	–	[b]	

[a] J.A. Skidmore, L.A. Coldren, E.L. Hu, J.L. Merz, K. Asakawa [*J. Vac. Sci. Technol. B (USA)* vol.6 (1988) p.1885–8]
[b] J.A. Skidmore, D.L. Green, D. Young, J.A. Olsen, E.L. Hu, L.A. Coldren, P.M. Petroff [*J. Vac. Sci. Technol. B (USA)* vol.9 (1991) p.3516–20]

actual rates may be much higher or lower than expected if etching is done at different ion fluxes. At high ion fluxes, appreciable heating may occur and thermally activated chemical reactions can yield much higher etch rates than with low flux conditions. Beam-current heating effects can occur with fluxes as low as 0.25 mA/cm^2. Etch rate dependence on extraction voltage, V, varies from $V^{5/2}$ under predominantly sputtering conditions to a sublinear dependence on V when chemical contributions are more important.

Rates in CAIBE and RBIBE depend on the uniformity of Cl-species across the wafer. Neutral fluxes can vary by up to an order of magnitude across a wafer if gas inputs are not optimally designed. Gas nozzle configuration, including angle and separation from the wafer, is critical for good wafer uniformity. Etch profiles can be especially sensitive to nozzle angle, with angled surfaces forming at temperatures below 100°C as features deepen and regions are shadowed from simultaneous reactive gas and ion flux. Reactive gas scattering from vertical walls can also produce trenching during "room temperature" etching.

Etch profiles for all ion-beam-based processes depend strongly on ion incidence angles and beam divergence. Normal incidence can produce angled sidewalls (overcutting) when sputtering is the dominant mechanism, but increasing the incidence angle slightly can produce vertical walls. Slightly tilting and rotating the sample can also increase verticality. Very small structures (9 nm quantum wires) have been etched using angled CAIBE following vertical etching of 80 nm quantum wires.

Beam divergence is more pronounced at lower ion energies. Some beam spread due to scattering from neutrals can occur for ion energies below 500 eV. However, high ion energies will produce electronic damage near the wafer surface. Damage of sidewalls is minimised by improved beam collimation. With well collimated beams, RIBE and CAIBE are extremely well suited for etching devices where sidewall damage must be minimised, assuming the greater damage on the horizontal etched surfaces can be tolerated. Sidewall etching by reflected primary ions or energetic secondary ion species can produce undercutting and damage. Trenching can result at feature edges when ions from poorly collimated beams are deflected from sidewalls. Wafer heating at high ion fluxes can increase undercutting due to thermally activated chemical processes even when the beam is well collimated. However, slight undercutting may be very desirable to remove damage from sidewalls in some applications.

For CAIBE, highest rates occur when an optimum balance is achieved between Cl_2 reaction with the surface to form a Cl layer and the ion-induced desorption of the products. Very smooth surfaces are possible, but greater initial surface cleanliness is required than with conventional RIE. Solvent cleaning just before loading wafers yields the smoothest surfaces.

Surface damage in CAIBE depends on the degree of chemical contributions to etching. While CAIBE with a Cl_2 flux $>1 \times 10^{18}/cm^2$ s with 500 eV Ar^+ produced diodes with ideality factors (n) equal to the control, an order of magnitude lower Cl_2 flux a few mm away on the same wafer yielded significant increases in n. Increasing the substrate temperature decreases the post-etch damage.

Surface stoichiometry depends on the balance between sputtering and chemical reaction through the Ar^+/Cl_2 ratio. Surface enrichment in As is greater under higher Cl_2 flux but decreases at higher beam energies.

5.10 GENERAL ISSUES FOR DRY ETCHING

5.10.1 Etch uniformity

To first order, uniformity is determined by the hardware design and quality. Most properly operating etch systems are capable of producing a uniform flux of reactants across the platen holding the wafer. However, if etch rates seem to vary in a reproducible way depending on how the wafer is located in the etch chamber, the hardware may not be performing up to specification and may need repair.

In ECR-HDPE systems, proper setting of the lower magnet, which primarily controls beam divergence, is important. The most uniform supply of reactants will be obtained when the lower magnet is set to produce a well collimated beam.

Even when the hardware is working ideally, the consumption of reagents will be different on the platen and the wafer. This can lead to either a greater supply and consequently faster rate near the wafer edge if the platen doesn't consume reactants or a lower supply and lower rates if it does. These types of etch non-uniformities are generally referred to as loading effects. If very high uniformity across the entire wafer is needed or if small pieces of a wafer are being individually processed, it may help to place sacrificial bulk GaAs wafers in contact with the device wafer to equalise reactant consumption rates.

A more local variant of loading effects is found when there is considerable variation in the size and spacing of devices on the wafer. Regions with a high resist-to-GaAs ratio (low exposed fraction of GaAs) may etch at different rates than those with low resist-to-GaAs ratios (high exposed fraction of GaAs). If fewer reactants are consumed etching resists, diffusion of reactants from resist regions into GaAs regions can locally increase the rate.

Etch uniformity may also depend on proper surface preconditioning before beginning to etch. Surface cleaning issues have been discussed in detail in Section 3.2. Without proper precleaning, some regions of the surface may be "micromasked" by surface chemical species that delay initiation of the etch at that location and thereby produce roughening.

Non-uniform oxides on Al- and Sb-containing materials pose the most risk for causing micromasking. This is especially true if the structure of the device involves more than one etch procedure, as in the case of a double-mesa like that shown in FIGURE 5.14. Differing thicknesses of oxide can form on exposed Al- and Sb-containing layers, leading to roughening during the second etch step. This can be prevented by a brief H-plasma treatment after returning the wafer to the plasma chamber but before beginning the second etch. Details of an effective pretreatment that yielded subnanometre RMS roughness for GaInAsSb and nanometre-scale roughness for AlGaAsSb have been reported [3]. The slightly greater roughness in the second case reflects the effect of Al versus Ga.

5.10.2 Damage from dry etching

While dry etching is to be preferred over wet etching for its better dimensional control, greater anisotropy, and generally more reproducible etch rates, the ion bombardment so essential to achieve these advantages simultaneously damages the near-surface region of the wafer. Therefore, in selecting etch conditions, the amount of damage that can be safely tolerated without excessively degrading electronic properties of the material must be carefully considered.

Degradation of electronic properties can result from two main mechanisms. The first is lattice damage resulting from direct energy transfer from the incident ion or energetic neutral. Disruption of the lattice can lead to the formation of point defects such as antisites (As_{Ga},Ga_{As}) and vacancies (V_{As},V_G). Ions can also produce changes in the surface stoichiometry, which can introduce surface states and change the Fermi level pinning position and even the near-surface doping type, as discussed in Chapter 3. The second is dopant passivation due to reaction of dopant atoms, especially Si, with hydrogen atoms generated in the plasma. Both problems can be reduced, although not necessarily eliminated, by post-etch thermal annealing.

Although damage will occur at any ion energy above the $\approx$40 eV atomic displacement threshold of GaAs, the damage produced at ion energies less than 100 eV is generally acceptable for most applications. Energies in excess of 100 eV are routinely encountered in RIE, RIBE and CAIBE, although HDPE processes readily permit low-energy bombardment for anisotropic profiles. Point-defect creation in the near surface can lead to carrier compensation up to 0.22 μm deep at 100-V bias. With bias voltages less than 50 V, undegraded optical properties and Schottky diode behaviour can be achieved. However, as ion energies are lowered below 100 eV to reduce damage, the increased chemical component of the etch can lead to excessive undercutting and loss of vertical profiles. While GaAs MESFETS etched with $SiCl_4$ between 100 and 300 eV have been described as "damage free", a 200% overetch using $SiCl_4/SF_6$ and an AlGaAs etch-stop layer decreased the source saturation current due to active layer depletion. More damage-sensitive HEMT devices, which operate at lower bias voltages, are completely damaged with a DC bias of 200 V and a 200% overetch. At voltages over 200 eV, deep-level electron traps form, but most of these can be annealed out. Ion energies over 300 eV may produce permanent degradation of electrical properties. While annealing reduces the total defects, it can also cause propagation of the remaining defects to greater depths.

Strategies to reduce ion damage can improve device performance. Confining the damaged region as close to the surface as possible is advantageous. Damage can be reduced by minimising the ion penetration depth by reducing ion energy or using heavier ions, which reduces the ion penetration at a given energy. Increasing the chemical contribution and consequently increasing the etch rate can help. Higher etch rates leave less apparent damage that tends toward a steady-state value for etch times greater than the damage depth/etch rate ratio. The best strategy may be removal of the damaged region at the end of the etch. A brief etch under conditions with low ion energies (<40 eV) can etch away the thin damaged region, leaving only high-quality material. This approach is easy to implement in an HDPE system by reducing the RF bias immediately before quenching the high-density plasma; it is difficult to do in conventional RIE systems where the plasma and bias are generated by the same RF power supply.

For very small etched devices where the sidewall area is an important characteristic of the device, ion-induced damage of the sidewall may be important. Even though top-surface and sidewall ion fluxes may be similar, the energy of ions striking the sidewall is generally lower with the impinging material consisting mainly of reflected primary ions, sputtered material and chemical reaction products from the adjacent flat surface. Sidewall damage can be minimised by low ion energy, highly collimated ion beams (which is more difficult at low ion energy) and a significant chemical contribution to the etch mechanism that will etch the sidewall enough to remove the damage but not enough to cause excessive undercutting. Unlike top-surface damage, which tends towards a steady-state value determined by the balance between ion penetration and removal of the damaged layer by etching, sidewall damage continues to increase with total etch time.

Changes in surface stoichiometry due to differential removal of As- and Ga-containing reaction products can change the location of the Fermi level pinning at the surface. While many studies report that n-type and p-type pin at the same midgap energy for GaAs(100), others report pinning of p-type material closer to the valence band. Dry etching produces an As-rich surface and As at the interface with the native oxide that forms after etching shifts the p-type pinning position closer to midgap. The implications for forming high-quality ohmic and Schottky contacts are clear.

Plasmas with H-containing source gases expose the wafer to a flux of H-atoms that readily diffuse into the wafer and passivate dopants. The depth of dopant passivation will depend on the diffusion characteristics of H in the material or combination of materials in a heterostructure. Deeper penetration can result from higher ion energies, higher wafer temperatures during etching (due to heating by ion bombardment), larger H-atom fractions in the plasma, longer exposure at longer etch times and the competition between diffusion rates and the rate of removal of the surface.

Thermal annealing can release the H and reactivate the dopants. The effectiveness of annealing depends on the material. While GaAs reactivates quite readily with a 5 min anneal at 360°C, AlGaAs materials may be more difficult to reactivate, especially for higher Al mole fractions. While the sheet resistance in $Al_{0.15}Ga_{0.85}As$ is restored with a 30 min anneal at 450°C, the same annealing conditions increase the sheet resistance in $Al_{0.3}Ga_{0.7}As$. The depth of carrier reduction increases from 170 nm in the unannealed material to 300 nm in the annealed material. This may be related to the DX centre, which becomes important for Al mole fractions in excess of 0.22.

5.10.3 Resists and their behaviour in dry etching processes

Device fabrication is not possible without the use of a resist to define the pattern by protecting certain regions from etching while permitting etching in other regions. Many general photoresist issues have been discussed in Section 4.4.1. This section will be more specifically addressed to resist issues important in dry etching. Resists can be divided into two general categories: 1) conventional photoresists (PR) based on organic polymers and 2) metallic and dielectric resists. While ideally one would want the resist to be unaffected by the GaAs etch process, in reality the resist will be at best physically sputtered and at worst chemically etched as well, sometimes at rates faster than the etch rate of GaAs. It is, therefore, very important to understand how a particular dry etch process is going to affect the resist and to select the resist type and thickness so it will not fail during device fabrication. A most important point to remember is that the smoothness (or roughness) of the resist before etching will determine the smoothness of the final feature wall. While this may not be of paramount importance for many electronic devices, for optoelectronic devices any edge roughness may lead to unacceptable light scattering at the roughened surface. It is always desirable to have as smooth a resist edge as possible.

The effect of ion bombardment of the resist must first be considered in selecting both the resist type and the thickness for a particular application. Even if the resist is not readily attacked by the chemistry employed to etch GaAs, it will be sputtered by the impinging ions. This can limit the critical dimension (smallest feature size) that can be achieved for a given required etch depth. Since thinner photoresists are required for smaller feature sizes, resist erosion places an upper limit on the achievable etch depth in the final device. Organic PRs sputter quite readily and can be rather quickly removed by high-energy ions, such as those used in RIBE processes. Metallic and dielectric resists are frequently required if appreciable etch depths are required. However, the ease of using standard photoresist makes it the most attractive first choice for many applications.

A matter of some concern with organic photoresists is the reaction of their exposed sidewalls with halogens like Cl to form partially halogenated compounds that have very different solubilities from the unaltered resist. This can make the altered sidewalls less soluble in acetone than the original photoresist and impede removal of all the resist after etching. This shows up as "fences" of altered photoresist along the upper edges of etched mesas. Removal may require a more aggressive solvent, such as NMP (N-methyl pyrolidone) and ethanolamine, and/or oxygen plasma cleaning.

The resist will also be heated during the process. The simplest effect of heating is an increase in the resist erosion rate. While RIE etching using BCl_3/Cl_2 appeared to produce only slightly faster erosion of a Novolac resist than an SiO_2 or Si_3N_4 resist below 100°C, the Novolac photoresist etched up to two orders of magnitude faster when substrate temperatures near 140°C were used. This contrast was most extreme at high RF powers, where additional heating of the wafer due to high ion energies occurs.

If photoresist (PR) is heated excessively, it can reticulate. Reticulation is the wrinkling of the resist that can occur when the surface has not been cured enough to withstand the volume change of the bulk of the resist as it becomes more dense and contracts as the temperature increases. Careful attention to pre-etch PR curing protocols can minimise this problem.

Photoresist can also flow if it is heated above its glass-transition temperature, T_g, which is between 100 and 125°C

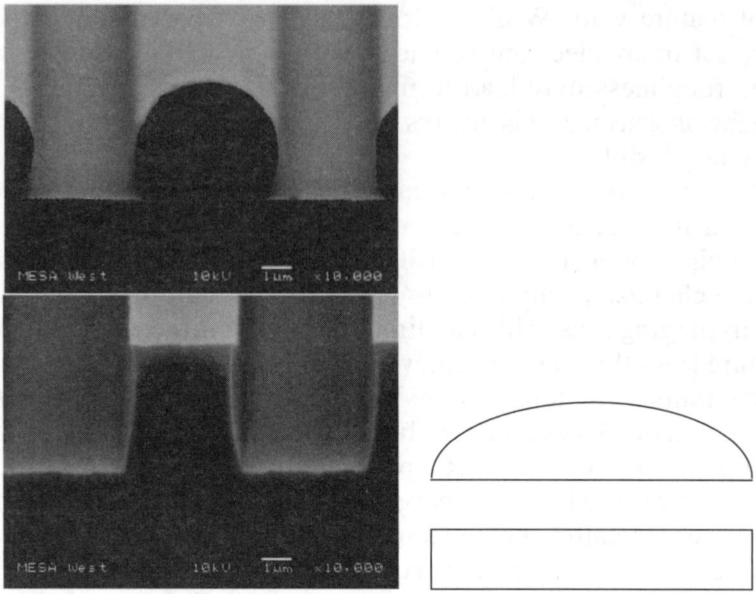

thick reflowed resist thin reflowed resist

FIGURE 5.13 Reflowed resist profiles.

for many Novolac resists and between 140 and 160°C for 4-polyhydroxystyrene (PHS). White light exposure has been found to reduce the glass-transition temperature of resist, so performing all resist handling under yellow lighting is advisable. The T_g for a particular resist is available from the resist manufacturer. When a resist exceeds T_g, it begins to flow to reduce its total surface energy by assuming a curved surface (FIGURE 5.13). While this phenomenon can be harnessed to controllably etch angled facets into GaAs by performing the reflow before beginning to etch, it can be a serious problem if it occurs unintentionally during an etching process because an angled profile may result when a vertical one was desired. The horizontal extent of the resist tends to remain constant, but the volume redistributes into an arched or even nearly spherical resist profile, depending on the initial height-to-width ratio.

Deliberate variation in the thickness of the resist can be used to advantage to achieve angled sidewalls with processes that otherwise produce vertical profiles. This may be especially desirable for a continuous connection between a metal contact on a mesa ledge with a metal bond pad on the adjacent wafer surface. To achieve this effect, the photoresist is first exposed and developed. It is then heated slightly above its glass transition temperature to induce some flow. The resist assumes a curved profile to minimise its surface energy. Controlled erosion of the resist as the wafer etches gradually exposes the underlying wafer regions, which then etch to produce the desired angled structures. The type of profile that results from a two-step etch using first a straight-wall photoresist and then a reflowed photoresist is shown in FIGURE 5.14. The angled wall of the second mesa allows good metal contacts to be made from the top of the angle-walled mesa to bond pads on the lower etched surface.

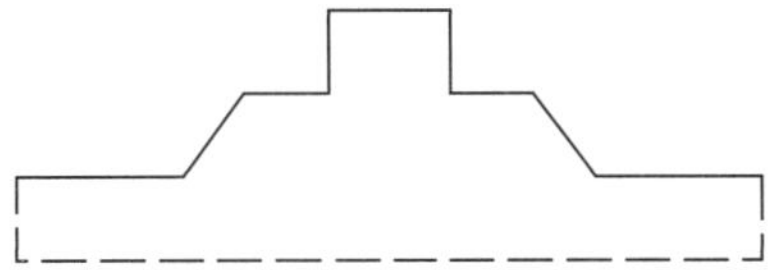

FIGURE 5.14 Angled-wall double mesa structure from two-step etch with first straight-wall resist and then reflowed resist.

To avoid thermal problems for longer etches, one or more cool-down steps may be included during the etch process. With source gases that require plasma dissociation to produce appreciable etching, this can simply involve stopping the plasma discharge while the gases continue to flow for a few minutes. A good starting time for determining what is required for adequate cooling is about 5 min. The final etch depth and profile will be largely determined by the time the plasma is on. For gases that can do appreciable etching without plasma activation, such a cool-down step might allow an undesirable amount of isotropic etching, resulting in degraded profiles.

The required thickness of a resist, especially a metal resist, will be defined largely by its sputtering rate. Metal resists or multilayer photoresist/dielectric resist stacks are used in processes such as RIBE for deep, narrow features where conventional photoresist cannot withstand the prolonged high-energy etching required.

One solution to the problem of excessively rapid photoresist erosion is the use of PR/dielectric stacks as trilevel or quadlevel resists. A quadlevel resist that works well for RIBE and ICP etching is shown in FIGURE 5.15. An organic liftoff layer, such as polymethylglutarimide (PMGI), is applied to the substrate. The main PR mask layer is then applied to the PMGI and extensively baked (60 min at 190°C followed by 60 min at 250°C). A half micrometre of a dielectric such as Si_xN_y is applied to this hard-baked PR. Another PR layer that is used for patterning the dielectric is applied on top of the stack.

The desired pattern is photolithographically defined in the top PR layer. This pattern is then etched into the Si_xN_y using an SF_6/O_2 RIE process. The pattern is then transferred into the hard-baked PR and liftoff layer using an ECR or ICP etch with O_2 and Ar. Following the completion of etching the III–V substrate (typically by RIBE), what remains of the resist stack can be removed by soaking in a solvent that releases the liftoff layer from the surface.

Nickel is often a good choice for a metal resist due to its relatively high resistance to sputtering and its tendency not to redeposit on the wafer surface when it is sputtered. Redeposition of sputtered material, either resist or low-volatility III–V products, can produce micromasking. The result is very rough etching, producing a "grassy" surface that often looks black or "burned" upon inspection.

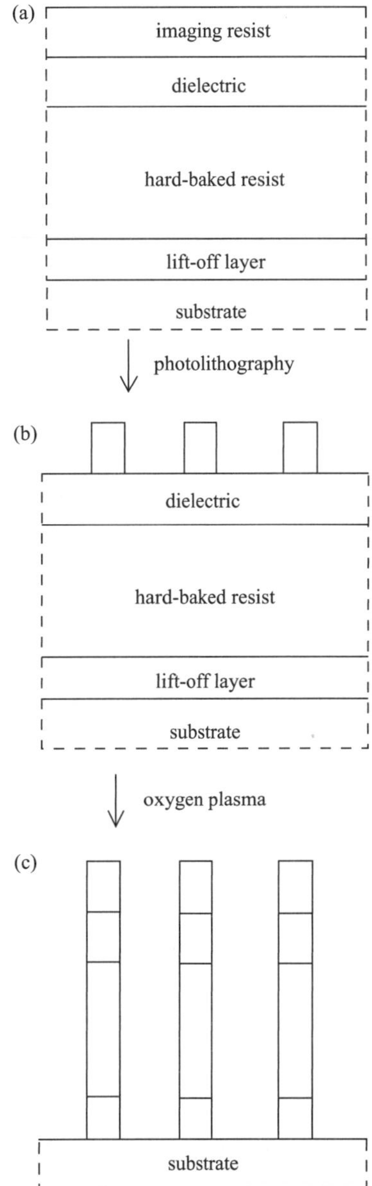

FIGURE 5.15 Quad-level mask.

FIGURE 5.16 Effect of sputtered Au micromasking.

When a device wafer is being subjected to more than one etch process with metallisation being applied between the etch steps, one must be especially careful to cover the metal with photoresist to avoid sputtering and redeposition of the metal onto adjacent surfaces. This is particularly true for Au-based metallisations. When Au is sputtered onto the nearby GaAs surfaces, it serves as a micromask and destroys the smooth surface that is generally obtained in dry etching (FIGURE 5.16). Careful attention to covering all exposed Au is essential to avoid ruining the device wafer due to this effect.

5.10.4 Advantages of Ar addition

Argon addition makes it easier to strike a plasma and therefore extends somewhat the range of gas mixtures, pressures and RF power that can be employed in RIE. It increases the sputtering component of any gas mixture, which increases the anisotropy and often serves to increase the smoothness of a slightly roughening etch, especially if micromasking is involved. Its optical emission can also serve as a calibration for semiquantitative applications of optical emission as a plasma diagnostic. These advantages of Ar addition apply in both RIE and HDPE processes. Addition of a few percent Ar can be sufficient to achieve most of these improvements.

5.10.5 Methods for end-point determination

The simplest method for determining the end point is to etch for a specific time assuming a known etch rate. While this is often the only available method, it is not a very reliable one. Day-to-day variation in the chamber wall conditions can lead to

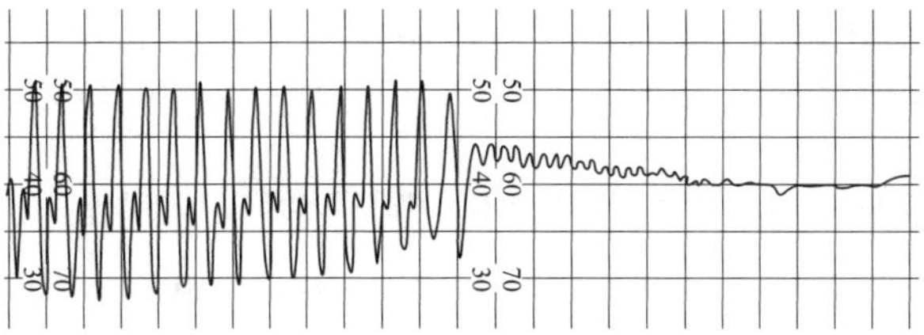

FIGURE 5.17 Optical reflectance spectrum showing the transitions between different heterolayers.

sufficiently different etch rates that variations in excess of 10% are routinely encountered. This is especially true when the etch system cannot be dedicated to a single process or single wafer layout. Even variations in the ratio of exposed GaAs to resist can produce significant rate variations in a dedicated system.

A better method for determining end point while etching heterostructures relies on the difference in optical reflectance from GaAs and III–V ternaries. Optical reflectance of a collimated light source, such as a laser, is monitored continuously. The resulting oscillations (FIGURE 5.17) indicate the passage of the etch front through the heterointerfaces. The physics of this is well understood and straightforward to model [4]. Modelling is not necessary to use this technique, however. By simply counting the oscillations, it is possible to determine when one reaches the specific interface where etching is to stop. If etching needs to proceed into the underlying GaAs substrate, where no interface oscillations will be observed, one can still obtain better etch control by calculating the etch rate through the heterostructure under current chamber conditions and timing the etch assuming that rate.

If an etch is roughening the wafer, this technique will give advance warning that it is occurring. As the surface begins to roughen, the amplitudes of the oscillations are damped. When damping is first observed to be starting, it is sometimes possible to "rescue" a sample before it is totally destroyed by the etch process, correct the source of roughening and try again. The reflectance spectrum of the aborted run provides information regarding the depth at which the process was stopped and allows one to resume monitoring the etch process during the second attempt.

There are some limitations to this technique. Although optical reflectance can reveal when the desired depth has been achieved at the point being measured, etch-rate uniformity across the wafer can lead to a variation in etch depths. In addition, the beam must be centred on exposed GaAs instead of resist. This can be difficult with wafers with a high ratio of resist to exposed GaAs (low exposed fraction of the GaAs surface). Longer-period oscillations due to the slower erosion of resist will be observed and can dominate the signal when there is little exposed GaAs.

5.10.6 Plasma diagnostics for trouble-shooting

While a tremendous amount of information about the chemical species in a plasma can be gleaned from optical emission spectroscopy (OES), interpretation of the spectra often requires time and expertise beyond that of the normal etch system operator. Quantitative interpretation of OES requires knowledge of the electron energy distribution function, the relative excitation cross-sections for the various plasma species at the range of energies present in the system, the interspecies relaxation processes and the individual emission spectra of the different excited-state species. In addition, typically less than 5% of the species in a plasma are in an excited state capable of emitting light, so the majority of the species that may be contributing to the etch process are not interrogated by OES.

There are, however, some relatively simple situations in which optical emission spectra can be extremely useful in identifying the source of an etching problem. The most significant of these is the identification of the presence of a leak in the system. While Cl-based plasmas appear violet, the presence of oxygen from a moderate leak will change the dominant colour to blue. Oxygen has characteristic emission peaks at 777.4 nm for O atoms and 486.0 for O^+ ions. The presence of air or water will be manifest as an oxygen peak in OES. Simple-to-use spectrometers with fibre optic inputs are sold for plasma monitoring and make leak detection simple.

5.10.7 Chamber cleaning issues

Since the composition of the plasma will be affected by reactions of radicals with the chamber wall, highly reproducible etching requires that the composition of the chamber walls be relatively consistent. To achieve this, chamber cleaning protocols are essential.

A simple and rather effective way to maintain a relatively consistent wall condition is the frequent use of oxygen plasmas to remove polymer buildup. The main sources of this polymer in Cl-based processes are products of photoresist erosion. Since photoresist is an organic polymer, alkyl fragments will be sputtered from the surface during etching. As these fragments enter the plasma above the wafer surface, they can be ionised and/or dissociated into highly reactive radicals. In the presence of Cl, some will become chlorinated as well. When these radicals encounter a chamber wall, polymer deposition will occur. The amount and chemical composition of the polymer will depend on the number and types of etches that have been performed in the chamber since the last cleaning. Although the specifics of this polymer will vary widely from chamber to chamber, its predominantly organic nature makes it susceptible to attack by oxygen atoms in the cleaning plasma. Additionally, the reaction of BCl_3 with oxygen leads to the formation of B_2O_3 deposits on the walls.

For optimum reproducibility, the oxygen clean would ideally be run between each etch. For practical purposes, this is generally not done. However, it is important to have a regular cleaning schedule that will depend on the specific use patterns of the etch system. If regular oxygen cleaning is not done, frequent mechanical scrubbing of the chamber walls may be needed to remove the polymer buildup. Even with frequent oxygen-plasma cleans, mechanical cleaning is occasionally necessary.

5.10.8 Effect of chamber materials

Etching plasmas are not merely in contact with the wafer to be etched; they are also interacting with the chamber walls and the cathode material upon which the wafer sits. This can sometimes alter the composition of the plasma in ways that affect plasma processing, especially if they serve as a source of oxygen to the plasma.

The effect of an HDP interacting with the plasma-generation region has been discussed in Section 3.3.6 for silane-based dielectric deposition plasmas. Oxygen can enter the plasma from the interaction of the plasma with the ceramic liner or the window of an ECR reactor. This might affect etching of Al-containing materials.

Another potentially unanticipated source of oxygen is the cathode. Cathode surfaces are made from a variety of materials, including anodised aluminium (Al_2O_3), quartz and graphite. Occasionally a Si coverplate is placed on the electrode. If an oxidised electrode is etched or appreciably sputtered, the release of oxygen into the vapour phase can lead to the formation of aluminium oxides. This may affect etch rates and surface smoothness if the oxide so formed serves as a micromask.

5.11 CONCLUSION

The proper balance of chemical etching and physical sputtering allows dry etch processes such as RIE, HDPE, RIBE and CAIBE to produce devices with small feature sizes and vertical profiles. Although some electronic damage results from the ion bombardment that permits fine features and profile control, these processes are and will remain the basis of most III–V device fabrication.

REFERENCES

[1] C.I.H. Ashby [in *Properties of GaAs* Eds. M.R. Brozel, G.E. Stillman (Institution of Electrical Engineers, London, 1995) ch.18, p.717–63]

[2] R.J. Shul, S.J. Pearton [*Handbook of Advanced Plasma Processing Techniques* (Springer, Germany, 2000)]

[3] G.M. Peake, R.J. Shul, C.I.H. Ashby, J.G. Cederberg, M.J. Hafich, R. Biefeld, M.N. Palmisiano [*J. Vac. Sci. Technol. B (USA)* vol.21 (2003) p.843–7]

[4] W.G. Breiland, K.P. Killeen [*J. Appl. Phys. (USA)* vol.78 (1995) p.6726–36]

Chapter 6

Ohmic contacts

6.1 CHAPTER SCOPE

This chapter will cover the physics, material science and practical aspects of metal/semiconductor interfaces that are fabricated for ohmic contacts to GaAs. The material presented in this chapter focuses on the generalised GaAs-based ohmic contact, its ideal properties and a wide range of implementations. Chapters 8 and 9 on FETs and HBTs also include sections on ohmic contacts where contacts are discussed in the context of specific types of high frequency transistors.

Section 6.2 introduces the necessary basics including a physical description of metal/semiconductor interfaces and the principles of the common metallurgical reactions. Section 6.3 covers the basic methods of fabrication and testing. Section 6.4 surveys the different types of n-type ohmic contacts to GaAs, their intended use and the motivation for engineering contacts with different properties. Section 6.5 surveys the different types of p-type ohmic contacts.

6.2 PRINCIPLES OF OHMIC CONTACTS

Metals and semiconductors come into close contact, and physical or chemical bonds are formed at an interface between them. Such contacts are of two generic types, Schottky and ohmic. Ohmic contacts are those that follow Ohm's law that current increases linearly with the applied voltage. This behaviour implies that a low barrier to current conduction exists at the metal/semiconductor interface in ohmic contacts.

In fact, almost all metal/semiconductor interfaces have barriers to mobile carrier conduction. The key to making an ohmic contact is to design these barriers to be low enough that current can cross the interface by thermionic emission or to be narrow enough that current can cross the interface by field emission. The barriers come about because electronic states at the metal/semiconductor interface cause band bending in the conduction and valence bands near the interface.

In an actual device, the resistance to current flow through the device is usually minimised and the resistance of the ohmic contact should be a small fraction of the total device resistance. In some optical devices such as laser diodes or solar cells, the area of the contacts is large and the specific contact resistance need not be reduced to state-of-the-art levels. In high-performance electronic devices such as FETs and HBTs, the situation is reversed. Contact areas are small and specific contact resistance should be as small as possible to ensure high performance.

6.2.1 Definitions

The resistance of a contact can be visualised by considering the semiconductor slab of length L and end area A shown in FIGURE 6.1. Current flows from the metal contact on the left to the contact on the right of the slab. The resistance of the semiconductor R_s is given by

$$R_s = \rho L/A \tag{6.1}$$

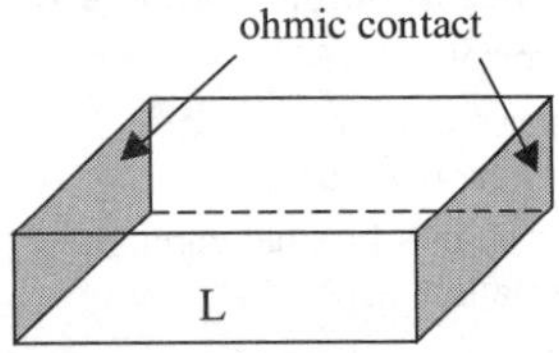

FIGURE 6.1 Semiconductor slab with ohmic contacts across two ends.

where ρ is the resistivity of the semiconductor. The total resistance, R, is given by adding a contact resistance, R_c, for each metal/semiconductor interface:

$$R = \rho L/A + 2R_c \tag{6.2}$$

The contact resistance is inversely proportional to the area of the contact, which makes it useful to define a specific contact resistance, r_c, such that

$$r_c = R_c A \tag{6.3}$$

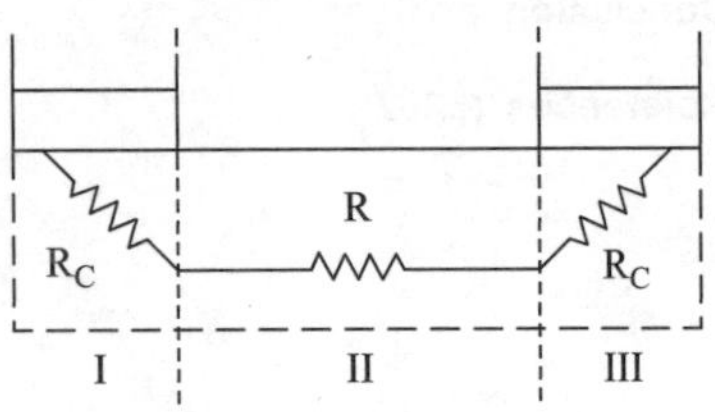

FIGURE 6.2 Resistances in a planar contact.

The specific contact resistance is the best metric for a device with current flow perpendicular to the contact, such as a laser diode and the emitter contact of an HBT. Planar devices such as GaAs FETs have non-perpendicular angles between the main current flow in the device and the contact, as illustrated in FIGURE 6.2. In such a case, the current flow is through a thin region near the surface of the semiconductor, both in the region of the device and in the region under the contact. For this geometry, R_c is defined to be the sum of the resistance through the metal/semiconductor interface and the resistance through the semiconductor in region I or III of FIGURE 6.2, while the current and resistance of the device are those that exist in region II. The contact resistance is a function of both the specific contact resistivity of the metal/semiconductor interface and the resistivity of the semiconductor under the contact.

An important aspect of the planar device geometry arises as the device currents pass from region II to region I or III. Current entering one of these regions finds the path of least resistance to the contact metal, resulting in greater current flow into the edge of the contact and lesser current flow further from the edge. The transfer length, L_t, is defined as the distance at which the current drops to 1/e of its value at the edge of the contact. Today's GaAs FETs have transfer lengths that may be only a few tenths of a micron, while the p-type base contacts of HBTs may have transfer lengths of several microns.

The transfer length L_t is described more completely in Section 6.2.2.

6.2.2 Physics of ohmic contact formation

A qualitative discussion of the physics of ohmic contacts is presented in this section to provide an understanding of the basic mechanism of ohmic contact formation. This presentation should be sufficient to explain and contrast the various types of metal recipes that are in common use for ohmic contact fabrication.

The properties of a metal/semiconductor interface of an ohmic contact are similar to the properties that characterise a semiconductor and the interface of its surface to air or vacuum. The conduction and valence bands of the semiconductor are flat in the bulk of the material in the absence of an electric field. These bands bend due to interface states at the metal/semiconductor interface, as the Fermi levels of the metal and semiconductor line up (FIGURE 6.3). The origin of the band bending is similar to that for semiconductor surfaces. At the interface, the atoms lose their four-fold bonding to other semiconductor atoms, which results in unpaired electrons. The energy of these unpaired electrons is higher than that for a fully formed bond. As a result, the overall energy can be reduced if the interface atoms rearrange themselves and form new bonds with each other or with the metal contact. In addition, defect structures such as missing atoms formed at the interface contribute to the electronic states at the interface. Surface or interface states are the two-dimensional analogue of the three-dimensional band structure of crystalline semiconductors. Like the band structure of a three-dimensional crystal, the electrons of a single surface or interface bond are not considered localised energy levels belonging to a particular bond, but they may be influenced by a two-dimensional periodic potential if it exists. Energy levels may also behave more as localised states if local disorder is high. These new states have energy levels that mostly lie in the forbidden gap between the conduction and valence bands. When the Fermi level is lined up between the semiconductor and the metal, the states in the gap at the interface set the Fermi level near midgap, while the n-type dopants set the Fermi level near the

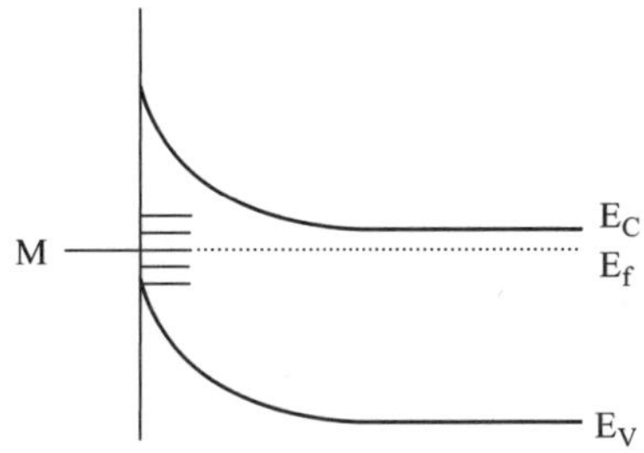

FIGURE 6.3 Energy bands of a semiconductor/metal interface.

conduction band in the bulk of the semiconductor. The bending of the conduction band at the metal/semiconductor interface sets the barrier to current flow across the interface, as in FIGURE 6.3.

Electrons may cross the potential or Schottky barrier by one of two mechanisms: thermionic emission or field emission (tunnelling). A combination of the two mechanisms is termed thermionic-field emission. In thermionic emission, a fraction of the electrons have sufficient thermal energy to cross the barrier. The fraction is related to the factor

$$\exp(-q\phi_b/kT) \tag{6.4}$$

and the current due to thermionic emission is given by

$$\begin{aligned} J &= (\text{const}) \exp(-q\phi_b/kT)[\exp(qV/nkT) - 1] \\ J &= J_s[\exp(qV/nkT) - 1] \end{aligned} \tag{6.5}$$

where ϕ_b is the barrier height, q is the electron charge, k is Boltzmann's constant, V is the applied voltage across the barrier and n is the ideality factor (discussed in Section 7.2.1). The purpose of an ohmic contact is to provide a low-resistance contact. EQN (6.5) indicates that low currents arise with low voltages unless the Schottky barrier is fairly low.

The second mechanism for current flow across a barrier, field emission, allows for quantum mechanical-tunnelling through the Schottky barrier. The equation governing current from field emission is given by

$$J = \exp(-q\phi_b/E_{00}) \tag{6.6}$$

where

$$E_{00} = q\hbar(N/\varepsilon m^*)^{1/2}/2 \tag{6.7}$$

and $\hbar$ is Planck's constant divided by 2π, ε is the dielectric constant, N is the doping concentration and m^* is the electron effective mass. Field emission works by electrons tunnelling through the Schottky barrier, whose width is roughly the same as the thickness of the depletion region. The tunnelling current increases exponentially with the doping concentration. Practically speaking, this current becomes significant enough to make low resistance contacts to GaAs when N exceeds 10^{19} cm^{-3}.

In reality, both thermionic and field emission can come into play for contacts at room temperature. This type of conduction is termed thermionic field emission and it works by electron tunnelling from thermally excited electrons that tunnel through a somewhat thinner barrier by virtue of the electron energy being somewhat higher than the Fermi level.

As seen from EQNS (6.5)–(6.7), lowering the barrier height and increasing the doping concentration are the two basic means for creating an ohmic contact to any semiconductor. Simple as this seems, GaAs presents a somewhat difficult case for barrier height lowering. Surface or interface state densities are fairly high for GaAs, compared with the SiO_2/Si interface, for example, which has the practical effect of pinning the Fermi level to midgap and constraining the barrier heights of most metal/GaAs contacts to a fairly narrow range (Section 7.4). If barrier-height-lowering is to be a practical method, the creation of lower bandgap materials or the creation of lower-barrier-height interfaces are the only available approaches. Some metallurgical approaches along this line will be presented in the next section. Alternatively, the strategy for designing an ohmic contact can rely on achieving high doping levels.

6.2.3 Metallurgy of ohmic contact formation

Although Ge/Au/Ni is the dominant contact metallurgy for n-type GaAs, literally hundreds of recipes exist for ohmic contact formation to GaAs and no mechanism is expected to explain all the details of the various ohmic contact processes. It is somewhat surprising, therefore, that most of the contacts to GaAs seem to follow the broad outlines of a general mechanism. Studying this mechanism can help the practising engineer understand the importance of the various elements of the recipe and relate them to the otherwise confusing folklore that is sometimes told about GaAs contacts. Prior to a description of this mechanism, a few historical highlights of GaAs ohmic contacts will be presented.

Sn was used as an early contact to GaAs, while GeAuNi became the most widely used contact after its report in 1967 [1]. Interestingly, Ge and Au were used in their eutectic composition, 88% Au and 12% Ge by weight, which is the composition of lowest melting point. Metal alloys with eutectic compositions have many uses, including the attachment of semiconductor chips to their package by eutectic melting and bonding. Ohmic contacts with low-bandgap surface regions obtained by epitaxial growth of InGaAs were first reported in 1981 [2] and many variations on this theme were later reported. Contacts based on solid-state regrowth using the Pd/Ge metal system were introduced in 1985 as a way of improving the reliability of ohmic contacts by understanding and controlling their metallurgical reactions [3]. Although these contacts were never widely adopted, the subsequent research on solid-state regrowth solidified our understanding of basic mechanisms in contact formation.

The metallurgy of ohmic contact formation to n-type GaAs will include some or all of the following steps: 1) an element of the

Some folklore about GaAs contacts (and reply). See text for more mechanistic explanation.

AuGe eutectic: Eutectic chosen so that contacts will melt at chosen temperature; molten alloy allows Ge to diffuse better into GaAs.

Reply: The mechanisms for Ge and other constituents to diffuse into GaAs can work well at temperatures well below the eutectic melting of GeAu.

Role of Ge: Ge acts as n-type dopant in GaAs.

Reply: Ge reactions with Ni are also important.

Role of Ni: Ni acts as a wetting agent so that the melted AuGe will not "ball up."

Reply: Ni plays an important role at temperatures well below the AuGe eutectic.

contact reacts with the native oxide present on the GaAs surface, 2) a low temperature process occurs where one element of the contact, the Ni in Ge/Au/Ni alloys, diffuses into GaAs and reacts to form intermediate X-GaAs complexes, "decomposing" the GaAs or otherwise facilitating interdiffusion between the contact and GaAs, 3) an element of the contact capable of doping GaAs diffuses into the GaAs aided by defects created in step 2 and dopes GaAs, 4) an element of the contact capable of forming a low bandgap alloy with GaAs diffuses into the GaAs, 5) Ga outdiffuses from GaAs and reacts with a contact element, often Au, 6) elements of the contact continue to react and form their thermodynamically favoured alloys until the reaction is complete or the thermal budget is exhausted, 7) excess As diffuses to the surface where it resides or vapourises.

Summary: 7 common steps of GaAs ohmic contact alloy reactions

1. An element of the contact reacts with the GaAs native oxide.
2. An element of the contact reacts with GaAs at low temperatures to form a X-GaAs complex.
3. An element of the contact diffuses into GaAs mediated by defects created in step 2 and dopes GaAs.
4. An element of the contact diffuses into GaAs and forms a low bandgap semiconductor phase.
5. Ga outdiffuses from the contact and reacts, usually with Au.
6. Elements of the contact react and form their thermodynamically preferred compounds.
7. As diffuses from the GaAs to the contact surface where it resides or vaporises.

The most commonly used contact is GaAs/Ge/Au/Ni. The metallurgy of this contact probably includes six of the steps outlined above. The Ge/Au/Ni contact will be described in terms of these steps of the metallurgy as a specific example of the reactions. In step 1, the Ni reacts with the native oxide. The native oxide is not easily characterised on GaAs and may vary in unrepeatable ways due to variations in processing, so it is useful for the contact to have a constituent such as Ni that will reliably react with different types of surface oxide. GeAu contacts without Ni are known to have poor morphology and large variability. The native oxide of GaAs may have regions that allow easy penetration of the GeAu and the alloy formation occurs rapidly in these areas. Other areas may never react because the GeAu does not penetrate the native oxide and poor morphology results.

In step 2, the Ni forms intermediate complexes with GaAs. The Ni-GaAs complexes later give way to more stable products such as AuGa intermetallic reaction products. The important role of the intermediate complexes is to disturb the crystal lattice to make way for diffusion of other atoms and further reactions. In the case of GeAuNi contacts, the absence of Ni would result in more thermal energy required for the Au and Ge reactions to get started. Once started they quickly get out of control leading to severe spiking and poor morphology. The Ni-based complexes are thought to form at temperatures below 100°C.

Once the Ni-GaAs complexes have disturbed the crystal lattice, the way is prepared for the diffusion of Ge into the GaAs in step 3. Normally, diffusion of dopants such as Ge in GaAs requires a substitutional-interstitial mechanism where the Ge can diffuse along interstitial sites until it finds a vacancy in the GaAs lattice. This type of mechanism has high activation energy and normally occurs to a significant degree at temperatures of 600°C and above. In a GaAs ohmic contact, the diffusion process is thought to occur at much lower temperatures because of the driving forces of the Ni-GaAs complex formation and the driving force of the next reaction which further disturbs the crystal lattice. The high activation energies of creating and filling Ga vacancies are overcome by complex reaction intermediates involving Ni-Ge-GaAs rather than by thermal effects that are necessary during substitutional-interstitial diffusion. The diffusion of Ge into GaAs at temperatures between 100 and 400°C leads to very high doping levels that make tunnelling contacts possible for doping levels above 10^{19}.

It is not presently believed that ohmic contact formation with Ge/Au/Ni in GaAs participates in step 4 through the formation of a low bandgap alloy, though NiAs phases were considered possible low-barrier-height phases in the past. The second part of the driving force for steps 2 and 3 comes from a step 5 reaction, the Au-Ga reaction. The Au-Ga alloy is more thermodynamically stable than any other compound between the five elements that can form. At temperatures at and above 250°C, this reaction is kinetically favoured and with sufficient time will continue until the supply of Au is exhausted.

In step 6, other reactions between the GaAs and the contact constituents continue. Some of these reactions form intermediate compounds such as NiAs while others such as NiGe are more stable.

The last step is the transport of As to the surface of the contact and its desorption from the contact. Among the five reacting elements, As is most stable in the elemental phase and leaves the GaAs. The significance of this step along with step 5 is the consumption of GaAs. Voids formed by the loss of As are formed and filled with AuGa, NiGe and other stable compounds. In the ideal case with good processing technique, the reaction occurs uniformly. Because the ideal case is usually only approximately achieved and because the reactions occur very fast, spikes form if there are any process non-uniformities. Visual inspection of an alloyed GaAs/Ge/Au/Ni contact will show a textured surface for a properly alloyed contact (FIGURE 6.4(a)) and a much greater morphology change for over-alloyed contacts (FIGURE 6.4(b)). Surface profilometry or atomic force microscopy measurements will show depressions in the areas where the contact has spiked and excess As has been consumed. Transmission electron microscopy directly images the regions where spiking has occurred and has been used to identify some of the reaction products.

Most contact metallurgies in use today follow some of the seven steps outlined in this section. Some of these steps have resulted in problematic characteristics for a GaAs ohmic contact and have stimulated further research. Much of the research has resulted

(a)

(b)

FIGURE 6.4 Optical micrograph of an alloyed GeAuNi ohmic contact: a) slight texture for properly alloyed contact, b) greater texture for over-alloyed contact.

in optimisation and better understanding of the mechanism of the GaAs/Ge/Au/Ni contact. Some alternative metallurgies have been developed and the more common ones will be reviewed in Section 6.3.

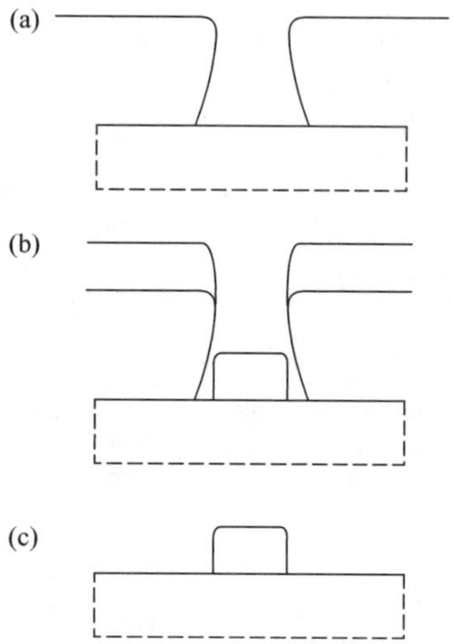

FIGURE 6.5 Illustration of the steps required for forming an ohmic contact.

6.3 FABRICATION AND TESTING OF OHMIC CONTACTS

6.3.1 Fabrication of ohmic contacts

The fabrication of an ohmic contact involves the steps of patterning the GaAs with resist or another suitable mask, preparing or cleaning the surface, depositing the metal, and alloying the metal, if needed. The most common process is to pattern the sample first and to use a liftoff process to produce the patterned contact. The basic steps are illustrated in FIGURE 6.5. A photoresist is applied to the wafer and spun typically between 1500 and 6000 rpm to uniformly coat the wafer. The resist thickness should be uniform so that good control of the final linewidth can be achieved. Photolithography is discussed in greater detail in Chapter 4. A glass or quartz mask is then aligned to the pattern on the wafer so that light can pass through the areas to be exposed and be blocked in the areas to be protected from the light. Any of the available lithography tools may be used, but the optical stepper is the tool of choice for most high-performance and high-yield processes in HBT and FET fabrication. Once the mask is aligned with the wafer, the resist can be exposed to the light and the pattern can be developed (FIGURE 6.5(a)). A re-entrant cross-sectional profile is desired so that the surface of the resist shadows the sidewalls and prevents the evaporated metal from coating them. Re-entrant profiles can be achieved with either positive or negative (image reversal) resists (Section 4.4.1). The choice of resists depends on the type of optical aligner used, the wavelength of the light and factors relating to the device morphology.

Prior to depositing the ohmic contact, a surface treatment may be used. This may include a plasma oxygen treatment to remove monolayers or less of organic material that remain after the resist develop process. The surface treatment may also include oxide removal or other steps. The ohmic metal is then evaporated over the whole wafer where it forms the contacts in the patterned areas and covers the resist over the rest of the wafer (FIGURE 6.5(b)). The unwanted metal is removed from the wafer when the resist is dissolved, typically by means of a solvent spray. At this point, the ohmic metal (FIGURE 6.5(c)) can be tested electrically if a non-alloyed ohmic contact process was employed. Otherwise, the contact will be alloyed so that the metallurgical reactions needed for ohmic contact formation can proceed.

The preceding description is an overview of the fabrication process. Several of the fabrication critical steps will now be described in more detail. The pre-clean step can be critical for achieving low electrical resistance in the contact. One common procedure used with Ge/Au/Ni contacts employs a plasma oxygen clean followed by an oxide removal step. The GaAs native oxide is about 1–2 nm thick. Exposure of a clean GaAs surface to the ambient results in the formation of a consistent native oxide. This native oxide is easily removed by dilute acids such as HF or HCl or bases such as an $NH_4OH : H_2O$ mixture (Section 3.2.2). During routine semiconductor processing the GaAs comes into contact with other chemicals such as hydrocarbons in a photoresist formulation that may leave residues on the surface of the GaAs. In such cases, both the formation of the native oxide and the reaction of a deposited ohmic metal are complicated by the presence of hydrocarbon or ionic residues. It is found that ohmic contacts can have excellent electrical characteristics but are not repeatable. Consequently, various attempts have been made to improve the repeatability of ohmic contacts. Several of these involve methods to remove the hydrocarbon residues.

Surface treatments for removing native oxides of GaAs were described in Section 3.2.2.

A common procedure to accomplish the hydrocarbon removal is to expose the patterned GaAs wafer to an oxygen plasma. Typically RF radiation at 13.56 MHz is used to create a plasma with O_2 and maybe a carrier gas such as Ar. Equipment specifically designed for this purpose is often available and one type is commonly referred to as an "asher". A popular geometry is the cylindrical shape with a powered metal cylinder enclosing an interior region that contains the plasma and the samples, which mount to grounded rods that are placed at positioned breaks in the powered electrode design. The operation of a barrel asher was described in Section 5.6 and FIGURE 5.4. This particular design is useful for hydrocarbon residue removal because the geometry leads to a lower plasma sheath voltage than the parallel electrode design. Such a lower self-biasing of the plasma is important for a general purpose plasma oxygen process because it is potentially used multiple times in a typical device fabrication and some of the steps will have requirements of low or no plasma damage.

A newer method of hydrocarbon removal uses UV-generated ozone exposure. Ozone can react with hydrocarbon residues by releasing molecular oxygen and an oxygen radical that quickly reacts with the hydrocarbon to form carbon oxides and other organic molecules such as alcohols and ketones. The reactive ozone is a neutral molecule which forms in the absence of a plasma and cannot cause the types of damage caused by the acceleration of charged particles. It is therefore ideally suited as a general purpose hydrocarbon removal tool. Commercial ozonation tools are among the least complex clean room equipment in that they only need a source of electrical power. In contrast to ashers, which are designed to remove several microns of photoresist per hour, ozonators typically remove resist at a rate of several monolayers per minute. They are especially valuable for stubborn submonolayer hydrocarbon removal.

The removal of hydrocarbon residues prior to ohmic contact deposition is a good way to ensure repeatability of the surface condition. In addition to removing hydrocarbon deposits that can block interface reactions, it also can create repeatable conditions for oxidising the GaAs surface. Although oxides themselves are potential impediments to interface reactions between the GaAs and the ohmic contact, they are easily removed prior to contact deposition and only partially regrow if samples are loaded within minutes (up to ten is a reasonable compromise between unrealistically short times and times long enough to fully regrow the oxide) to the deposition system. Common methods of native oxide removal by acids such as HCl and HF or bases such as NH_4OH also work for the plasma or UV-ozone oxides. A further benefit of incorporating an oxide removal step in the ohmic contact process is that any low-level impurities that settle on the GaAs surface are dissolved with the oxide.

Following the surface preparation methods comes the actual contact deposition. Evaporation is the technique of choice for most metal depositions on GaAs because it provides a method of high purity deposition with minimal damage to the GaAs. This is especially so for Au-based contacts because an evaporation process with liftoff patterning works well with Au, while etch processes are limited to ion milling. Evaporation is carried out in a high-vacuum chamber with pressure in the mid to low 10^{-7} torr pressure range. This pressure level is required so that the residual gases in the vacuum chamber will not be a source of contamination as the deposition proceeds. As previously noted, electron beam evaporation results in minimal damage to most semiconductors compared to the main alternative, sputtering. With electron-beam deposition, no energetic ions are produced which can be accelerated to the sample surface and cause lattice damage. The main concern for damage with electron-beam deposition is the effect of energetic electrons. For cases where this type of damage is a concern, thermal evaporation is used. For GaAs ohmic contacts, lattice damage is not usually a concern because lattice disruption is an intended consequence of the contact reaction.

A variety of evaporation equipment is available depending on the degree of throughput and automation needed. For high-volume production, extremely large evaporators are available with the capability of depositing on dozens of wafers per run. These large vacuum systems are equipped with heavy duty vacuum pumps that can bring about the required vacuum conditions in considerably less than 1 h. On the other end of the spectrum, research tools have much smaller vacuum chambers and pumps and may require several hours for pump down after sample loading. A brief description of the steps in an electron beam deposition follows.

An electron beam typically accelerated to 10 kV is directed at one of the hearths or crucibles containing the high-purity solid. The energetic electrons enter the solid where collisions and scattering events with valence and core electrons from the solid result in energy transfer from the accelerated electron to the solid source. Ultimately, the source heats up in the localised area where the electron beam is focused and begins to melt and evaporate. It becomes a source of atomic flux throughout the chamber. The acoustic wave properties of a quartz crystal are used as a flux monitor to provide feedback to the electron beam source and to stabilise the flux to a preset value. Once the flux is stable, a shutter is opened to allow the sample to receive a thin film deposition. The process is ended when the desired film thickness is reached and is then repeated for any other depositions that are needed in a multimetal stack. A thermal evaporation follows a similar procedure, except that instead of an electron beam being used to heat up the source material, electrical current is passed through a refractory metal holder that contains a small amount of the source material.

The various components of the ohmic contact are typically deposited layer by layer. For example, deposition of a Ge/Au/Ni ohmic contact will proceed with the deposition of the Ge layer first, followed by the Au layer, and finish with the deposition of the Ni layer. On occasion two constituents will be co-deposited. This occurs frequently with Au and Ge due to the fact that an alloy of 12% by weight of Ge and 88% of Au forms the eutectic composition and melts at a lower temperature (361°C) than other alloy compositions. After deposition is completed the shutter is closed and the evaporator is vented for sample removal.

Source material deposition is completed by the removal of the resist material and the unwanted metal through a liftoff process. At this point the contact formation is completed by means of a thermal annealing step, if required. This step is often termed an ohmic alloy for contacts that undergo an alloy process. For Ge/Au/Ni contacts the thermal anneal is typically carried out near 400°C for 10 s to several minutes. Usually the shorter times require slightly higher temperatures than the longer times.

A rapid thermal annealer (RTA) is commonly used for this step, although furnaces and strip heaters are also used. An RTA is composed of several necessary components such as an inert chamber (often quartz) which will contain the samples, a heater such as an infrared lamp, a network of plumbing for inert gas introduction and pumping, and a control system. It is important to exclude oxygen during the anneal. Oxygen can cause several deleterious effects. First, it can oxidise the exposed GaAs surfaces. Because this oxidation will be occurring at elevated temperatures, the As-oxides are volatile and the process can continue to the point of consuming a considerable amount of As from the GaAs. Second, oxygen can react with the contact materials. For the Ge/Au/Ni ohmic contact, Ge and Ni oxides would seriously inhibit the intended reactions. The chamber containing the samples is usually purged with a high purity inert gas to drive out the oxygen. Some equipment is designed for pump and purge cycles to speed up this process while other equipment is designed for vacuum annealing.

Other considerations also come into play for the RTA process. One of these is the method of sample placement, which will affect the thermal loading. One common type of annealer uses a Si wafer mounted to quartz rods attached to the door of the chamber with a thermocouple embedded in the wafer. In such an arrangement, the sample is mounted directly on top of the Si wafer. Another method involves placing the sample in a specially designed susceptor, as illustrated in FIGURE 6.6. An example of a sample susceptor for RTA is a SiC disk that is coated with graphite for chemical inertness and as a radiation absorbing film. A central circular area is recessed a distance slightly deeper than the typical sample thickness and is the region where the sample will be placed. A second disk with a matched diameter to the recessed region acts as a cover to the sample. This type of sample arrangement adds thermal loading to the sample and more uniformity to the temperature profile, and makes for a more accurate temperature calibration. However, this method is less amenable to automated handling of the wafers and may be disadvantageous in production environments that require high throughput.

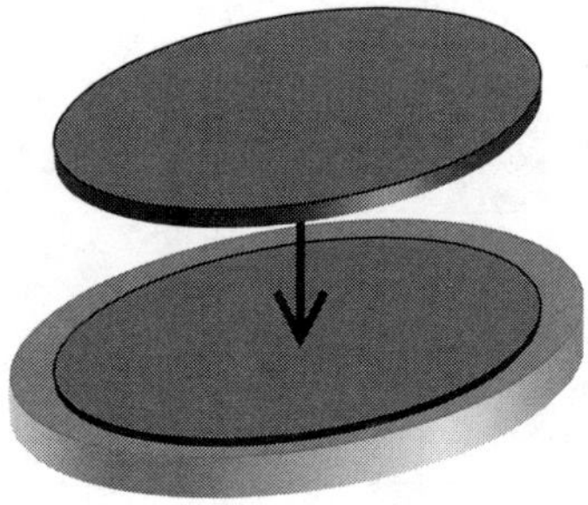

FIGURE 6.6 Illustration of a sample susceptor for an RTA.

Another consideration relates to the efficiency of the heating process. To reach 400°C, the lamps do not require very much power and will emit more infrared than visible radiation, much of it below the GaAs bandgap and not efficiently absorbed by GaAs. However, the metals in the ohmic metal do absorb the infrared radiation effectively. In such a situation, the temperature calibration becomes difficult to achieve, as one must take into account the quantity and type of thin film deposited on the wafer. Obviously, the results can be ohmic-pattern dependent. Because of the difficulty of accounting for the pattern in the temperature calibration, the second type of sample placement within an IR absorbing susceptor is often preferable.

6.3.2 Measurements of ohmic contacts

The purpose of an ohmic contact is to provide a linear relationship between current and applied voltage in a semiconductor device. It is usually desirable to have low access resistance from the device to the outside world. Consequently the measurement of ohmic contacts to accurately determine their small but finite resistance is an important part of semiconductor engineering and one which can be the subject of detailed treatments. For the purpose of this chapter, only two practical methods of measurement will be introduced. More detailed treatment of other methods and other models can be found in [4]. Both methods treated in this chapter will be based on the planar contact common to FETs and the base and collector of HBTs. The two methods differ in the test structure chosen, either a linear contact array or a circular contact test structure.

From a linear contact array illustrated in FIGURE 6.7 with resistances defined in FIGURE 6.2, the contact resistance R_c is given by

$$R = 2R_c + R_{sh}L/w \tag{6.8}$$

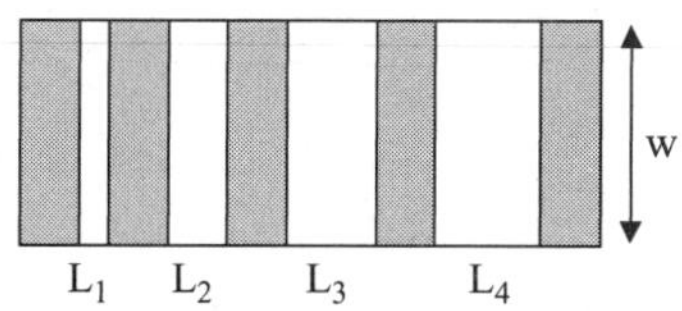

FIGURE 6.7 Illustration of linear contact array.

where R_{sh} is the sheet resistance of the semiconductor between the contacts, L is the length between the contacts and w is the width of the contacts. In practice, the linear array contains contact pairs with differing separations. The resistance of a contact pair is measured and plotted as a function of L, as illustrated in FIGURE 6.8 for a contact width of 100 μm. The slope of the resistance gives the sheet resistance (201 Ω/square after normalising length/width of

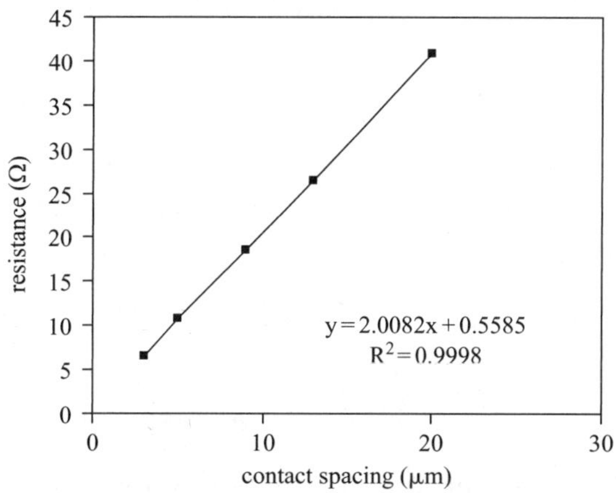

FIGURE 6.8 Plot of resistance versus contact separation to extract contact resistance.

the contact), while the intercept gives twice the contact resistance ($R_c = 0.03\ \Omega$-mm in this example). The contact resistance R_c is the most important value for planar devices such as FETs as it directly relates to the parasitic resistances encountered in the device. Furthermore, it is simple to measure and it requires no model to extract. However, it is unsatisfactory from the standpoint of the device metallurgy because it contains elements of semiconductor resistance in addition to interfacial resistance as seen in the definition of FIGURE 6.2. One would prefer a more direct link between the chemistry of the interface and its electrical properties. For this reason, the contact resistivity r_c of EQN (6.3) is more appropriate and other models have been developed to extract r_c from a planar contact.

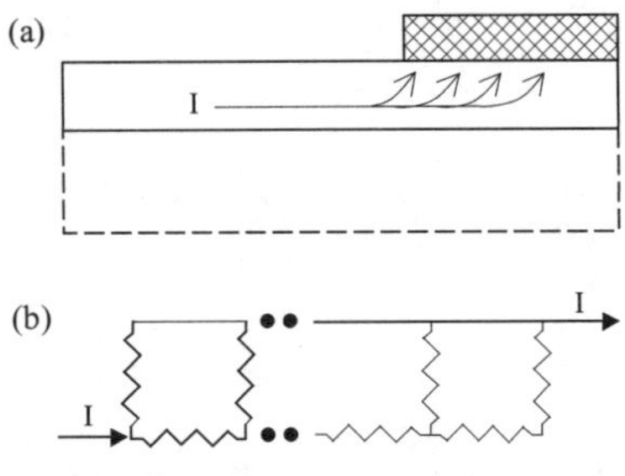

FIGURE 6.9 Plot of resistance versus contact separation to extract contact resistance.

The geometry of a planar contact is shown in more detail in FIGURE 6.9(a). In the planar contact model, it is assumed that a semiconducting region is sandwiched between a semi-insulating substrate and the contact metal. The resistance of the semiconductor-contact combination can be modelled by a resistive transmission line with uniform sheet resistance R_s, as illustrated in FIGURE 6.9(b) and as modelled by Berger [5]. Standard microwave methods are used to determine the voltage and current distributions along the transmission line:

$$\begin{aligned} V(x) &= V_1 \cosh(\gamma x) - I_1 Z \sinh(\gamma x) \\ I(x) &= I_1 \cosh(\gamma x) - V_1 Z \sinh(\gamma x) \end{aligned} \tag{6.9}$$

where Z is the characteristic impedance of the transmission line and γ is the propagation constant. The equations for Z and γ are given by [4]:

$$\begin{aligned} Z &= R/G^{1/2} = w^{-1}(r_c R_{sh})^{1/2}(1 + j\omega C r_c)^{-1/2} \\ \gamma &= (RG)^{1/2} = (R_{sh}/r_c)^{1/2}(1 + j\omega C r_c)^{-1/2} \end{aligned} \tag{6.10}$$

The analysis can be restricted to the direct current case ($\omega = 0$). If the contact is assumed to be infinite in length, then $Z = R_c$ and the equations simplify to

$$\begin{aligned} R_c &= w^{-1}(r_c R_{sh})^{1/2} \\ \gamma &= 1/L_t = (R_{sh}/r_c)^{1/2} \end{aligned} \tag{6.11}$$

where L_t is the transfer length. The definition of a transfer length is motivated by the fact that in a contact of infinite length, most of the current will flow from the semiconductor to the contact over a finite length near the edge of the contact. This length, the transfer length, is proportional to the square root of the contact resistance and inversely proportional to the square root of the sheet resistance.

The relationships contained in EQNS (6.11) are used in combination with the measured contact resistance of EQN (6.8) to derive contact resistivity and transfer length for GaAs-based ohmic contacts. However, it is important to realise that the sheet resistance of EQNS (6.11) is the sheet resistance under the contact, which is likely to be different from the measured sheet resistance of the semiconductor between the contacts. The sheet resistance under the contact is likely to be lower due to the fact that most alloyed or sintered contacts introduce a heavily doped region directly under the contact. In addition, spiking of the contact can further degrade the assumption of a uniform conducting region under the contact. In fact, there is no good way to measure the sheet resistance under an ohmic contact. Despite this fact, many researchers prefer to use EQNS (6.11) under the assumption that the sheet resistance under the contact is the same as that of the semiconductor between the contacts. Unfortunately, there is no good understanding of the error introduced by this assumption. Because of the limitations in these assumptions, it is prudent to be sceptical of any claims of contact resistivity less than $10^{-6}\ \Omega\text{-cm}^2$, as measured from a planar contact (as is the case for the example in FIGURE 6.8, where r_c is calculated as $4 \times 10^{-8}\ \Omega\text{-cm}^2$).

A method does exist in principle to measure the sheet resistance under an ohmic contact and is described by Williams [4]. The method depends on the experimental analysis of the data in FIGURE 6.8 and on the measurement of the end resistance of the contact geometry illustrated in FIGURE 6.10. A current I is forced through the first two contacts. The voltage measured at the end of the second contact (x = d) divided by the current at the beginning of the second contact is the end resistance R_E, which is given by:

$$R_C/R_E = \cosh(d/L_t) \qquad (6.12)$$

In principle, the measurement of R_E is straightforward, but for GaAs semiconducting regions with low resistivity, the length of the contact should not be greatly larger than the transfer length L_t for this method to work. In practice, L_t is often less than 1 μm for good GaAs contacts and the end resistance method is problematic.

Of the many ohmic contact test structures available, the linear contact array is most commonly used with the planar contacts of FETs and HBTs. However, one other useful structure is worth discussing. The circular TLM [6], illustrated in FIGURE 6.11, gives much of the same information as the linear contact array,

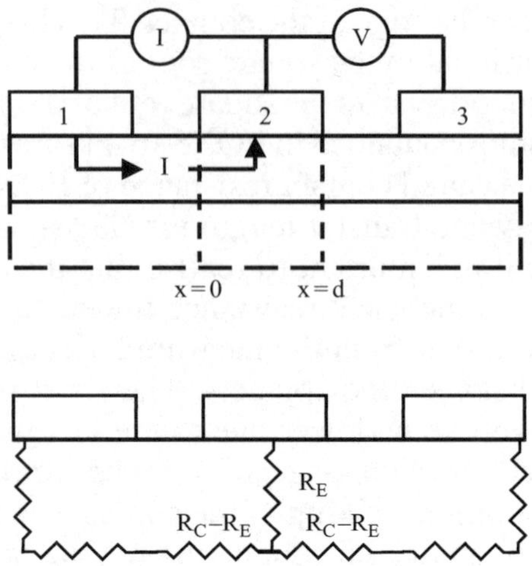

FIGURE 6.10 The end resistance method using a perfect current source and high-impedance voltage monitor.

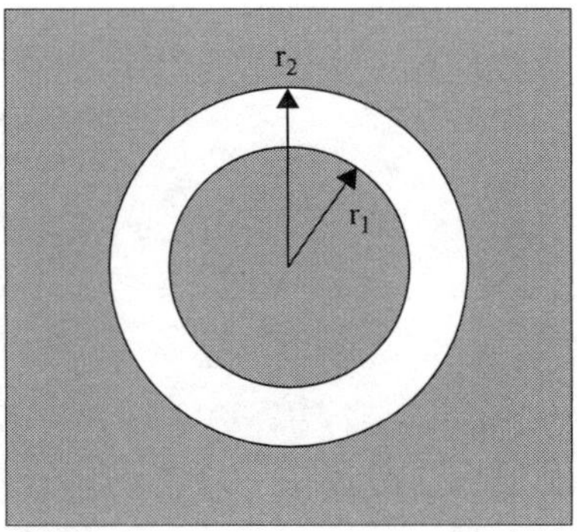

FIGURE 6.11 Illustration of the circular TLM.

but does so without the need for an isolation step (or a patterned implant). The circular TLM also takes up more space than the linear contact array for a comparable contact pad dimension. For this reason it may not be appropriate as an in-process diagnostic, but rather is most useful for rapid turnaround in contact optimisation experiments.

The geometry of a circular TLM makes the resistance calculation slightly more complicated, but still straightforward. If r_1 is the inner radius and r_2 is the outer radius, and d the difference between the two, then the resistance is given by

$$R = R_{sh}/2\pi[L_t(r_2^{-1} + (r_2 - d)^{-1} + \ln[r_2(r_2 - d)^{-1}])] \quad (6.13)$$

where other terms have been defined previously. EQN (6.13) is a simplification of a more exact equation containing modified Bessel functions which applies as long as the radii are greater than L_t by a factor of four or more.

Practical aspects of the measurement of ohmic contacts are key to achieving consistent results. Measurements of adjacent contacts in a linear contact array are generally made with a setup that can be as simple as a probe station connected to a current source and a voltmeter, although semiconductor parameter analysers are much more flexible to set up and control. A width of 100 μm for the contact array is typical. This width is large enough to easily probe each pad with two probes. Spacings between contacts need to be small enough that the resistance between contacts is not a large fraction of the expected intercept (illustrated in FIGURE 6.8) and are generally made small enough to comfortably print with medium-resolution optical lithography. Errors in the actual versus intended spacing will directly limit the accuracy of the measurement. For this reason, the spacings are not made so small that the lithography is not reproducible. Commonly, the smallest gap is in the range of 2–4 μm.

The resistance of the probes must be taken into account. The intercept of the resistance data from a linear contact array is in the range of several ohms for a 100 μm contact width for a state-of-the art contact. A probe contacting the pad may have a resistance that varies from 1–4 Ω and introduces errors comparable to the value of the resistance being measured. Elements of probe resistance include its material composition (tungsten tips penetrate pads well but are more resistive) and changes in the pressure on the pad each time it is placed in contact with the pad. Also, its location on the pad will cause variability in the resistance if the pad metal is somewhat resistive. Contact pads with an Au overcoat, usually a separately deposited pad metal, do not generally suffer from this latter problem. A separate step for the pad metal avoids forming excessive amounts of AuGa at the surface of the pad. A four-point probe measurement illustrated in FIGURE 6.12 will eliminate the probe resistance and lend greater accuracy to the ohmic contact resistance measurement. Two of the probes are used to force a known current through the adjacent contacts. The remaining two probes are used to sense the voltage across the contact pads. Voltage measurements are less sensitive to probe resistance or other probe contact variability issues.

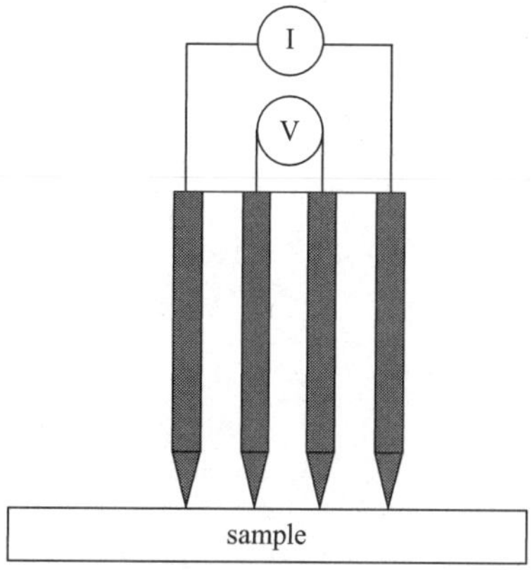

FIGURE 6.12 Illustration of the four-point probe method.

6.4 OHMIC CONTACTS TO n-TYPE GaAs

A wide variety of contacts to n-type GaAs have been investigated over the years. Of these, the GeAuNi contact is now the most widely used. Much of the reason may be historical, as the GeAuNi contact was first reported in 1967 [1] and was found over the years to be a great improvement over the earlier In-based ohmic contacts. When GaAs manufacturing first took off with light emitting diodes followed by MESFETs in the 1970s and early 1980s, the GeAuNi contact was the best contact available. Manufacturing always brings the need to establish reliability and repeatable processes. GeAuNi ohmic metals satisfactorily passed these hurdles and became difficult to dislodge in spite of the numerous perceived drawbacks. Nevertheless, other ohmic contacts are now in use in both research institutions and manufacturing facilities in some niche applications. This section will review the use of GeAuNi contacts and some newer approaches to addressing the limitations of GeAuNi.

6.4.1 GeAuNi ohmic contacts

A common ohmic contact recipe: 26 nm Ge, 54 nm Au, 15 nm Ni annealed at 400–420 °C for 0.5–3 min. Sheet resistance of 2 Ω/square can be expected.

The reaction of the GeAuNi contact was used as an example of the metallurgical reactions described in Section 6.2.3. Proper attention to those principles and to good fabrication procedures as described in Section 6.3.1 can lead to reliable and repeatable contacts. Good surface preparation is critical as is optimising the alloying temperature and time for the particular metal layers chosen. No one optimum recipe exists. Rather, within the framework of the reactions involved, several are possible. The Ge and Au thicknesses are almost always chosen so that the composition in a perfectly mixed alloy is near the eutectic composition. The Ni thickness is often chosen to be between 15 and 30% of the sum of the Ge and Au thicknesses. The total thickness of the layers does not seem to matter very much, as long as the contact resistance is optimised as a function of the alloy temperature and time. Often the Au and Ge are co-evaporated so that they can melt as the eutectic temperature of the alloy (361°C) is reached, but this can be impractical with electron beam evaporators and layer by layer depositions can be just as effective. Electrical contact resistivities less than 10^{-6} Ω-cm^2 are routinely achieved.

Some variations of the basic recipe give equally good results. For example, Ni has been recommended as a first layer to achieve more consistency, uniformity and less spiking [7]. The authors of this work undoubtedly explored this idea with great care and thoroughness. However, they did not publish any work comparing the Ni-first contact with no special surface preparation against the standard contact with careful surface preparation, as outlined in Section 6.3.1. When the Ni-first contact was evaluated by other groups, it was found that the recipe works as advertised, as long as the thickness of the initial Ni layer is kept at or below a certain thickness of near 5 nm. For groups that have good manufacturing control of such thin layers, the Ni-first contact may be a good choice. For other groups, the standard recipe with extra attention to surface treatment is preferred. In general, one will find that many variations of the basic recipe will give excellent results. The key is to perform a careful optimisation of contact resistance using the parameters and experimental conditions suited to a given laboratory.

Ohmic contact with thick Au overcoat: 26 nm Ge, 54 nm Au, 15 nm Ni, 200 nm Au annealed at 380–400 °C for 20–40 s. Sheet resistance of 0.3 Ω/square can be expected.

Other permutations of a GeAuNi contact recipe are common as well. In addition to minor variations in either the total thickness or the thickness of individual layers, an extra Au overcoat is often used. GeAuNi with the thicknesses quoted in this section will leave the metal with a resistance of about 2 Ω/square, which is too high for most devices. An extra Au layer must be added in a subsequent step, usually an interconnect step. Alternatively, one may choose to add the extra Au as part of the ohmic contact process in order to minimise the amount of processing needed. The use of an extra Au overcoat is advantageous for the measurement of the contact resistance. Without the extra Au, the contact resistance measurement can be sensitive to the placement of the probes on the pads and also to the pad pressure. With the extra Au, the metal resistance is lowered by a factor of ten and the resistance through the metal pad and the probe contact resistance becomes negligible.

One concern with the addition of extra Au is the potential for greater reactions with Ga along with excess tendency to spike during the alloying process. The presence of the Ni layers acts somewhat as a barrier to Au diffusion during the alloying process. However, at high enough temperatures, the Ni is reacting with Ge and no longer acts as a diffusion barrier. As a result, the Au must be controlled by limiting the thermal budget of the reaction. As for any contact recipe, the contact resistance and morphology must be carefully optimised with regard to the anneal conditions. Generally speaking, the alloying process uses slightly lower temperature and times for contacts with a thick Au overlayer.

Thermal budget refers to the maximum temperature (or temperature-time product) that a process can handle. It is used to define the bounds of a process and to create an optimum process sequence.

The GeAuNi contacts have shown adequate reliability for many applications. Nevertheless, there have been some concerns for other applications. The temperature stability of GeAuNi limits the temperature of any post-processing steps and reduces the flexibility for processing GaAs. Temperature stability also limits the use of GaAs for high-temperature electronics. Other concerns for GeAuNi contacts are its use in extremely thin layers such as the base of a pnp heterojunction bipolar transistor (Chapter 9) because of its propensity to spike and because reliability in extremely thin layers is unproven. For some of these reasons, alternative contacts have been developed.

6.4.2 Limited Au contacts for improved thermal stability

Much effort has been expended to attain ohmic contacts with improved thermal stability. Almost all of these approaches eliminate or severely limit the amount of Au in the contact in order to remove the temperature restriction of the Au-Ga reaction. Perhaps the most comprehensive effort has been directed towards contacts based on solid-phase regrowth. The PdGe ohmic contact was reported in 1985 and was the first of a number of contacts that fall under the general solid-phase regrowth mechanism [8]. In this contact the Pd plays a similar role to the Ni in the GeAuNi contact. At temperatures near 100°C, Pd reacts with the GaAs to form Pd_x-GaAs complexes. This reaction serves to disturb the GaAs lattice for later Ge incorporation as an n-type dopant. Likewise, the Pd forms more stable compounds with the Ge, which acts as a driving force for the removal of Pd from GaAs during a continuing temperature ramp and the regrowth of the GaAs with the incorporation of a Ge dopant. The Pd-Ge alloy is stable up to temperatures near 600°C, giving this contact more thermal stability than the GeAuNi contact. The term solid-phase regrowth was used to describe this reaction for two reasons. First, no eutectic melting takes place as part of the alloying reaction, and therefore all reactions are solid-phase. Second, the intermediate reaction of Pd with GaAs is only temporary and the GaAs regrows as the Pd is removed from the GaAs during the Pd-Ge reactions. Because of the absence of a Ga consuming reaction, the PdGe contacts are suitable for shallow contacts with excellent morphology and no spiking. Excellent resistivity comparable to GeAuNi contacts has been reported. Other ohmic contacts based on solid-phase regrowth have been studied as well, but the PdGe contact shows the best electrical properties of these. In spite of the excellent thermal stability and good morphology, the PdGe contact is not as widely used as the GeAuNi contact, but is an excellent choice for applications that require higher temperature operation, extremely shallow contacts or greater edge acuity. The main significant drawback of the PdGe contact is its high resistivity, depending on the thickness. This drawback can be remedied by the application of Au-containing interconnect.

Other Au-free contacts provide greater thermal stability. Simple removal of the Au from GeAuNi also results in a contact stable up to 600°C. This contact is a close analogue of PdGe, but one with slightly higher contact resistivity and a narrower optimum composition. As such it is not very useful and would not even rate a mention were it not for its improvement when small amounts of Au are added. With as little as 3 nm of Au, the contact resistivity is improved to within a factor of two of the GeAuNi contact without adding so much Au that the morphology becomes poor. This contact illustrates the principles of using the known metallurgical mechanisms to engineer a contact with desired properties.

Other non-Au contacts worth noting are the GeAlNi and the Cu_3Ge contacts. The Al plays a similar role to the Au in GeAuNi. However, the AlGe eutectic temperature (424°C) is more than 60° higher than that for AuGe, giving the contact higher thermal stability. In addition to the higher eutectic temperature, stable phases such as Al_3Ni and NiGe compounds are formed within the contact structure after sintering at 500°C [9]. Another promising non-Au contact to n-type GaAs using Cu_3Ge exhibits a planar and abrupt interface and excellent contact resistivity of 6×10^{-7} Ω-cm^2 to moderately-doped GaAs (1×10^{17} cm^{-3}) [10]. Ohmic contacts were observed with both excess Cu and excess Ge from a stoichiometric Cu_3Ge phase, suggesting that epitaxial Ge need not be a primary reason for the observed ohmic behaviour. In addition, Ga was observed by SIMS in the $\varepsilon 1$-Cu_3Ge phase suggesting to the authors that Ga outdiffusion promotes Ge incorporation in GaAs as an n-type dopant. These contacts can be considered if one has concerns about other contacts in use. However, one might be concerned about reliability of devices with active regions close to the ohmic contacts since Cu diffusion is very rapid in GaAs.

6.4.3 Ohmic contacts to heavily doped surfaces

When the doping level increases above $10^{19}\,cm^{-3}$, formation of non-alloyed ohmic contacts through a tunnelling mechanism typically occurs and this will provide greater flexibility in the choice of metal for the contact. Common n-type dopants have a doping concentration limitation of 2–4 $\times$ $10^{18}\,cm^{-3}$ in GaAs by ion implantation and most commonly used epitaxy methods, but there have been some novel methods to overcome this limitation. In one such example Sn segregation to the surface of an n^+ GaAs layer during MBE growth provided a sufficiently high surface concentration to allow a non-alloyed contact based on TiPtAu, Al or TiW metals [11]. Non-alloyed ohmic contacts using $(NH_4)_2S$ surface passivation followed by thermal desorption in vacuum have been demonstrated, and it was suggested that the S incorporated below the surface as an n-type dopant, raising the surface doping level to $7 \times 10^{18}\,cm^{-3}$ [12]. This topic is presented in more detail in Section 3.3.4.

6.4.4 Refractory metals and contacts based on reducing the surface bandgap

InAs with a bandgap of <0.4 eV and a surface Fermi level that is pinned in the conduction band is a much better surface layer than GaAs for an n-type ohmic contact. Accordingly, InAs or InGaAs has often been used as an epitaxial surface contact layer for non-alloyed ohmic contacts, when the device allowed such a structure. The emitter surface of HBTs sometimes incorporates epitaxially grown InGaAs layers for non-alloyed contacts to refractory metals such as WSi [13].

Surface InGaAs is also a motivating factor in ohmic metal structures containing In such as InW and NiInW. In-containing contacts without the use of a refractory metal have in the past typically produced contacts with high contact resistivities of about $10^{-5}\,\Omega\text{-cm}^2$ or approximately one order of magnitude worse than GeAuNi contacts. The poor results for the Au-, Ag- and Pt-In contacts are due to uncontrollable, low-temperature, liquid-phase reactions with GaAs. The use of a refractory metal with In has greatly improved this situation by producing contacts with resistivities that are nearly identical to GeNiAu contacts for InW [14] and essentially identical by adding Si to the NiInW contact [15]. A number of other refractory In-containing contacts have been reported as well. The refractory metal must be inert with the GaAs at higher temperatures than conventional ohmic contacts, since the InGaAs phases are formed and observed by TEM in the refractory contacts only after alloying above 600°C. Not surprisingly, the thermal stability of these contacts at 400°C is excellent. The role of the refractory metal is to act as a barrier for a fairly thin In film to maximise the coverage of InGaAs phases at the GaAs interface. Films 3 nm thick were optimum for InW [14] with thinner films not producing enough InGaAs and thicker films resulting in poor thermal stability due to In-rich In(Ga,As) phases.

6.5 OHMIC CONTACTS TO p-TYPE GaAs

For Au-based ohmic contacts to p-GaAs, many of the same considerations of GeAuNi n-type contacts are also relevant. The metallurgical reactions outlined in Section 6.2.3 are mostly applicable to p-type contacts and mainly need to be modified by the choice of elements needed for doping the GaAs. Though the principles of Au-based and non-Au-based contacts to p-type GaAs are similar to those for n-type GaAs in most respects, several distinct differences exist. Metal contacts to p-type GaAs form lower barrier heights, making thermionic-field emission a relatively easier process for ohmic contact formation. Also, it is somewhat easier to dope p-type GaAs above 10^{19} cm^{-3} for the formation of non-alloyed ohmic contacts. Finally, the elements chosen to dope the GaAs p-type and those chosen to promote an interface bandgap lowering reaction are different for p-type contacts.

AuZn [16], AuBe [17] and TiPtAu [18] are some of the more commonly utilised p-type contacts. These contacts lack the low-temperature lattice disrupting constituents common to n-type GaAs contacts. These contacts were developed two to three decades ago. Had the reaction mechanisms been better known at the time, we might have seen AuZnNi or AuBeNi contacts developed. Instead, the contacts were probably developed under the assumption that a p-type dopant was needed to act in the place of the Ge for an AuGe contact. Since no eutectic melting takes place with AuBe, third element components were probably not thought necessary. Small percentages of Zn or Be (e.g. 0.5–2% Be was found to be optimum) were adequate for contacts in the low 10^{-6} Ω-cm^2 resistivity range. In these contacts, the Au provides the role of a lattice disrupting element. No surface oxide reacting element is present, so the methods of surface preparation described in Section 6.3.1 are critical. Also, the excess Be tends to diffuse to the surface of the contact, where it oxidises and can form an insulating barrier to a subsequent metallisation step or cause difficulties in probing. Either the use of minimal amounts of Be in the contact, an overcoat layer such as TiAu or a removal of the surface Be-oxide is called for. The latter process can be accomplished with an HF dip. AuBe contacts are usually thermally evaporated because electron beam evaporation of the alloy quickly depletes the Be from the surface of the alloy and causes problems for repeating the contact composition in subsequent runs.

High p-type doping of GaAs is straightforward with epitaxy, diffusion and ion implantation for a number of dopants including Be, C, Mg and Zn. Non-alloyed ohmic contacts to these heavily doped GaAs surfaces are then chosen based on other contact properties required by the given application. Non-alloyed contacts to p-GaAs are the norm in most GaAs devices. TiPtAu contacts or variants on these compositions are common for the base layer of npn HBTs, where the doping is desired to be high to minimise the base sheet resistance. Virtually any metal will make a non-alloyed contact to the base layer of a highly doped npn HBT. Other considerations determine the optimal composition. One popular choice is the PtTiPtAu contact [19] with 15, 10, 15, 60 nm thicknesses of the layers. A composition of this type is chosen for good adhesion, and the promotion of a uniform, shallow-reacted contact (Section 6.4.2). Non-alloyed W and WSi contacts to p^+ regions based on shallow Zn and Mg implants are other types of non-alloyed contacts that have been reported. These contacts have approximately $1–2 \times 10^{-6}\ \Omega\text{-cm}^2$ contact resistivity.

As is the case for n-type contacts to GaAs, a wide range of non-Au metallisations exist as well. Many types of non-Au contacts with improved morphology and/or temperature stability have been reported for p-type GaAs. An example is NiSi(Mg) [20] which is the p-type analogue of the n-type solid phase regrowth contact using NiSi. Contact resistivity $<10^{-6}\ \Omega\text{-cm}^2$ has been achieved. Surface bandgap lowering contacts using NiInW(Mg) have been reported [21] with contact resistance of 0.8 Ω-mm. Contacts utilising both solid-phase regrowth and surface bandgap lowering using PdSb(Mn) [22] with contact resistivity $<10^{-6}\ \Omega\text{-cm}^2$ are other examples.

6.6 CONCLUSION

Ohmic contacts to GaAs technology have been widely studied. In part, this is due to the search for "greener pastures". The n-type GeAuNi contact was discovered early in the history of GaAs and has proven to be quite reliable for a wide variety of applications. Thanks to decades of ohmic contact research, many alternatives are available for those who desire greater thermal stability, shallower contacts or a contact better suited to the special requirements of one's process. Alternative contacts include those based on solid-phase regrowth, refractory metals, bandgap engineering and methods of heavily doping the GaAs surface. Research has also yielded a better picture of the basic reactions occurring at the GaAs/contact interface and has given insight for better ways of achieving contact reproducibility.

Many choices also exist for p-type contacts to GaAs. In this case, no single contact metallurgy dominates as in the case for n-type contacts. GaAs devices requiring p-type contacts usually can provide high enough doping for making non-alloyed ohmic contacts. GaAs devices such as HBTs require shallow, thermally stable contacts; all GaAs devices require repeatable processes.

REFERENCES

[1] N. Braslau, J.B. Gunn, J.L. Staples [*Solid-State Electron. (UK)* vol.10 (1967) p.372]
[2] J.M. Woodall, J.L. Freeouf, G.D. Pettit, T.N. Jackson, P. Kircher [*J. Vac. Sci. Technol. (USA)* vol.19 (1981) p.626]
[3] E.D. Marshall, W.X. Chen, C.S. Wu, S.S. Lau, T.F. Kuech [*Appl. Phys. Lett. (USA)* vol.47 (1985) p.298]
[4] R.E. Williams [*Gallium Arsenide Processing Techniques, Second Edition* (Artech House, Norwood, MA, 1990)]
[5] H.H. Berger [*Solid-State Electron.* vol.15 (1972) p.145]
[6] G.K. Reeves [*Solid-State Electron.* vol.23 (1980) p.487]
[7] Y.-C. Shih, M. Murakami, W.H. Price, E.L. Wilkie, A.C. Callegari [*J. Appl. Phys. (USA)* vol.62 (1987) p.582)]
[8] L.C. Wang [*Mater. Res. Soc. Symp. Proc. (USA)* vol.319 (1994) p.93]
[9] R. Zuleeg, P.E. Friebertshauser, J.M. Stephens, S.H. Watanabe [*IEEE Electron Device Lett. (USA)* vol.7 (1986) p.603]
[10] M.O. Aboelfotoh, C.L. Lin, J.M. Woodall [*Appl. Phys. Lett. (USA)* vol.65 (1994) p.3245]
[11] F. Ren, A.Y. Cho, D.L. Sivco, S.J. Pearton, C.R. Abernathy [*Electron. Lett. (UK)* vol.30 (1994) p.912]
[12] V. Fischer, T.-J. Kim, P.H. Holloway, E. Ristolainen, D. Schoenfeld [*J. Vac. Sci. Technol. B (USA)* vol.12 (1994) p.1419]
[13] M.L. Lovejoy et al. [*Thin Solid Films (Switzerland)* vol.253 (1994) p.496]
[14] H.-J. Kim, M. Murakami, W.H. Price, M. Norcott [*J. Appl. Phys. (USA)* vol.67 (1990) p.4183]
[15] Y.-C. Shih, M. Murakami, W.H. Price [*J. Appl. Phys. (USA)* vol.65 (1989) p.3539]
[16] Y. Lu, T.S. Kalkur, C.A. Paz de Araujo [*J. Electrochem. Soc. (USA)* vol.136 (1989) p.3123]

[17] I. Ishida, S. Wako, S. Ushio [*Thin Solid Films (Switzerland)* vol.39 (1976) p.227]
[18] G. Stareev [*Appl. Phys. Lett. (USA)* vol.62 (1993) p.2801]
[19] E. Nebaur, M. Mai, J. Wurfl, W. Osterle [*Semiconductor Science and Technology* vol.15 (2000) p.818]
[20] C.C. Han et al. [*J. Appl. Phys. (USA)* vol.68 (1990) p.5714]
[21] P.-E. Hallali, M. Murakami, W.H. Price, M.H. Norcott [*J. Appl. Phys. (USA)* vol.70 (1991) p.7443]
[22] C.C. Han, X.Z. Wang, S.S. Lau, R.M. Potemski, M.A. Tischler, T.F. Kuech [*Appl. Phys. Lett. (USA)* vol.58 (1991) p.1617]

Chapter 7

Schottky contacts

7.1 CHAPTER SCOPE

Schottky contact is the term used to describe metal/semiconductor interfaces that display rectifying or diode-like behaviour. The material presented in this chapter will enable the reader to understand the basic physics of Schottky contacts, methods of forming them and ways that they form gate contacts to GaAs FETs.

The physics of Schottky contacts and their interfacial properties are given in Section 7.2. Their formation and electrical testing are discussed in Sections 7.3 and 7.4, respectively. Reliability is discussed in Section 7.5.

7.2 PHYSICS AND CHARACTERISATION OF SCHOTTKY CONTACTS

7.2.1 Physics of Schottky contacts

The physical description of a Schottky contact begins in a similar way to the description of the metal/semiconductor interface for ohmic contacts in Section 6.2. Ga or As atoms have tetrahedral symmetry in the GaAs semiconductor. Each Ga atom is surrounded by four equivalent As atoms, i.e. equally spaced, equal bond lengths, equal bond angle (109.5°). Each As atom can be described in the same way. At a free surface, the atoms lose their four-fold bonding to other semiconductor atoms, which results in unpaired electrons. The energy of these unpaired electrons is higher than that for a fully formed bond. As a result, the overall energy level can be reduced if the surface atoms rearrange themselves and form new bonds with each other (Section 3.3.1). At an interface with a metal, similar atomic rearrangements happen, but with new bonds with metal atoms also being possible. In addition, defect structures such as missing bonds formed at the interface contribute to the electronic states at the interface. Surface or interface states are the two-dimensional analogue of the three-dimensional band structure of crystalline semiconductors. Like the band structure of a three-dimensional crystal, the electrons of a single surface or interface bond are not considered localised energy levels belonging to a particular bond, but they may be influenced by a two-dimensional periodic potential, if it exists. Energy levels may also behave more as localised states if local disorder is high. These new states have energy levels that mostly lie in the forbidden gap between the conduction and valence bands. When the Fermi level is lined up between the semiconductor and the metal, the states in the gap at the interface set the Fermi level near midgap, while the n-type dopants set the Fermi level near the conduction band in the bulk of the semiconductor. The bending of the conduction band at the metal/semiconductor interface sets the barrier to current flow across the interface, as illustrated in FIGURE 7.1. The Schottky barrier is defined as the energy difference between the metal Fermi level and the semiconductor conduction band energy at the metal/semiconductor interface.

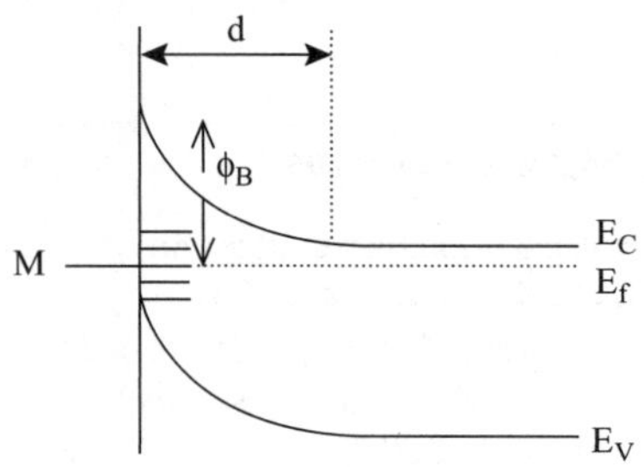

FIGURE 7.1 Energy bands of a semiconductor/metal interface.

For ohmic contacts the Schottky barrier is an undesirable impediment to free current flow, but it is a desirable barrier for Schottky contacts to FETs. In Section 8.2.1 it will be seen that the FET acts as a voltage-controlled current source and the Schottky gate is the medium for the voltage input to the device. It is highly desirable that Schottky contacts conduct no current and act only as a capacitive input to the transistor. Although the Schottky contact does not completely meet this ideal, it offers a reasonable approximation over a certain voltage range. The current equations of a generalised metal/semiconductor interface were presented in Chapter 6 and are repeated below:

$$\begin{aligned} J &= (\text{const})\exp(-q\phi_b/kT)[\exp(qV/nkT) - 1] \\ J &= J_s[\exp(qV/nkT) - 1] \end{aligned} \tag{7.1}$$

where ϕ_b is the barrier height, q is the electron charge, k is Boltzmann's constant, V is the applied voltage across the barrier and n is the ideality factor. An ideal diode has $n = 1$, and n usually ranges between 1 and 2 for Ga(Al)As-based Schottky contacts. The first term in the exponent dominates over the second for qV a few times greater than kT and describes the forward conduction of the Schottky diode. The second term in the exponent describes the reverse leakage current.

Current conduction for moderately doped semiconductors mostly occurs by the fraction of electrons that are sufficiently energetic to traverse the Schottky barrier. This current mechanism is termed thermionic emission and is well described by using $n = 1$ in EQN (7.1). More highly-doped semiconductors begin to add a tunnelling component to current conduction, as described in Section 6.2.2, and do not make very good Schottky contacts. Even with moderate to low-doped semiconductors, other current mechanisms, probably related to defects or interface states, exist and are described by ideality factors that are usually greater than 1. These other factors can be influenced by processing conditions also.

The interface states responsible for the conduction band bending in FIGURE 7.1 deplete the carriers in the near surface region of the semiconductor. This depletion region is also affected by the applied voltage,

$$w = [2\varepsilon(V_{bi} - V_a)/qN]^{1/2} \qquad (7.2)$$

where ε is the dielectric constant, q is the electron charge, N is the doping density, V_{bi} is the built-in voltage of the Schottky contact and V_a is the applied voltage of the Schottky contact. The built-in voltage represents the depletion region that exists with no externally-applied voltage. Positive applied voltage will reduce the depletion region, while negative applied voltage will increase it. EQN (7.2) is derived by solving the Poisson equation for a constant doping and is useful for estimating depletion widths for certain structures. Analytical modelling tools are a more general way of analysing more complex structures. Other assumptions used in deriving EQN (7.2) break down when $(V_{bi} - V_a)$ approaches zero and when the electric field represented by $(V_{bi} - V_a)/w$ approaches the breakdown field for GaAs. These limits also represent the useful operating range of Schottky diodes.

The capacitance (C) of a Schottky diode is another useful characteristic. Its equation is given by the relation

$$C = A(q\varepsilon N/2V)^{1/2} \qquad (7.3)$$

or

$$1/C^2 = 2V/q\varepsilon NA^2$$

where A is the area of the metal/semiconductor interface and the other terms are as before. No assumptions of constant doping were used to derive EQN (7.3).

The second form of EQN (7.3) is particularly useful since C is easily measured, can be manipulated to the form of EQN (7.3) and is used to extract the doping as a function of voltage. The doping and voltage can then be used with EQN (7.3) to extract the doping as a function of depletion depth. This method is illustrated in FIGURE 7.2, where C is measured as a function of voltage (FIGURE 7.2(a)), then $1/C^2$ is plotted as a function of voltage (FIGURE 7.2(b)). The slope of this plot yields the doping concentration as a function of voltage, which is plotted as N versus depth in FIGURE 7.2(c).

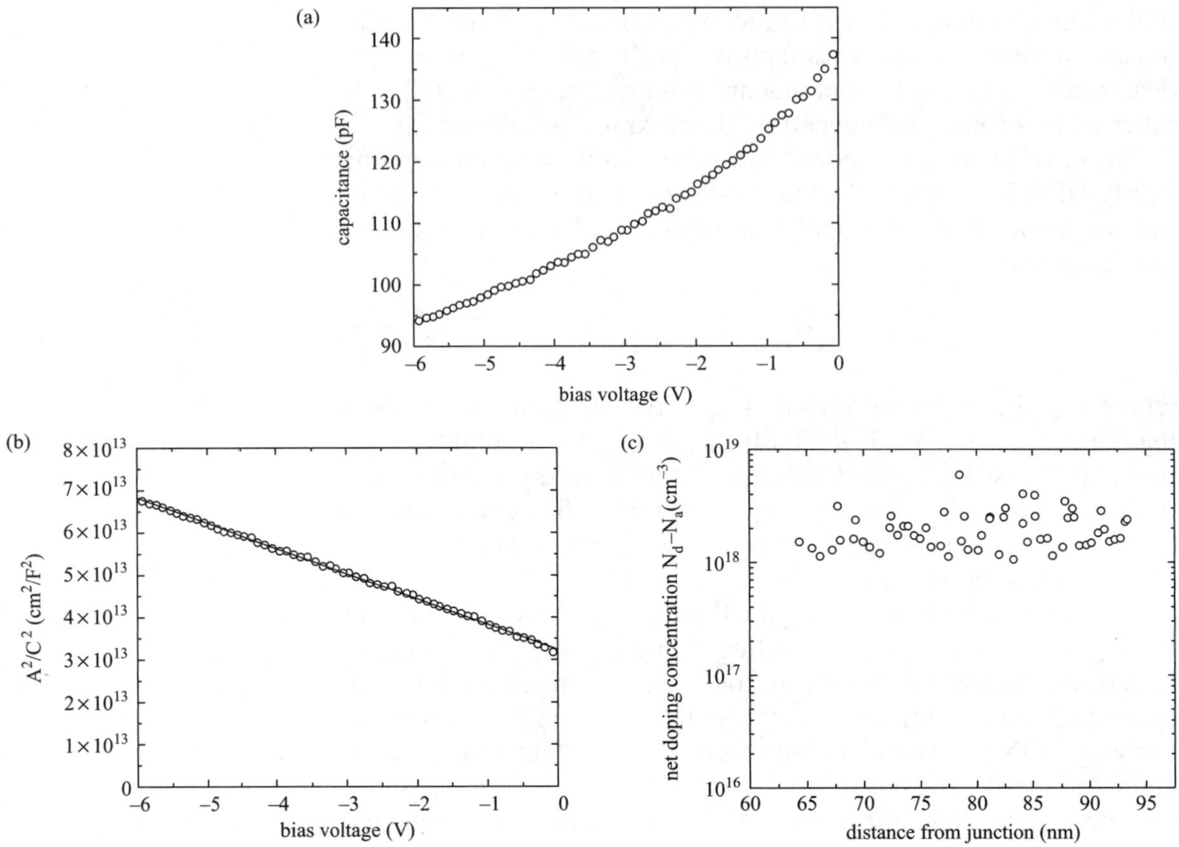

FIGURE 7.2 Capacitance-voltage and doping-depth plots.

Direct capacitance measurements require fairly large areas for the Schottky contacts, of the order of 10^4 μm^2 (depending on the carrier concentration), in order to measure capacitances in the tens to hundreds of picofarads. This area represents contacts with 2–3 orders of magnitude larger areas than typical Schottky gates for FETs so testing is carried out on special test structures (large diodes) created during FET fabrication or with techniques such as mercury probes (see Section 2.5.4). Special techniques (based on S-parameter measurements – see Section 8.2.1) are used to extract the much smaller capacitances of individual gate fingers, and these techniques are not normally used to determine the doping in FETs.

Other examples of Schottky contact characterisation will be briefly discussed. Current-voltage (I-V) plots can be taken for Schottky diodes of all different sizes, including the gates in actual FETs. When EQN (7.1) is expressed in logarithmic form, the result is

$$\ln(J) = \ln(J_S) + qV/nkT \tag{7.4}$$

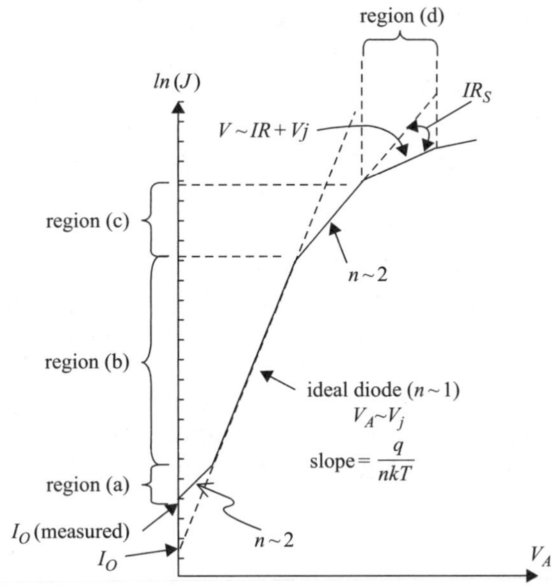

FIGURE 7.3 Ln(J) versus V plot for an idealised Schottky contact.

where J_s can be given by

$$J_s = A^{**}T^2 \exp(-q\phi_b/kT) \tag{7.5}$$

and A^{**} is a constant (the Richardson constant), ϕ_b is the barrier height and the other terms were previously defined. A plot of ln(J) versus voltage gives some useful information about a Schottky contact. FIGURE 7.3 shows an idealised plot for a Schottky contact. Four distinctive regions of this plot stand out. Region (a) is for the first several multiples of kT (0–100 mV). In this region, the assumptions behind EQN (7.4) are not yet valid and no log-linear relation exists yet. Generation-recombination current may dominate in some cases. In region (b), ln(J) versus V is linear over a range of voltages spanning many decades of current. In this region, the slope of the curve equals q/nkT and the ideality factor n can be extracted from the slope of the plot evaluated according to EQN (7.4). High-quality Schottky diodes often have ideality factors near 1. In regions (c) and (d), the diode begins to turn on and the current is high enough that the resistive effects of GaAs conduction become important. Analysis of the ideality factor is important in determining the quality of a Schottky contact. Examples of this analysis will be given in sections with topics varying from metallurgy of Schottky contacts (Section 7.2.2), recessed gates of FETs (Section 7.3.1) and refractory gates for FETs (Section 7.3.2). In addition, analysis of pn junction ideality factors (Section 9.2.1) proceeds in a similar way.

EQNS (7.4) and (7.5) may also be analysed to extract ϕ_b. The Schottky barrier height is important in GaAs FETs because it directly affects the zero bias depletion width and variability in its value can, therefore, cause variability in FET currents (refer to Section 8.2). From FIGURE 7.3 and EQN (7.4), the intercept gives $\ln(J_S)$. From EQN (7.5), the Schottky barrier height can be extracted as long as A^{**} is known. For GaAs, the commonly accepted value for A^{**} is 8.6 (compared with a free electron value of 120). Because of the exponential relationship between ϕ_b and A^{**}, relatively large errors in A^{**} will result in only small uncertainties in ϕ_b.

Schottky barrier extraction techniques other than the I-V method are sometimes used as well. These techniques include the capacitance-voltage or the activation energy methods. The reader is referred to references [1–3] to delve further into these methods.

7.2.2 Interfacial properties of Schottky contacts

Free surfaces are very reactive, as explained in Section 3.3.1. Although many attempts have been made to study the pristine GaAs surface and the way it forms an intrinsic Schottky barrier with a wide host of metals, it is safe to say that the fundamental mechanisms are poorly understood, at least in comparison to ohmic contacts. What is understood is that defects are easily created in the Schottky barrier formation. Theories describing the role of defects in Schottky barrier formation are summarised in Section 3.3.1.1. In one very telling experiment repeated in a number of different laboratories, researchers took a clean GaAs surface in ultra-high vacuum, cooled the sample, coated it with a solidified noble gas such as Ar and then deposited a metal, which physisorbed on the Ar. When the sample was slightly warmed, the Ar vaporised and the thin metal film contacted the GaAs surface without having the energy of bombardment. Insight about the intrinsic low-temperature properties of Schottky contacts was gained. Midgap states were present even under those benign conditions. Most known Schottky metals will interact with the GaAs surface and create midgap defect states at fairly low temperatures. These interfacial interactions are consistent with the physical picture of surfaces and interfaces presented in Section 7.2.1. Though a wide variety of metals have been studied, most have remarkably similar Schottky barriers that vary within only about 0.2 eV (Section 7.4). This situation contrasts greatly with Si Schottky barriers, where the work function of the gate metal affects the Fermi level in the semiconductor to a greater extent than for GaAs.

Some considerations in the choice of Schottky metals include the metal adhesion, resistivity, metal reactivity and long-term reliability. No single metal meets all of these diverse requirements and so stacks of layered metals are often used. Al perhaps comes the closest; it has low resistivity, reasonable adhesion and reasonable stability. One of its drawbacks is its propensity to react with Au (which could be a bond pad overlay or interconnect metal) under high current density to form a high-resistivity phase known colloquially as "purple plague". This process, which occurs gradually under current densities reasonably expected in normal operation of GaAs circuits, is a reliability concern. Certainly, preventive measures in the form of diffusion barriers are possible for purple plague, so Al can be a perfectly suitable Schottky metal and is one commonly used contact.

High metal resistivity in micro-circuits will cause undesired IR voltage drops. In the gates of FETs, high resistivity will also degrade high-frequency performance. For low-resistivity gates, four metal choices exist: Au, Al, Cu and Ag. The latter two metals are rarely used for GaAs Schottky contacts due to their reactivity, though in the case of Cu, it is not clear that suitable diffusion barriers would not prevent this.

One important interfacial property of the Schottky metal is its adhesion. Cr and Ti have particularly good adhesion to GaAs. So does Al. These metals, along with the refractory metals such as WSi, are common choices for Schottky contacts. Au has poor adhesion to GaAs. It is also excluded for the GaAs/metal interface due to its reactivity with GaAs, as will be elaborated on in some of the material that follows.

Most metals will react to some extent with GaAs at sufficiently high temperatures. The key is to match the temperature requirements of the device processing (or its application, if that is higher) with the thermal constraints of the GaAs/metal interface. We will review the interfacial reactions for a number of commonly used contacts.

Among the layered metal stacks, TiPtAu is one of the most widely used for low-temperature processes (gate after ohmic metal; see Chapter 8). Its use and the rationale for its use are often the source of some misunderstanding, so we will review them here. We will begin by describing the simpler GaAs/Pt, GaAs/Ti and GaAs/Au interfacial reactions before describing the GaAs/Ti/Pt/Au reactions.

A GaAs/Pt contact reacts in tens of minutes at temperatures near 300°C, but will show relatively stable Schottky characteristics up to 500°C for short annealing times [4,5]. The interface of GaAs/Pt forms Pt-arsenide complexes, principally $PtAs_2$, while the Ga diffuses through this boundary layer into the Pt. The reaction proceeds with a $t^{1/2}$ dependence at temperatures between 300 and 400°C, indicating that the Ga diffusion is the rate limiting step.

The reaction proceeds with an activation energy of 1.6 eV and consumes about 10 nm of GaAs in 1 h at 300°C. The Schottky barrier height increases by about 0.05 V for the GaAs/$PtAs_2$ interface. The diodes show a gradual increase in ideality factor from 1.1 to 1.2 after temperature treatment.

GaAs/Ti does not react significantly for moderate times (tens of minutes) until about 400°C [6]. At these temperatures or higher, the reaction proceeds with a strong similarity to the GaAs/Pt reaction. Ga diffuses into Ti and Ti-As compounds form at the original GaAs/Ti interface, consuming GaAs in the process. The activation energy for the reaction is 1.7 eV. The contact remains Schottky and the Schottky barrier height increases slightly.

The GaAs/Au interface reacts extensively at 250°C in about 1 h. Ga diffuses through the Au and accumulates at the surface of the Au. The Schottky barrier drops from about 0.9 to 0.6 eV. $AuGa_2$ is the stable phase.

On the other hand, a GaAs/Pt/Au contact will become leaky at approximately 300°C and will show ohmic-like behaviour for short anneals at 400°C. The reason for the very different electrical properties of Pt/GaAs interfaces in the Au-containing and Au-free contacts is not due to gross differences in the interfacial reactions, but rather due to minor differences in vacancy-related effects. An example of these differences will be presented later in this section.

Often, the role of Pt in the GaAs/Ti/Pt/Au contact is described as a diffusion barrier that keeps Au and Ga apart. This reputation may have arisen from reliability studies on Si-based beam lead diodes in the 1960s where TiPtAu is deposited on SiO_2. It is certainly not a perfect barrier as metallic intermixing occurs near 250°C for hundreds of hours [7]. One problem is that Pt, like most evaporated metals, is typically polycrystalline, permitting fast diffusion along grain boundaries until their available sites become saturated. TiPtAu will exhibit resistance increase by thermal aging in tens of hours at 250°C due to slight Pt intermixing with Au when deposited on SiO_2. The situation is more complex when Ti interfacial reactions with GaAs are considered. Overall, no gross metallurgical differences exist in GaAs/Ti reactions with or without Au-containing overlayers. Subtle differences affecting electrical properties do exist, and though not thoroughly understood, they probably involve small amounts of Au at the GaAs interface. One might argue that Ti is a better diffusion barrier than Pt because it maintains the integrity of the interface to GaAs up to a higher temperature. Even better diffusion barriers than Pt or Ti are amorphously deposited refractory metals.

At temperatures below 350°C for short or intermediate times, no obvious interfacial reaction occurs, but electrical characteristics indicate carrier reduction in n-type GaAs, consistent with Ga outdiffusion. An activation energy of 1.4 eV is associated with the electrical degradation [8]. Thus, GaAs/Ti/Pt/Au interfacial reactions show slight, if any, metallurgical differences with non-Au contacts, but can have large electrical differences. The evidence for Pt as an effective diffusion barrier is not compelling.

Metal reactivity with Ga or As is what drives the interfacial reactions with GaAs. Examples of M/GaAs interfaces that react at relatively low temperatures have been given (Au, Pt). A few generalisations about M/GaAs reactions can give some insight, though they are no substitute for a yet-to-exist complete picture. Stable end-reaction phases exist for many metals, which provide a powerful driving force for interfacial reactions with GaAs. Interfacial intermixing occurs readily at low temperatures for many metals, affecting electrical properties, but also providing reaction pathways. Reactions beyond the thin interfacial layers depend on diffusion of M, Ga or As. Ga generally diffuses faster than As, as seen in the GaAs/Pt, GaAs/Ti reactions. Metals that diffuse faster than Ga, generally those with weaker M-M bonds, react at relatively low temperatures. M-M bond strength, in turn, generally correlates with the heat of vapourisation of the metal. Due to the lower mobility of As, M-As phases tend to form at the original GaAs interface and the M-Ga phases form above the M-As phases. For metals that do not diffuse readily at low temperatures (<400°C), GaAs decomposition with Ga and As outdiffusion becomes important at higher temperatures. This outdiffusion is suppressed to the greatest extent by the use of amorphous refractory metals. Complete characterisation of M-GaAs reactions with intermediate and final phases is complex and less easily generalisable.

Among refractory Schottky metal contacts, Mo, Ta and W are stable at temperatures below 700°C. Ta interfacial reactions occur at 700°C in seconds, while Mo and W interfacial reactions occur between 750 and 800°C. All of these metals show strong polycrystalline signals in X-ray diffraction.

WSi and WSiN can be deposited as amorphous metals for optimum composition. The optimum composition of tungsten silicide is $W_{0.45}Si$. For lower tungsten fractions, the film will either contain crystalline W phases or recrystallise at lower-than-optimum temperatures. For higher-than-optimum W composition, the film recrystallises to form WSi_2. For the optimum composition, WSi remains amorphous up to about 850°C. WSiN also remains amorphous up to 850°C and its recrystallisation is less strongly dependent on composition. Other refractory metal alloys such as WN and TiW are sometimes used as well. However, these films are at least partially crystalline as deposited. At high enough temperatures, usually 900°C and above, GaAs decomposition becomes the dominant degradation effect, regardless of the Schottky contact used. WSi is also used as a diffusion barrier for Au-based contacts even in many low-temperature processes.

Aluminium-containing refractory metals such as AlMo alloys are sometimes used. These alloys can be stable up to 800–850 °C. Interfacial reactions with Al at these temperatures can be desirable to increase the Schottky barrier height.

Many more types of interfacial Schottky reactions have been characterised, but those presented here are representative of important, widely used contacts.

7.3 FABRICATION OF SCHOTTKY CONTACTS

7.3.1 Basic recessed gate fabrication

Field effect transistors can have either recessed or self-aligned gates. In this section, we will describe a general process for making recessed-gate structures.

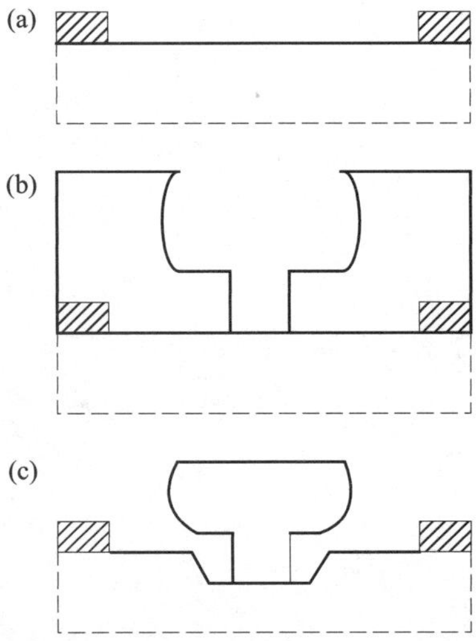

FIGURE 7.4 Illustration of a recessed-gate process.

Gate recessing requires a method for determining the etch depth. Direct measurement is not easy or quick, so process engineers usually resort to an electrical method of measuring the source to drain current to determine the proper etch depth. From a structure such as that of FIGURE 7.4(a), the initial source-drain current is measured. Then gate lithography patterning follows using either optical lithography or electron beam lithography (FIGURE 7.4(b)). An initial recessed-gate etch is timed so that it will not reach completion. The sample is removed from the etch solution and retested. The etch and test process is iterated until a target source-drain current is reached. Once the final etch target is reached, the wafer is ready to have gate metal deposited (FIGURE 7.4(c)).

Prior to the recess etch, a surface treatment may be used. This may include a plasma oxygen treatment to remove monolayers or less of organic material that remain after the resist develop process. The surface treatment may also include oxide removal or other steps. After recess etching to the proper depth, the Schottky metal is then deposited over the whole wafer where it forms the contacts in the patterned areas and covers the resist over the rest of the wafer. The unwanted metal is removed from the wafer when the resist is dissolved, often by means of a solvent spray. As an alternative to solvent sprays, tape pull processes, where an adhesive tape adheres to the uppermost metal that covers the resist, are used. The unwanted metal and resist are easily pulled away from the wafer during this process and the wafer is cleaned to remove unwanted organic residues. At this point, the Schottky metal can be inspected and the transistor can be tested.

Most recessed-gate structures result in a fully testable FET after the liftoff is completed. The FETs can be tested by the procedures described in Section 2.5.4. Specific transistor tests are described in Section 8.2.1. The Schottky contact can be tested as a special structure or as part of the FET test. The FET can be biased as a Schottky diode to measure the forward and reverse currents, Schottky barrier height and ideality factor. Of the tests described in Section 7.1.1, only the capacitance measurements require a test structure other than the FET. These diode structures should have an area of at least 10^4 μm^2 for direct capacitance measurements. Generally, one uses baseline measurements of Schottky diodes for comparison. Increases in the ideality factor are generally a sign of degradation as are increases in the reverse leakage current, increases in the generation-recombination current or any change in Schottky barrier height. These parameter changes are usually an indication of deviations in material growth or processing.

Several of the issues involved with the recessed-gate fabrication steps will be described in more detail. The equipment and many of the issues involved with metal evaporation are the same as for ohmic contacts. Refer to Section 6.3.1 for these. The recess etch reproducibility is one of the greatest challenges for making FETs. The geometry of the recessed channel must also be very uniform. One will easily lose control of the manufacturing tolerances of a wafer if these processes cannot be kept uniform. Finally, the metal deposition process can also be critical. The details of the deposition will depend on which gate structure is chosen (Section 7.3.3).

For the purpose of the ensuing discussion, we will assume that a GaAs FET uses a thin (several tens of nm) n^+ doped cap layer to lower source resistance. A recessed-gate etch will remove this layer and also part of an underlying GaAs or AlGaAs layer. See Chapter 8 for other epitaxial layer structure choices. Reproducibility of the surface and etch chemistry is necessary for obtaining good results in recessed gate etching, as discussed in Chapter 3. The use of a first-step SiN passivation is one way of achieving this outcome.

One must pay careful attention to the GaAs surface prior to etch initiation. If the GaAs surface is not protected early in the process sequence with SiN, it can pick up many kinds of impurities that come from the air, photoresists, developers or other process steps. A resist process will coat the GaAs surface with organic material, which is intended to develop away in the area of a gate recess. After an alignment and develop process, the resist will be mostly removed. However, submonolayer surface coatings remain if the surface bonding situation makes it energetically favourable for them to do so. Due to lithography, thin film depositions, rapid thermal annealing (during ohmic contact formation) and any rework processes that are necessary, the surface condition in the vicinity of a gate recess may be anything but predictable. Therefore, the process engineer should institute procedures that will return the surface of the GaAs to a well-defined state by means that are very similar to the surface preparation described prior to ohmic contact formation in Section 6.3.1 and elaborated on in Section 3.2.2. In the case of the ohmic contact, the goal is to create a surface that will not inhibit the ohmic alloy reactions that are necessary for achieving a good ohmic contact. In the case of recess etching, one wants the GaAs surface to wet with the chemical etchant promptly, to start the etching without any sort of time delay and to etch at the rate expected without roughening. For more details, refer to Section 4.4 to review some of the chemically necessary controls that should be put into place any time wet chemistry is needed. The procedures for surface reproducibility include a plasma oxygen (or other method of producing reactive oxygen) treatment for removal of carbonaceous surface species followed by a GaAs oxide removal step for returning the GaAs surface to a native state. The surface oxygen treatment must be vigorous enough to oxidise stubborn carbonaceous surface occupants yet gentle enough to not significantly alter resist profiles.

One approach to circumvent the uniformity issues surrounding the gate recess etch would be to use selective etches to form the gate recess. For example, a phosphoric acid, hydrogen peroxide aqueous mixture might be used for a non-selective GaAs MESFET gate recess and for the doped GaAs layer in an AlGaAs/InGaAs PHEMT. However, one might also use citric acid : hydrogen peroxide : water because it does not etch AlGaAs. By stopping on AlGaAs, the selective etch will eliminate depth uniformity issues that arise with non-selective etches. The only problem with this approach is that some types of devices cannot tolerate lateral undercutting of the recess etch very well. For example, power amplifiers show increased RF dispersion effects and long term degradation with laterally undercut gates (Sections 8.2.2 and 8.8). For other applications, lateral undercut may be less of an issue.

When using wet recessed-gate etches, one must take care to avoid electrochemical etching. If the gate lithography pattern is misaligned and too close to an ohmic metal, a galvanic reaction can be set up from liquid electrolyte touching the ohmic metal (or its laterally reacted extent) and extremely fast etching can occur (Section 4.7). Even if the gate metal is properly aligned, electrochemical etching can occur if ohmic metal is exposed nearby (the distance to an ohmic metal test pad intended for recessed gate monitoring!). For this reason, recessed gate test structures are protected with photoresist and the test engineer scratches through photoresist with the electrical probes to monitor the recessed gate etch progress.

A dry etch is a possibility as long as one can avoid plasma damage. Plasma damage is often recoverable at least to a certain extent by heating the sample to 400°C or above. Most GaAs recessed-gate structures cannot be heated beyond 300°C. Those that can, should not be heated past what the ohmic contact can handle without degradation (near 400°C for GeAuNi). Generally, one must use DC self biases below 30–40 V to avoid surface damage (Section 5.10.2). Because of the low DC biases, a minimally contaminated surface starting condition is a necessity, as is careful attention to the starting surface condition in order not to cause etch initiation delays or roughening of the surface. Wet cleans after dry etching are generally necessary to remove etch residues. An advantage of dry etching is potentially greater uniformity over wet etching. Dry etch is also the main option for etching through SiN, and low self-biases are also required to avoid damaging the GaAs.

7.3.2 Self-aligned Schottky gates

A Schottky gate can be used as an implant mask for self-aligning source and drain regions with the gate of an FET (Section 8.2.4). In a self-aligned process, the heavily doped source and drain regions extend to the gate edge by means of these implants. The gate must also remain in place during the implant activation step so that a realignment will not be necessary. Refractory metals such as those described in Section 7.2.2 have the attributes needed for self-aligned gates. The self-aligned gate process is the mainstream technology for Si MOSFETs and many of the ideas for GaAs self-alignment are adapted from Si technology. These ideas and technology have great appeal because this planar process is considered more repeatable, manufacturable and scalable to high integration levels. It is more suited for digital but not microwave applications, though exceptions do exist.

Two basic types of self-aligned processes exist and are illustrated in FIGURE 7.5. A heavily doped n^+ region can extend to the edge of the gate (FIGURE 7.5(a)). Alternatively, an intermediate doped region is placed at the gate edge as a buffer between the gate and the n^+ region (FIGURE 7.5(b)). The first type of self-aligned process leads to the highest performance FETs, but the breakdown voltage is low. For digital applications with high integration levels, low voltages are preferred to limit power consumption and this approach is viable. However, in mixed-mode or specialised digital applications, higher breakdown voltage is desired. The intermediately doped region aids in increasing the gate-drain breakdown voltage. This intermediately-doped region

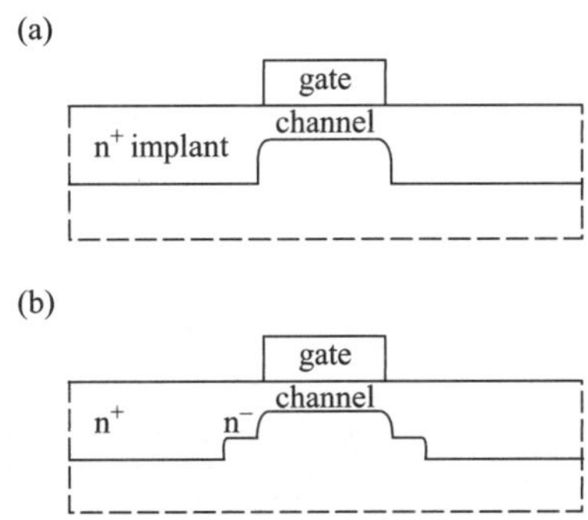

FIGURE 7.5 Types of self-aligned gate processes. In (a) the implanted n^+ region extends to the edge of the gate. In (b) the n^+ region is separated from the edge of the gate.

is mainly useful on the drain side of the gate, but most methods of fabrication are symmetric about the gate and create an intermediate doping on the source side as well. FETs with intermediately doped regions are usually called LDD FETs, for lightly-doped drain, even though the source may have a lightly-doped region as well. A variation of the LDD process extends the lightly-doped region further towards the drain but not the source.

Si is the most common dopant because of good activation efficiency and low diffusion. After the implant, the sample must be annealed to activate the implant. Either a furnace or, more commonly, a rapid thermal annealer is used to activate the implant. With rapid thermal annealing, a typical time and temperature of 850°C for 30 s may be used. Of course, temperature calibrations may vary from one laboratory to another. The anneal may be carried out with a dielectric encapsulant to protect the GaAs from thermal decomposition. If no dielectric is used, As overpressure must be used to protect the GaAs. During the anneal, it is important to exclude all sources of oxygen, especially for capless anneals. Oxidised GaAs quickly desorbs, effectively etching the GaAs.

In a furnace, As overpressure is provided by sublimation of solid As within the furnace. In an RTA, a special technique with a sample susceptor such as that described in Section 6.3.1 is used. The method consists of impregnating the inside of the sample susceptor with a thin film of solid As, so that upon annealing it provides the required gaseous As overpressure into the small space that exists between a GaAs wafer in contact with the susceptor lid. One method of providing the impregnated As is to anneal a mechanical GaAs wafer inside the susceptor to a somewhat higher temperature than the implant activation process requires. For example, during an 850°C, 30 s anneal process, the susceptor will provide sufficient As overpressure when it was previously annealed at 950°C for 5 min with high-purity GaAs.

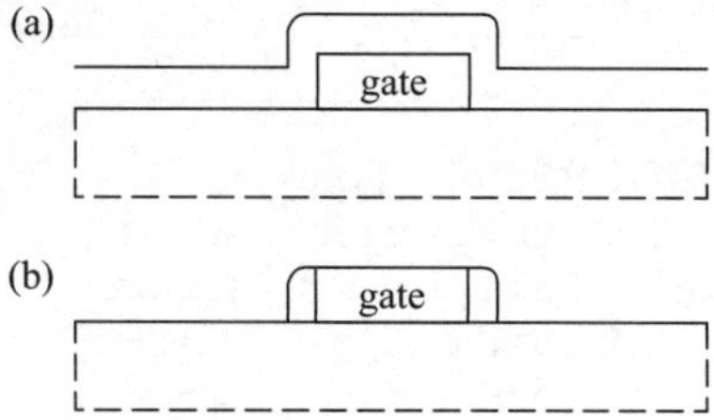

FIGURE 7.6 A dielectric sidewall process for spacing implants.

If a dielectric encapsulant is used, it must both be mechanically robust and maintain the surface integrity of the GaAs. This latter point will be elaborated on in Section 8.3 when ion implantation for FET channel doping is discussed. When used as an encapsulant for n^+ source and drain implants, the encapsulant must also avoid stressing the gate edge during the high-temperature excursion. This latter requirement is tricky to achieve and many process engineers prefer to use capless anneals instead.

The most common way to implement the LDD process is by the use of dielectric sidewalls (FIGURE 7.6). The LDD implant is performed after gate patterning. Then a dielectric is deposited over the gate. An anisotropic etchback using RIE is used to create dielectric sidewalls. Then the n^+ implant is performed and this heavily-doped region is blocked from the gate edge by the sidewall. The spacing of the n^+ implant to the gate edge is determined by the thickness of the sidewall, which in turn is proportional to the deposited dielectric thickness. Optimisation of the n^+ and LDD implants and their spacing from the gate edge form the heart of self-aligned gate technology and FET performance.

The last key element of self-aligned gate technology is the refractory Schottky contact. The thermal and interfacial characteristics were presented in Section 7.2.2. The most promising choices from the standpoint of interfacial stability are $WSi_{0.45}$ and WSiN because of their stable amorphous interface with GaAs [9]. Many other choices offer acceptable electrical characteristics and may, in fact, be used if higher Schottky barriers are desired.

7.3.3 Schottky gate structures

There exists a large number of Schottky gate structures for FETs. We will present a number of these as examples.

The basic recessed-gate structure was illustrated in FIGURE 7.4 (Section 7.3.1). A double recess is sometimes used to increase the breakdown voltage for power-amplifier or other high-voltage applications. In this structure, illustrated in FIGURE 7.7, a first recess etch is used to remove a highly-doped cap layer, so that this layer will not be subject to high electric fields near the drain edge. After another photoresist alignment, a second recess etch is performed to define the groove for the Schottky contact. In a double recess structure, only a slight second recess is used and a small gate-edge-to-recess-channel gap will arise. Such a structure can minimise hot electron effects that are described in Section 8.8.

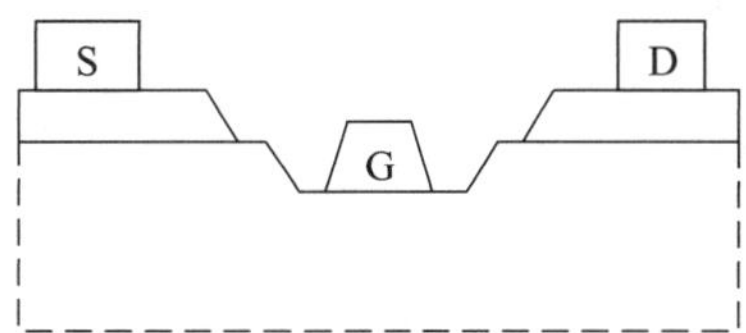

FIGURE 7.7 A double-recess structure for FETs.

Microwave devices also require low gate resistance for high performance. Such a device will have a small footprint and a wider top. The small footprint defines the Schottky contact and the depletion region of the FET, while the wider top provides a larger cross-section for increased gate conductance. Such structures are known as "mushroom gates" and are commonly implemented for microwave devices. One method of fabricating these is outlined and illustrated in FIGURE 7.8. A dual layer electron-beam resist is deposited on a wafer. The lower layer is a less sensitive, high-resolution resist such as poly methyl methacrylate (PMMA). The upper layer is a more sensitive, lower-resolution resist, such as methyl acrylic acid (MAA). The stack is written with a high dose for the narrow footprint and a lower dose for the liftoff profile of the upper part of the gate. After developing the resist, a recessed-gate etch is usually performed. An alternative to writing both layers by electron-beam lithography is to use optical lithography (after exposing and developing the electron beam resist) for the

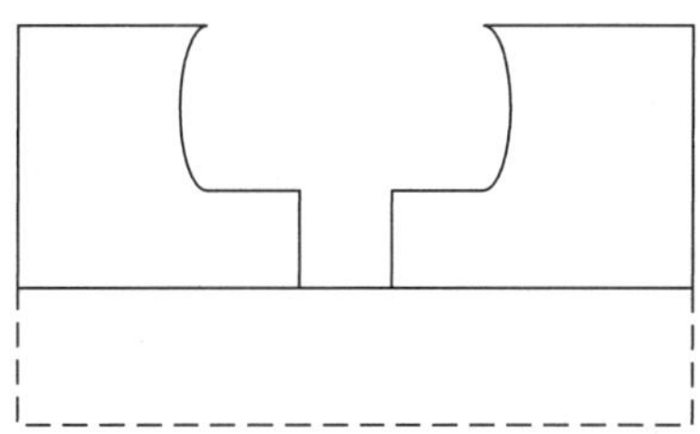

FIGURE 7.8 Illustration of an electron beam resist process for forming a mushroom gate.

upper resist. One can use thicker resist in this way for evaporating thicker gate metal than is generally available with electron beam resists.

In some applications, the gap between the gate and the gate-recess edge needs to be minimised or eliminated. One way to achieve this is to sputter the gate metal into the recessed channel. However, a special patterning technique must be used because a conventional photoresist will not lift off the metal well if the metal is deposited to fill the gaps. Instead, a tri-layer masking process such as that of FIGURE 7.9 can be used. When the side gap is eliminated, a depletion region at the sides of the gate adds to the gate capacitance. This parasitic capacitance reduces high-frequency performance, especially for short gate lengths.

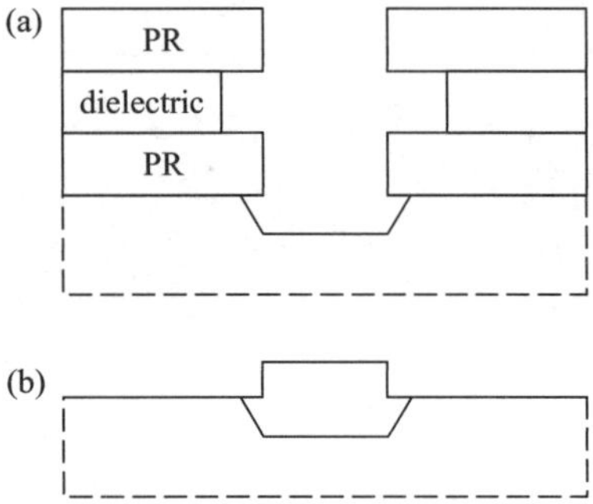

FIGURE 7.9 Illustration of a gapless gate deposition.

As an alternative to a recessed gate, a reacted gate may be used. A metal stack such as Au/Pt/Ti/Pt/GaAs is the usual choice. The thickness of Pt is chosen to react to the proper depth at temperatures below which the Ti reacts with the Pt-intermetallics or the GaAs. This process eliminates the gap between the gate and the GaAs that is common in recessed-gate etches. However, it is very difficult to control the Pt thickness, uniformity and reaction depth. In addition, a depletion region at the sides of the gate adds to the gate capacitance and is problematic for very short gates. This process is illustrated in FIGURE 7.10.

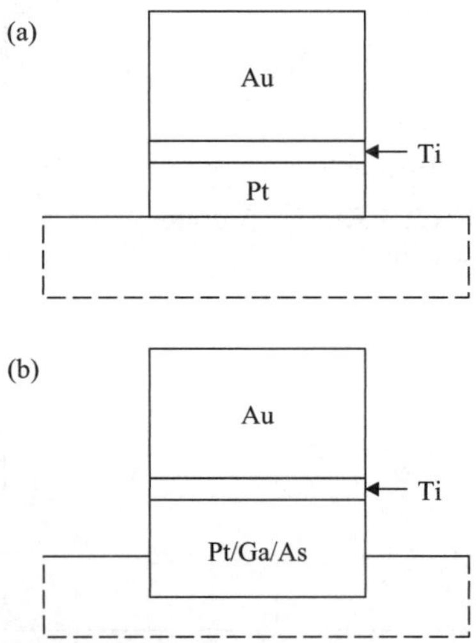

FIGURE 7.10 Illustration of a reacted gate.

The most straightforward way to make short-length, high-performance gates is to use electron beam lithography. Another way is to use deep-UV lithography with phase-shift masking techniques. The equipment for fine-line lithography steppers is very expensive and not available to all researchers. Consequently, other fine-line lithography techniques have been developed. Several of these will be described here.

In the first technique, a dielectric such as SiO_2 is deposited. An opening is formed in the dielectric with conventional lithography and reactive ion etching. Then a second dielectric, such as SiN, is deposited over the first dielectric and the opening. This process is illustrated in FIGURE 7.11(a). After another anisotropic reactive ion etch, a dielectric sidewall is formed at each edge of the original dielectric opening and the opening is effectively narrowed in a self-aligned manner (FIGURE 7.11(b)). The use of SiO_2 and SiN in the order shown is useful because SiN etches faster than SiO_2 in F-based reactive ion etches, but other dielectrics can be used as well. After the formation of the narrowed opening, a recessed-gate etch and metal-deposition process may be used (FIGURE 7.11(c)). Many choices are available for the gate metal process. One advantage of this type of process is that the gate metal is easily widened above the dielectric for low gate resistance. The main drawback of this technique is the difficulty in controlling the final gate length and controlling its uniformity across a wafer. A variation of this process would be to use high-resolution lithography for the dielectric opening without dielectric sidewalls as an alternative method to making a mushroom gate.

An advantage of the previous process is that one can have greater flexibility to deposit refractory metals as the interfacial layer when recessed-gate processes are used. This may be desirable for greater interfacial stability, as discussed in Section 7.2.2, or to eliminate hydrogen poisoning associated with Ti and Pt, as discussed in Section 7.5.

Another method for shrinking a gate beyond the capability of the lithography is the use of angled evaporation. After the photolithography process the wafer is tilted when placed in the evaporator. The shadowing of the resist results in metal deposition over only a part of the opening, as illustrated in FIGURE 7.12. The angled evaporation does not lend itself to an easy way to produce a mushroom gate and a bilayer gate technique (described later in this section) will be useful. Such short gatelengths will suffer from high gate resistance, limiting performance. The main attractiveness of angled evaporation is for the demonstration of hero results when expensive, high-resolution lithography tools are not readily available.

Another method for shrinking gates is the so-called Y-gate illustrated in FIGURE 7.13. A photoresist is developed with a re-entrant angle for liftoff. Next, a low-temperature dielectric is deposited over the resist. Low temperature is needed so that the resist will not reflow and change its profile. The dielectric thickness is chosen to narrow the opening of the resist. Next, an anisotropic reactive ion etch is performed to expose the GaAs. The vertical etch rate is faster than the lateral etch rate, so the dielectric opening remains small compared to the resist opening. Then a recessed-gate etch may be performed, followed by metal deposition and liftoff. This method will result in a narrow footprint and wide top, similar to a mushroom gate. Like other methods of gate shrinking, this process can be difficult to control.

A last example of novel gate shrinkage is illustrated in FIGURE 7.14. A conventional dielectric is formed by lithography and RIE. SiO_2 is a good choice because it etches relatively slowly in F-containing plasmas. A metal such as WSi is deposited, patterned and formed over one edge of the dielectric by RIE. A metal sidewall is formed by anisotropic etchback using RIE. The dielectric is then removed with an HF-based wet etch. The length of this type of gate is more reproducible than other gate-shrink processes because the sidewall dimension is closely correlated with the deposited thickness. A drawback of this process is that no

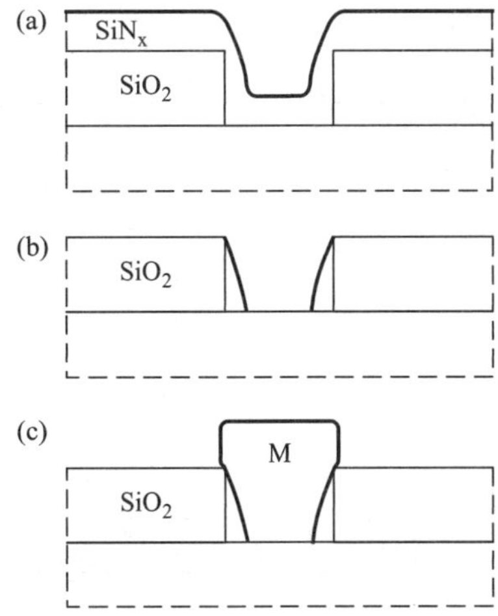

FIGURE 7.11 Illustration of a sidewall-based gate shrink process.

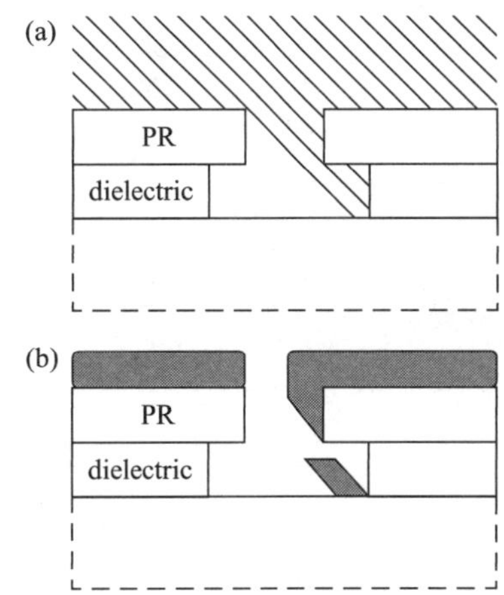

FIGURE 7.12 Illustration of angled evaporation as a gate shrink process.

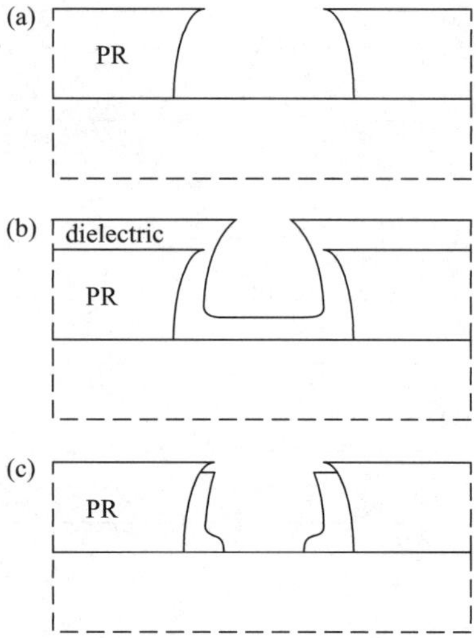

FIGURE 7.13 Illustration of Y-gate process.

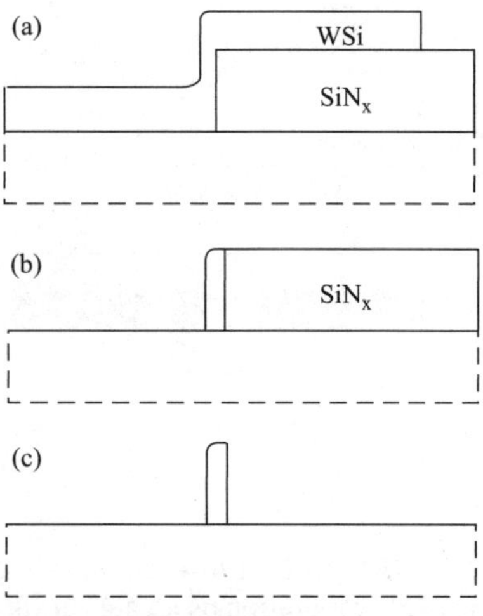

FIGURE 7.14 Illustration of sidewall gate process.

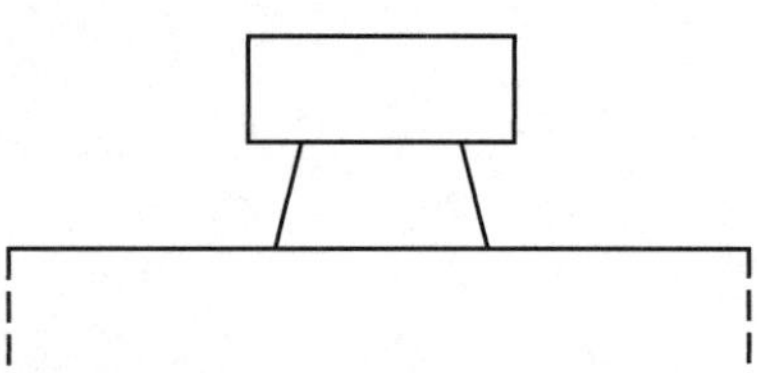

FIGURE 7.15 Illustration of bilayer gate process.

recess etching is possible. The process may be used for refractory-gate self-aligned FETs, but such short gates become very difficult to manage in a self-aligned process at gate lengths below about 0.4 μm. The encroachment of the self-aligned implants under the gate creates a parallel channel under the gate that is difficult to control. Also, the gate resistance of WSi becomes quite high when the gate is narrowed. A structure of this type will require some sort of bilayer gate, as described in the next set of examples.

A bilayer gate process is commonly used in self-aligned processes. The lower metal is chosen for its Schottky contact properties, while the upper metal is chosen to provide a lower gate resistance. A common choice is W/WSi, as illustrated in FIGURE 7.15. Both metals can be defined by F-based reactive ion etching. The W resistivity is about ten-fold lower than that of $WSi_{0.45}$. The WSi is sometimes undercut to provide a controlled spacing from a drain implant relative to the gate edge. Although liftoff is used to define an Au/Ti gate for recessed-gate structures, the Ti may also be undercut by means of an F-based reactive ion etch of the Ti, as for the W/WSi process. Other combinations of metals are also possible. This procedure is sometimes used as a method for tailoring the breakdown voltage.

Another type of bilayer gate process is illustrated in FIGURE 7.16. After a refractory gate (or angled evaporated gate) is formed first as the lower gate, a dielectric is deposited over the gate. This step is followed by a photoresist planarisation, which is used in a subsequent etchback process. Often a CH_4/O_2 gas mixture is used with the gas ratio and other plasma conditions adjusted for equirate etching of the photoresist and the dielectric. The etchback proceeds until the lower gate is exposed. Then after another photolithography step, the upper gate, usually Au-based, is deposited by evaporation in a lift off process.

A variation of the last two bilayer gate processes is illustrated in FIGURE 7.17. A bilayer structure using SiO_2/WSi is formed first by photolithography and RIE (FIGURE 7.17(a)). Then a second photolithography step is used to form an opening about the SiO_2/WSi structure with the thickness of the resist chosen appropriately for the next step (FIGURE 7.17(b)). The resist is then reflowed to close the gap between the resist and the bilayer structure (FIGURE 7.17(c)). The resist thickness is chosen so that the dielectric protrudes above the reflowed resist. The dielectric is then removed with an HF-based wet etch (FIGURE 7.17(d)), leaving a re-entrant profile. In the last steps of the bilayer gate process, an Au-based metal is deposited by evaporation for the final liftoff step (FIGURES 7.17(e)–(f)).

As can be seen by these examples, many creative processes can be used to achieve a given result.

7.4 ELECTRICAL CHARACTERISTICS OF GaAs SCHOTTKY CONTACTS

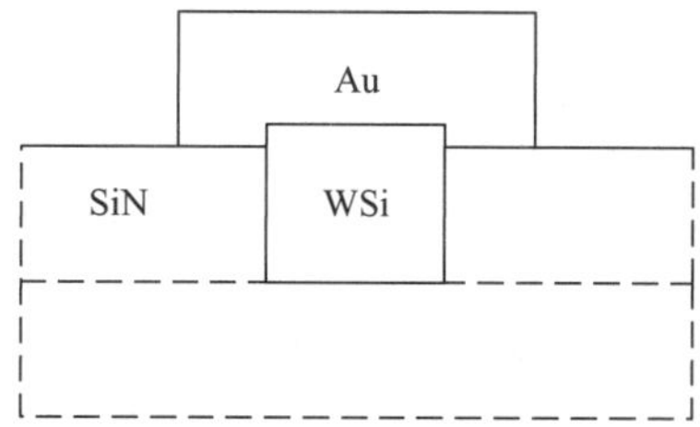

FIGURE 7.16 Illustration of bilayer gate process that utilises dielectric etchback.

Interface states play a major role in the electrical properties of Schottky contacts, as seen in Section 7.2.1. In turn, the treatment of the GaAs surface prior to Schottky metal deposition will play a part in determining these interface states. Even with clean surfaces, interface reactions of metal contacts with GaAs are the rule rather than the exception. Although the interface reactions can vary widely with the metal contact, they tend, in most cases, to produce interface states and densities that fill to approximately midgap. For these reasons, electrical properties of Schottky contacts with GaAs vary little in comparison with certain electrical parameters (such as metal work function, deemed important for contacts to other semiconductors (e.g. Si)).

More than 40 elemental metals have been reported as Schottky contacts to GaAs. Even more metal alloy combinations have been studied. In one heroic study, 43 elemental metal Schottky contacts were studied under the same experimental procedure for consistency of the GaAs, the GaAs cleans and the metal deposition procedure [10]. As has been discussed, consistency of the GaAs surface history prior to metal deposition is a necessary requirement for any hope of obtaining interface consistency. The GaAs was grown by MOCVD and doped to $1 \times 10^{16}\,\text{cm}^{-3}$. All depositions were carried out in an ultra-high vacuum system by electron-beam deposition or resistive heating. The results are summarised in TABLE 7.1. The Schottky barrier height, measured by the I-V

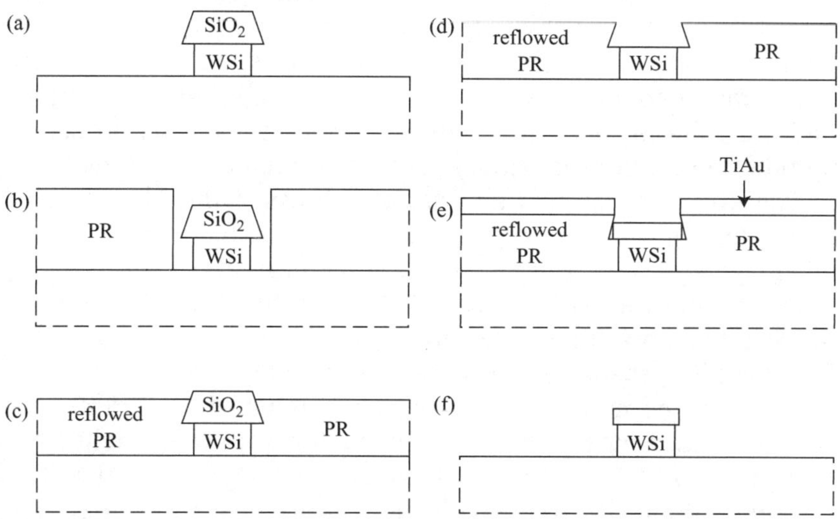

FIGURE 7.17 Another bilayer gate process utilising a photoresist reflow technique.

TABLE 7.1 Schottky contacts reported by Myburg et al. [10].

Contact	ϕ_B (I-V)	Contact	ϕ_B (I-V)
Mg	0.62	In	0.67
Al	0.77	Sn	0.71
Sc	0.71	Sb	0.97
Ti	0.83	Hf	0.81
V	0.80	Ta	0.78
Cr	0.80	W	0.79
Mn	0.82	Re	0.87
Fe	0.83	Ir	0.90
Co	0.83	Pt	0.99
Ni	0.83	Au	0.92
Cu	0.99	Pb	0.86
Zn	0.81	Bi	0.88
Ga	0.59	Pr	0.76
Y	0.71	Nd	0.76
Zr	0.77	Sm	0.75
Nb	0.77	Gd	0.75
Mo	0.87	Tb	0.75
Ru	0.86	Ho	0.74
Rh	0.89	Er	0.74
Pd	0.93	Tm	0.72
Ag	0.99	Yb	0.67
Cd	0.82		

method, varied only from 0.6 to 1.0 eV for all metals studied. For 39 of the metals, the variation was only from 0.7 to 1.0 eV. For the 14 metals most commonly used as single layer contacts to GaAs devices, ϕ_B ranges from 0.77 to 0.99 eV. The ideality factor varied from 1.01 to 1.03, a sign of high-quality contacts. Only a small correlation was found to exist between the Schottky barrier height and the metal work function, and only for a selected group of higher melting point metals. Compilations of Schottky barrier properties by other researchers generally arrive at similar conclusions.

The implications for FETs are as follows. Other things being equal, the higher Schottky barrier height is usually preferred, as it will generally produce lower leakage currents. In addition, it allows higher forward bias and higher currents for enhancement-mode transistors (those that are normally off at zero gate bias, Section 8.2.1) that are widely used in digital circuits.

It should be noted that the barrier heights in MESFETs will often be somewhat lower and ideality factors somewhat higher than reported in TABLE 7.1 because the channel doping of MESFETs is higher than $1 \times 10^{16}\,\text{cm}^{-3}$. In addition, surface conditions are never as ideal for FET processing as for Schottky barrier experiments and measurements in other laboratories can vary somewhat from those reported in TABLE 7.1. HEMTs and HFETs use higher bandgap materials (AlGaAs and InGaP) layers for the Schottky interface. Nevertheless, the values in TABLE 7.1 offer useful guidelines.

7.5 RELIABILITY OF GaAs SCHOTTKY CONTACTS

Issues that affect the reliability of GaAs-based Schottky contacts can be categorised in three ways: interfacial reactions, degradation of the interface and the underlying GaAs, and hydrogen poisoning of GaAs FETs. Other FET reliability issues (Section 8.8) may also involve Schottky contacts in more complex ways.

Interfacial reactions of the Schottky contact with GaAs are usually termed "gate sinking", to denote the consumption of GaAs and the associated movement of the Schottky interface. M-As or M-Ga intermetallics can form a new interface with GaAs. The consumption of GaAs results in electrical degradation of the FET, through threshold voltage shifts and current reduction. Some common interfacial reactions that occur in a short time period, generally several hours or less, were described in Section 7.2.2. Many of these same mechanisms will occur more slowly at lower temperatures according to the rate equation $R = R_o \exp(E_a/kT)$, but these need to be verified for specific cases. FET degradation (Section 8.8) is usually accelerated by either temperature or high electric fields, but it is generally accepted that gate sinking is purely thermally accelerated.

Most gate-sinking reliability studies have been carried out for TiPtAu gates, and we will summarise some of these here. Stress tests are usually thermally accelerated at temperatures above 225°C and often approaching 400°C. Gate sinking is positively identified by a physical test such as TEM or X-ray diffraction (Section 2.5) that reveals a degraded interface or new intermetallic reaction phases. Once a mechanism is known for a given metal system, electrical tests on Schottky diodes or FETs can be used to detect failure. As previously noted, gate sinking will result in a reduction in drain current for an FET, and it will result in a change in either the Schottky barrier height (ϕ_b), the ideality factor (n) or the doping level for a Schottky diode. TiPtAu gates have been observed to fail with an activation energy of 1.3–1.7 eV. Higher stress-test temperatures naturally produce shorter average times until device failure. The lower temperatures require longer stress times but give greater confidence that another mechanism with lower activation energy will not be the limiting one. It is possible that the range of activation energies represents differences in gate metal processes. However, some studies indicate that the higher activation energies represent gate sinking and the activation energy near 1.3 eV represents electrical degradation by carrier reduction [8]. The likely explanation for the carrier reduction is Ga outdiffusion during the initial interfacial degradation. Recall from Section 7.2.2 that Ga diffusion into metals is a commonly observed early step in M-GaAs interfacial reactions.

An extrapolated lifetime depends on the assumed operating junction (gate) temperature for the FET, which will usually be considerably higher than the ambient for a power FET and closer to ambient for a low-noise FET. A junction temperature of 125–150 °C is often assumed. Gate sinking mechanisms often result in extrapolated median lifetimes greater than a million hours, which means that they will not be a limiting factor in FET reliability. Based on this mechanism, TiPtAu gates are considered reliable by these criteria even though more stable alternatives exist (Section 7.2.2).

One critical process condition is the thickness of the layers. Ti and Pt thicknesses of 50–100 nm have been commonly used, but Ti as thin as 30 nm is considered reliable [11] and will subject the FET channel to lower stress.

In light of their excellent reliability against gate sinking, properly processed TiPtAu gates would not be considered a serious reliability concern were it not for hydrogen poisoning. Due to the use of hermetically sealed packages for high reliability space applications, hydrogen can become trapped in the package. Kovar material in the package is one source of the hydrogen. This hydrogen causes degradation, as described below, but many other sources of hydrogen exist, mainly from processing, and are difficult to control [12]. These include plasma processes such as PECVD (Section 3.3.6) and hydrogen ion implantation.

Hydrogen can take part in several processes that change Schottky barriers and FETs. It can passivate donors by bonding with unpaired electrons that would otherwise be conduction band carriers. A similar process works to passivate acceptors. Passivation of donors in FETs will reduce free carriers and cause threshold voltage shifts and a reduction in other electrical parameters such as the drain current. Pt or Pd are well-known catalysts for dissociating molecular hydrogen. When used as gate metal constituents they can dissociate molecular hydrogen that builds up in hermetic packages. The atomic hydrogen can then passivate donors in the channel, causing positive threshold voltage shifts and a reduction in channel current.

Another type of hydrogen poisoning has been observed in FETs without involving the Pt in the gate. TiH was observed by Auger spectroscopy (Section 2.5.2) in these FETs with hydrogen poisoning and was believed to be the cause of the degradation. TiH volume expansion and associated piezoelectric threshold voltage shifts were believed to be the cause of the hydrogen poisoning [13]. Since GaAs is a piezoelectric material, stress in the gate causes a change in the electric field in the GaAs, causing shifts in the threshold voltage of the FET. Hydrogen poisoning in TiAu gates generally causes a lower degree of degradation than Pt- or Pd-containing gates. In order to eliminate hydrogen poisoning, it is necessary to place a hydrogen getter in the hermetic package or to eliminate the Ti and the Pt from the gate-metal stack. At this time, it is not clear if hydrogen poisoning is a serious concern in non-hermetic packages. Nevertheless, many manufacturers are switching (or considering it) to WSi-based contacts, even for low-temperature processes.

7.6 CONCLUSION

Schottky contacts are one of the most widely studied aspects of GaAs technology. In part, this is due to the fact that no single Schottky contact satisfies all of the requirements for an ideal Schottky contact to GaAs. Another factor driving more research has been the desire for higher Schottky barrier heights. Yet another factor has been the poorly understood nature and reproducibility of the metal/GaAs interface. In spite of the existence of contacts with greater interface stability, TiPtAu exhibits more than sufficient reliability against gate sinking in GaAs-based FETs. However, TiPtAu should be used with hydrogen getters for hermetic packages or be replaced with non-Ti and non-Pt materials. Many choices of refractory contacts can be used for high-temperature self-aligned FETs. Some of the best are $WSi_{0.45}$ and WSiN because these remain amorphous to sufficiently high temperatures. A wide variety of processing choices are available to tailor the gate structure to the desired application using available equipment.

REFERENCES

[1] S.M. Sze [*Physics of Semiconductor Devices* (John Wiley and Sons, New York, 1981)]

[2] R. Williams [*Modern GaAs Processing Methods* (Artech House, Boston, 1990)]

[3] C.Y. Chang, F. Kai [*GaAs High-Speed Devices: Physics, Technology, and Circuit Applications* (John Wiley and Sons, New York, 1994)]

[4] D.J. Coleman Jr., W.R. Wisseman, D.W. Shaw [*Appl. Phys. Lett. (USA)* vol.24 (1974) p.355]

[5] C. Fontaine, T. Okumura, K.N. Tu [*J. Appl. Phys. (USA)* vol.54 (1983) p.1404]

[6] K.B. Kim, M. Kniffin, R. Sinclair, C.R. Helms [*J. Vac. Sci. Technol. A (USA)* vol.6 (1988) p.1473]
[7] Y. Kitaura, T. Hashimoto, T. Inoue, K. Ishida, N. Uchitomi [*J. Vac. Sci. Technol. B (USA)* vol.12 (1994) p.2985]
[8] J.-L. Lee, J.K. Mun, B.-T. Lee [*J. Appl. Phys. (USA)* vol.82 (1997) p.5011]
[9] E. Nebauer, U. Merkel, J. Wurfl [*Semicond. Sci. Technol. (UK)* vol.12 (1997) p.1072]
[10] G. Myburg, F.D. Auret, W.E. Meyer, C.W. Louw, M.J. van Staden [*Thin Solid Films (Switzerland)* vol.325 (1998) p.181]
[11] B.K. Seghal, B. Bhattacharyam, S. Vinayak, R. Gulati [*Thin Solid Films (Switzerland)* vol.330 (1998) p.146]
[12] S.J. Pearton [*Int. J. Mod. Phys. B (Singapore)* vol.30 (1994) p.1247]
[13] R.R.Blanchard, A. Cornet, J.A. del Alamo [*IEEE Electron Device Lett. (USA)* vol.21 (2000) p.424]

Chapter 8

Field effect transistors

8.1 CHAPTER SCOPE

This chapter will cover most of the basic GaAs-related processing steps for GaAs field effect transistors, including most of the front-end steps. Active steps are those that relate to the fabrication of the GaAs field effect transistor and do not include the interconnections or the integration of passive elements. FET fabrication and operation require an understanding of the GaAs semiconductor surface and its interface to metal junctions for etching, doping, contact formation and some particular aspects of performance optimisation. This chapter will draw heavily on material presented in the preceding chapters.

The basics of field effect transistors and their different types are given in Section 8.2 [1–3]. That section starts with a tutorial and continues to relate FET operation to the processing methods, device geometries and materials. It also highlights the main reliability issues and relates them to FET structures, processing and materials. Sections 8.3, 8.4 and 8.5 present the details of the main active processing methods: ion implantation, ohmic and Schottky metal formation, recessed gate etching and passivation.

8.2 FIELD EFFECT TRANSISTOR BASICS

The field effect transistor was proposed long before it was demonstrated in 1952. In the 1920s, semiconductors were poorly understood because of poorly developed materials technology. Two of the greatest materials problems of that time were a lack of purity and a lack of understanding of the semiconductor surface and its oxides. Both of these areas saw great progress in the 1950s with the development of zone refining methods of purification, among other important developments. These developments were followed by great progress in the development and understanding of transistors, which ushered in the modern age of microelectronics that has so radically changed our society in the latter half of the twentieth century.

8.2.1 Field effect transistor tutorial

Our discussion of FETs begins with a simplified tutorial on the operation and characterisation of field effect transistors. Some knowledge of semiconductor physics is necessary, but the transistor action is described without any prior knowledge assumed. A more complete treatment of FET operation is given in references [1–3].

The field effect transistor can be thought of as a voltage controlled resistor. The transistor consists of three contact electrodes and a semiconductor channel. The FET is known as a majority carrier device, which means that transport is carried out by the type of carriers, electrons or holes, which comprise the majority of the carriers in the semiconductor. Either electrons or holes can traverse the semiconductor channel, although electrons do so with greater speeds. The three contacts are represented by the nomenclature of source, gate and drain, which are schematically illustrated in FIGURE 8.1. An electric potential is applied between the source and the drain, which provides an impetus for carrier transport, or current, to the drain along the semiconducting channel of the device. The source and drain contacts are ohmic contacts and provide low-resistance paths across the metal/semiconductor interface. Carriers for transport in the channel are provided by the electrical dopants that are located in the channel region and are generally absent outside the channel. The gate potential can modulate the carriers in the channel and, therefore, the magnitude of the current from negligible levels to the maximum level. In this way, the input voltage to the gate can cause the transistor to operate as a switch in digital circuits, or a signal amplifier in analogue circuits. Ideally, no current flows into the gate, because this results in a parasitic current which does not promote switching or signal amplification.

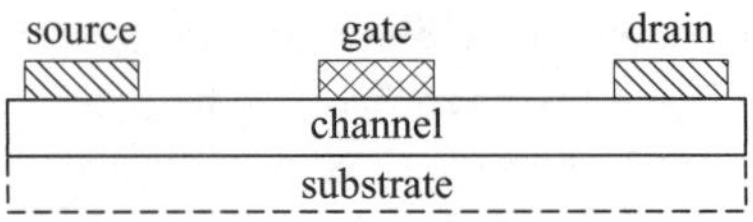

FIGURE 8.1 Schematic cross-section of a field effect transistor.

There are two main types of gate, the metal-oxide-semiconductor (MOS) gate, which forms the basis of the dominant CMOS Si technology, and the metal-semiconductor, or Schottky gate, which is used for GaAs technology. Transistor action in GaAs will be described in terms of the effect of a voltage applied to a Schottky contact. The detailed physical description of a MOS gate or other types of heterostructures with Schottky gates is somewhat different, but the end effect of depleting or adding carriers with the applied gate voltage is similar.

The basics of the Schottky contact are addressed in Chapter 7. The key concept for understanding transistor operation is that the potential of the Schottky contact (built-in potential plus applied voltage) creates a depletion region, so named because free carriers are depleted from this region and do not participate in the transport.

The depletion region can be made shallower by a positive forward bias for an n-type FET up to the value where the Schottky barrier is forward biased to turn on its diode. The depletion region can be increased in depth by negatively biasing an n-type FET. When the applied bias is sufficient to extend the depletion region completely across the doped channel region of the FET, the device is "pinched off" and current flow from the source to the drain is effectively stopped by being reduced to levels many orders of magnitude below the on-state currents. This voltage is known as the threshold voltage of the FET. Its value is an important parameter in the design of devices for specific applications and is given by the following expression for a MESFET with a uniform doping profile:

$$V_{th} = (-qN_D a^2/2\varepsilon) + V_{bi} \tag{8.1}$$

where q is the electron charge, N_D is the donor density in the channel, a is the thickness of the doped channel, ε is the dielectric constant of the semiconductor and V_{bi} is the built-in voltage. EQN (8.1) is derived from the expression for the depletion region of a Schottky contact (Chapter 7). HFETs, HEMTs and MESFETs with non-uniform channels will have other expressions for the threshold voltage, which need not concern us here.

The transistor operation of an FET can be better understood by examining some of the basic DC measurements. The most common DC measurements are the drain I-V characteristics, illustrated in FIGURE 8.2. The drain current is plotted against the drain voltage for a family of gate voltages which start with the

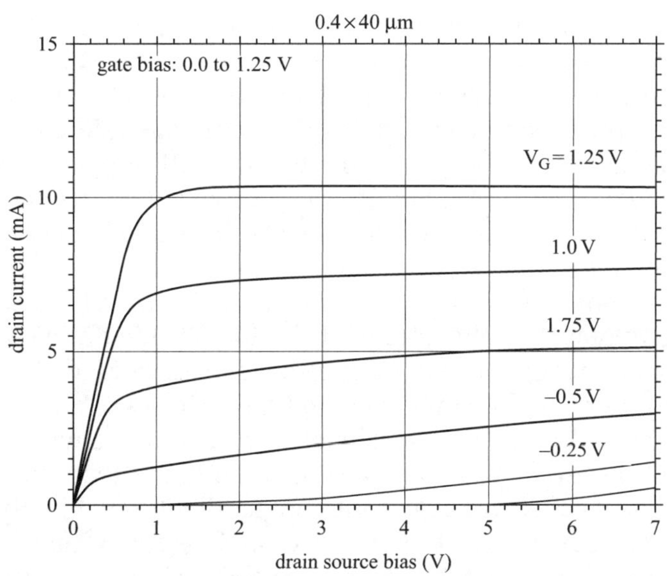

FIGURE 8.2 Drain I-V characteristics of an FET.

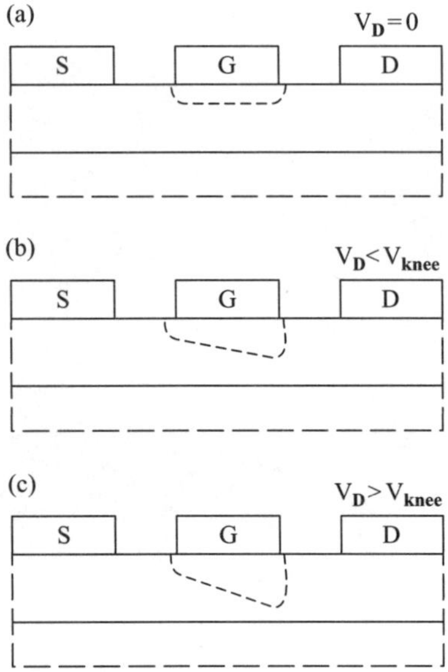

FIGURE 8.3 Depletion region of a gate at fixed gate bias and under various drain biases.

most positive (usually defined by the onset of gate current and corresponding to the highest drain currents). The gate voltage is stepped negative until pinchoff is reached. For a given gate voltage, the drain current increases initially with increasing drain voltage, reaches saturation at the drain voltage that defines the knee (the transition from linear to saturation currents) and is fairly steady for larger drain voltages. In the region of linear increase, the applied drain voltage leads to higher drain current through the following relation:

$$I_D = q\mu V_D \tag{8.2}$$

where q is the electron charge and μ is the electron mobility. As the drain bias is increased, electrons travel across the gate region with higher velocity. The gate depletion region is altered by the presence of the gate voltage, as illustrated in FIGURE 8.3. FIGURE 8.3(a) illustrates a gate depletion region for a transistor with no drain bias, while FIGURES 8.3(b) and 8.3(c) illustrate an FET with a drain bias below the knee voltage and above the knee voltage, respectively. The drain end of the gate experiences a more negative potential as a result of the drain bias, which results in a larger depletion region. The knee voltage results in pinchoff at the drain end of the gate. At this bias, the rest of the gate is screened from further drain bias increase and does not modulate significantly with higher drain voltages. Although the channel is pinched off in a small region along the drain side of the gate in this simple model, two-dimensional calculations show that a small undepleted channel exists, accounting for the non-zero current flow. Higher drain bias does not further deplete the channel and the current saturates. The depletion region can continue to extend towards the drain with further drain bias increase. This effect is considered in designing the crystal growth and in choosing processing steps in terms of how the gate edge region is doped in epitaxially-grown structures or self-aligned structures. This model of MESFET action is the Shockley model and generally describes long-channel devices [4]. In short-channel devices, velocity saturation also plays an important role.

Apart from knee-voltage determination, the drain I-V characteristics give other insights into the FET. Ideally, the drain current in the saturation region is flat. In short-channel devices, the drain current may increase with drain bias and special designs are employed to minimise this. In large-periphery devices, a decrease in drain current with drain voltage could be a thermal effect due to self-heating; this can be managed with proper diagnosis. Finally, kinks in the drain current as a function of drain bias can be an indication of surface or trap-related anomalies that affect performance or reliability.

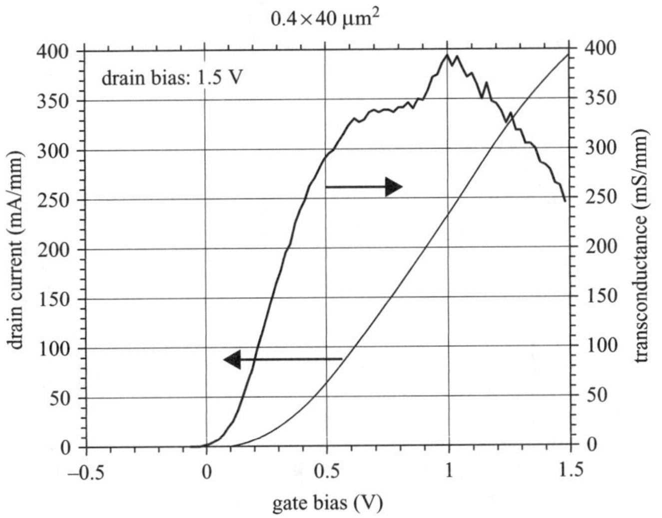

FIGURE 8.4 Plot of drain current and transconductance versus gate voltage.

Another common DC measurement is illustrated in FIGURE 8.4. The drain current is plotted against the gate voltage. The slope of the drain current gives the transconductance, g_m, of the device. High transconductance is important as it is directly related to the speed of the device by the relation

$$f_t = g_m/2\pi C_{gs} \tag{8.3}$$

where f_t is known as the cutoff frequency (described shortly) and C_{gs} is the gate-to-source capacitance. High transconductance also helps to increase the power density and maximum power output of a microwave amplifier. It can also minimise the transistor size required to switch a given current in a high speed digital circuit. The I_D–V_G plot is also the most useful way to determine the threshold voltage of an FET and to examine the subthreshold current characteristics. Sub-threshold current analysis is most useful if it is viewed as a log plot, as in FIGURE 8.5. One strives to maximise the logarithmic slope and minimise the lowest current level.

The measurements of FIGURES 8.2 and 8.4 have been described here as DC measurements, but they may also be made with short-current pulses (as short as 100 ns) to distinguish between thermal and trapping effects. Trapping effects play a large role in many aspects of FET performance and degradation, as will be described in Sections 8.2.3 and 8.8. Pulsed I-V measurements are critical in analysing these effects.

GaAs FETs are most commonly used in high-performance applications, such as amplifiers that involve high frequency, low

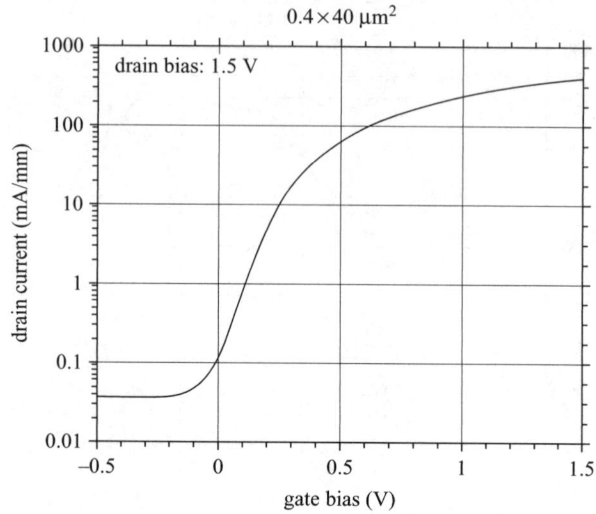

FIGURE 8.5 Logarithmic plot of drain current versus gate voltage.

noise or power with high linearity. For devices and circuits, time-dependent current-voltage characterisation is sufficient to completely describe a circuit. This method of characterisation works well if the frequency is low enough that the device (circuit) dimensions are small compared to the wavelength (the inverse frequency) at the frequency of interest. At microwave frequencies ($\lambda = 1$ cm at 30 GHz), this assumption breaks down and it becomes impossible to get accurate time-dependent I-V data. The high-frequency performance is usually characterised with scattering parameter (S-parameter) testing. An S-parameter test is one where the intensity of scattered and reflected waves for an incident (usually high-frequency) signal is measured at each port for an arbitrary n-port network. The FET is represented as a 2-port network for this type of characterisation, with ports at the gate and drain. The S-parameters can be converted to several other domains, where they can be related to voltages and currents in the device. The HBT (Chapter 9) is characterised with ports on the base and collector. For more details on high-frequency characterisation, the reader is referred to [1].

Equations are readily available to transform S-parameter data to other forms that allow one to derive high-frequency parameters of interest. The y-parameters transform the currents in a device to voltages. The z-parameters transform voltages in a device to currents and h-(hybrid) parameters perform a mixed (current, voltage) transformation. When this transformation is carried out for S-parameters to h-parameters, h21 describes the current gain of a device. When plotted as a function of frequency as in FIGURE 8.6, its extrapolation to unity current gain defines

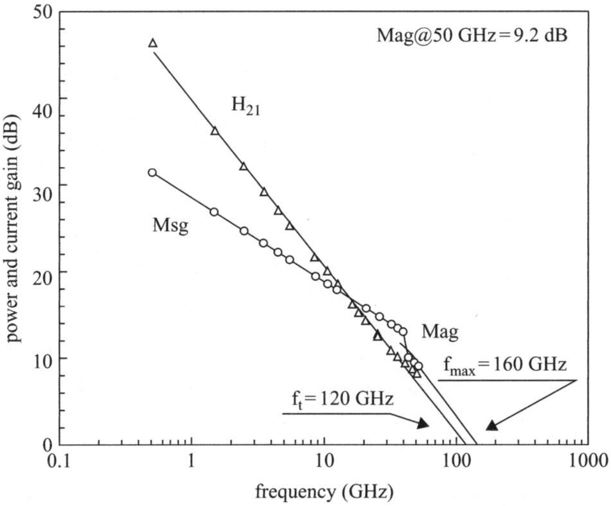

FIGURE 8.6 Example plots used to extract f_t and f_{max} in a FET. (*IEEE Microwave and Guided Wave Lett.* vol.9 (1999) p.28–30, reprinted with permission.)

the f_t of a device. The f_{max} is defined as the extrapolation to unity of the unilateral power gain, U, or alternatively as the maximum stable gain. Likewise, equations are readily available to transform S-parameter data to U. Extrapolation to unity gain is done using a 20 dB/decade slope, rather than a direct extrapolation of data as seen in FIGURE 8.6. Many device researchers focus on high-frequency measurements such as f_t and f_{max}, but other high-frequency measurements such as amplifier gain and noise figure are important as well.

FETs are often electrically described as a circuit in order to write equations that describe their physical or electrical behaviour. Such a description is termed an "equivalent circuit" and is used to predict the behaviour of circuits built from FETs. An example of an equivalent circuit description of an FET is shown in FIGURE 8.7. In this example, some of the equivalent circuit elements have an obvious correlation to a physical part of the transistor. For example, the source and drain ohmic contacts are represented by the resistors, R_S and R_D. However, in other cases, no direct physical correlation is made. Equivalent circuits used for the prediction of circuit operation should accurately model the transistor with a minimum of components so that circuits with high complexity can have manageable computational requirements. On the other hand equivalent circuits whose intent is to accurately model the physics of the FET often tend towards more complex models with as many physically based parameters as is practical. Equivalent circuit models for FETs are usually built from a combination of DC, pulsed and S-parameter measurements.

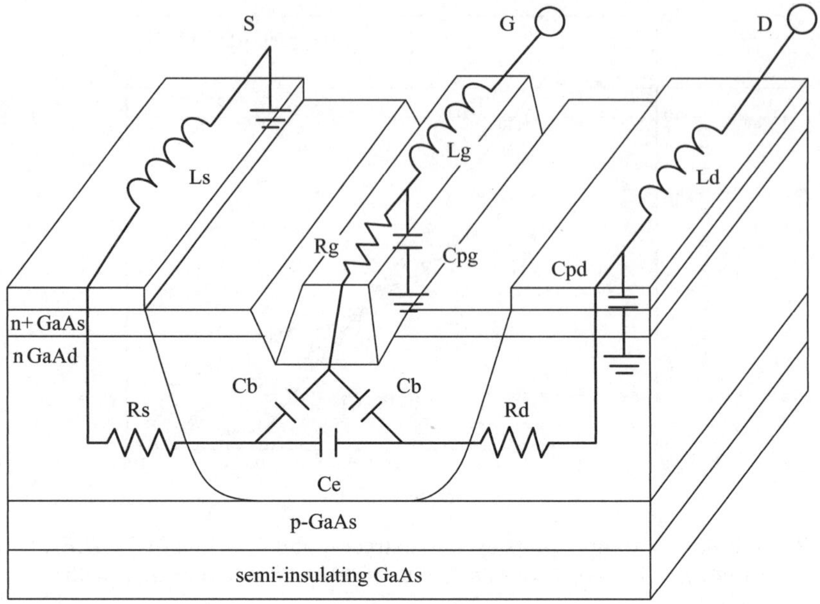

FIGURE 8.7 Example of an equivalent circuit description of an FET. (*IEEE Trans. Microwave Theory Tech.* vol.49 (2001) p.1410–18, reprinted with permission.)

8.2.2 Field effect transistor performance and reliability issues

FETs have performance and reliability issues that are often interrelated and so are presented here in the same section. Ideally, FET dimensions and materials properties would determine performance limits. Dimensions would systematically be reduced through improvements in processing equipment. Even using lithography tools two to three generations behind the Si world, GaAs technologies compete with current Si devices in certain measures of performance such as linearity and power-added efficiency. Materials choices impact performance and reliability through their inherent physical properties and complex interactions. Materials scientists have been engineering GaAs-based materials for several decades now, and much has been learned.

Thermal effects are extremely important considerations for high-performance FETs. The performance of an FET generally reaches its highest level at biases that give the highest transconductance (refer to FIGURE 8.4). The power dissipation of the transistor (voltage-current product) leads to heat flow away from the transistor as well as self-heating of the device. Accordingly, a given current level will lead to greater self-heating at higher drain bias. This is seen in the I-V characteristics when they are taken to higher drain bias, as in the slight droop of the top curve near

7.0 V in FIGURE 8.2. The performance of an FET as measured by f_t and f_{max} will generally decline as a function of drain bias due to self-heating effects. The peak in f_t is often near a drain bias of 1.0 V for GaAs FETs.

Not surprisingly, FET degradation is accelerated by temperature and is impacted by the self-heating effects that lead to an increase in the junction temperature of an FET. Calculation of the junction temperature depends on how the power is thermally dissipated. The following equation describes the relation between the temperature rise and the dissipated power for an FET in thermal contact with a perfect heat sink, i.e. one that does not experience a temperature rise in response to a thermal load:

Note: The term junction temperature comes from HBT terminology and is widely used by FET engineers as well. Strictly speaking, "channel" or "Schottky" temperature would be more accurate descriptions of FETs.

$$\Delta T = R_{th} I_D V_{DS} \tag{8.4}$$

ΔT is the junction-temperature rise over the heatsink temperature, R_{th} is the thermal resistance of the semiconductor and I_{DS} and V_{DS} are the drain current and drain voltage. The thermal resistance R_{th} is dependent on the geometry and the thermal conductivities of the materials in contact with the FET.

GaAs is a poor thermal conductor compared to Si (by a factor of three) and to some of the metals commonly used as interconnects, packaging materials and heatsinks. Junction temperatures for GaAs FET power amplifiers can commonly attain and even exceed 150°C, although smaller devices in other types of analogue and digital circuits can operate at lower temperatures. High junction temperatures have important implications for device, processing and reliability engineers. First, electron transport degrades with increasing temperature. The thermal energy of common lattice vibrations (phonons) increases with temperature and causes more electron scattering. Both electron mobility and saturated velocity decrease with increasing temperature. These factors negatively affect the transit time of FETs and therefore their high-frequency performance.

As an example of the potential for thermal effects to affect performance, consider a high-linearity FET power amplifier. The output power at a given frequency needs to be linearly proportional to the input power and meets this requirement at low input levels. At high enough input levels, the dissipated power of the device causes significant self-heating, which in turn degrades electron transport. The gain of the device drops and the output power is no longer linearly related to the input. Unless the heat can be managed better to control the junction temperature, the linear output power limit will have been reached.

High junction temperatures also have the potential to accelerate the aging of some of the materials of the FET. As discussed in

Chapter 6, reactions of GaAs with Au, Ni and other ohmic contact constituents occur within minutes and sometimes within seconds at temperatures of 300°C and above. It is not unrealistic to expect that some of these reactions can occur at temperatures of 150–200°C after a sufficient time. The same is true for thermal acceleration of the metallurgical reactions of some Schottky contacts.

Thermal-management solutions are sought to keep device temperatures below this upper junction temperature limit. Some common techniques in practice are the thinning of the GaAs substrate, sometimes to as thin as 50 μm, and judicious spreading of the gate fingers.

Another important performance issue relates to discrepancies found between low- and high-frequency measurements. Well-optimised devices perform in good agreement with physically-based models that set performance expectations for material parameters such as saturated velocity and mobility. These parameters are related through well-understood physical derivations to the current, transconductance and transit time in an FET. However, devices often show deviation from expected values of some of these parameters at high frequencies, an effect known as frequency dispersion. Output conductance (the slope of the drain I-V curve in the saturation region) and transconductance of all GaAs FETs show frequency dispersion to some degree. Those FETs and their processes or materials that display low degrees of frequency dispersion have obvious advantages for the companies that sell them.

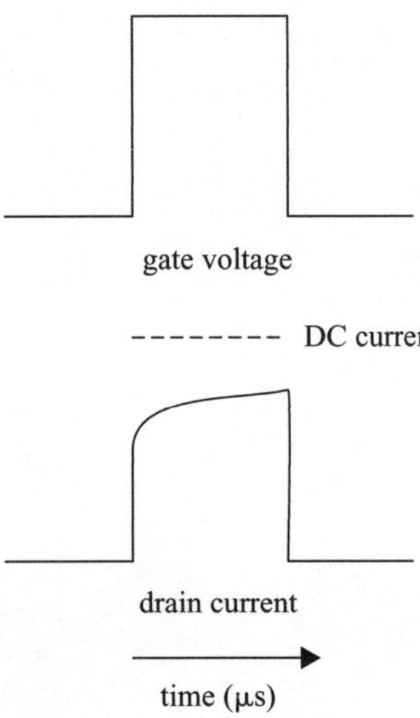

FIGURE 8.8 The current in an FET may not follow the voltage pulse on a fast timescale. This effect is known as "gate lag".

A performance issue probably related to frequency dispersion is known as "gate lag", which is illustrated in FIGURE 8.8. An arbitrary combination of gate voltage and drain voltage will imply a given drain current according to an FET's expected characteristics, such as those illustrated in FIGURE 8.2. However, when such voltages are applied to an FET with a fast pulser and viewed on a fast oscilloscope, the current response does not reach the expected DC current on a timescale that is very fast compared with the electron velocity. For example, an FET that is biased in the "off" state receives a fast pulse to the gate biasing the FET to the "on" state. A fraction of the expected DC current is measured on a short timescale (1 μs, for example), while the remainder of the current takes a much longer time (milliseconds, for example) to appear. The time lag in recovering to the expected DC current level is called "gate lag" if the gate is pulsed and "drain lag" if the drain is pulsed.

Electron traps are acknowledged to be responsible for many of these transient effects. Recall that a trap is a midgap state that is not mobile. Free electrons will respond rapidly to a change in electric field and the free-electron currents will adjust on a timescale of the order of the transit time of an electron (picoseconds). However, the traps will respond to the new conditions with much slower time constants because of the need to overcome an activation energy for the carrier to "de-trap" and become mobile. Also, fixed charge associated with the traps creates depletion regions that reduce the number of mobile carriers.

We shall now describe a few ways that traps negatively affect device performance and applications. Specific FET manifestations of some trapping effects will be discussed in Section 8.8. The crux of the problem is that both analogue and digital circuits can have rapid changes in bias conditions that can trap electrons during one part of their operating cycle and slowly release them in another part of the cycle. If large enough in magnitude, the effects can lead to unpredictable logic errors in digital circuits. In a microwave amplifier with drive conditions large enough to fill trap states in part of the cycle, the circuit operates under a continual state of trapped charge with the RF currents, and therefore power, significantly lower than those implied by DC measurements. Many high-performance metrics scale directly with the extent of frequency dispersion, and gate lag can be a useful screening test for bias conditions relevant to the application.

Many of these trap-related effects can be reduced or exacerbated by common process choices and must be considered by process engineers. Gate geometries and passivation are examples of these process choices and are described in Sections 7.3 and 8.7. The trap-related performance issues become degradation and reliability issues (Section 8.8) when long-term operation can cause permanent changes in the trapping effects and lead to a performance decline.

One big factor in FET performance is parasitic capacitance between device elements and the substrate. These effects are minimal in semi-insulating GaAs and other compound semiconductors, but are an important factor in Si-based technology. A semi-insulating substrate is needed for low-loss microwave integrated circuits on GaAs and reduced parasitics in high-speed digital circuits. The differences between GaAs and Si substrates are a result of both the physical properties and the crystal growth. GaAs has a higher bandgap than Si and consequently has a lower intrinsic carrier concentration. Both Si and GaAs can incorporate impurities at a parts-per-billion level that dope the material and can lead to uncontrolled substrate resistivity and doping type. Si technology deals with this by using substrates with controlled doping and by constructing pn junctions for backside isolation from the substrate. GaAs technology deals with the background doping issue by incorporating deep levels in the crystal growth to compensate the background doping. The most common deep level in GaAs is a crystal defect known as EL2, which is incorporated in concentrations near $10^{16}\,cm^{-3}$ and can vary over an order of magnitude depending on the crystal growth method.

Non-ideal scaling of FETs also affects performance. Ideally, device geometry is optimised for an FET and increases in performance occur stepwise in succeeding generations of process technology through proportionally reducing all of the key equivalent circuit values (mostly resistances and capacitances) that affect FET performance. Scaling the FET in this manner usually requires a proportional reduction in all device dimensions and generally requires increases in effective doping or charge density. Strictly proportionate scaling of GaAs FETs is no longer possible because many of the material, processing and physical limits are being approached, especially for the vertical dimensions (e.g. channel thickness) and dopant concentrations. Consequently, generational performance increases now require complex tradeoffs between device design and processing. Non-ideal effects arise from shortening the gate length of the device without proportionally reducing all equivalent circuit parameters; these are termed "short-channel" effects. Improved processing strategies are needed to address short-channel effects.

Much more can be written about performance, especially in the context of specific applications that may involve circuit speed, noise performance, power-added efficiency and linearity. Some of these topics are specifically addressed in sections where process choices affect them. Others are not addressed in this book.

8.2.3 Field effect transistor structures and materials

Two basic types of transistors are made with GaAs materials: heterostructure field effect transistors (HFETs) and metal-semiconductor field effect transistors (MESFETs). MESFETs utilise a Schottky contact for the gate and the semiconductor portion of the device is entirely GaAs. Heterostructure FETs contain one or more heterojunctions in the material structure, as in AlGaAs/GaAs high-mobility electron transistors (HEMT), for example. A HEMT is a special case of an HFET, where the donor ions reside in a high-bandgap layer (most often AlGaAs or InGaP) and the mobile carriers reside in the adjacent low-bandgap layer (most often GaAs or InGaAs), as illustrated in FIGURE 8.9. Often the low-bandgap material is also doped or other adjustments are made to the classic HEMT structure, and these devices are generally referred to as HFETs.

The classification of FET structures can be more complex, but we will generally refer to two basic types: recessed-gate structures and planar structures. The geometries, types of structures

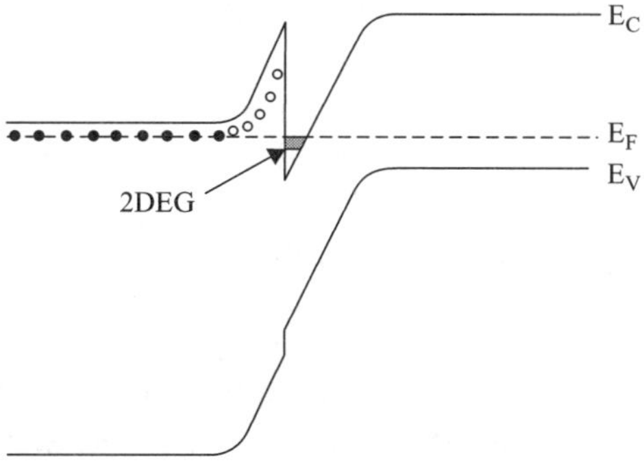

FIGURE 8.9 Illustration of the conduction and valence bands of a two-dimensional electron gas used in HEMTs.

and common methods of processing these gates are presented in Chapter 7.

Typical HEMT and MESFET structures are illustrated in FIGURE 8.10. The active layers tend to be very thin, which places constraints on the manufacturing method for forming the structure. Ion implantation has historically been the most cost-effective method of forming the doped regions for GaAs MESFETs (FIGURE 8.10(a)) and has contributed to their entry in high-volume commercial markets. However, ion-implantation doping is a less precise method for forming doped regions of FETs and is less applicable to very thin active layers for high-performance devices. Historically, epitaxy has been used for high-performance specialty markets (military, test equipment) and ion implantation for high-volume commercial markets. In recent years, the production of epitaxial substrates has become much more cost-effective and the performance/cost tradeoff balance has tilted in favour of epitaxial substrates even in many commercial markets.

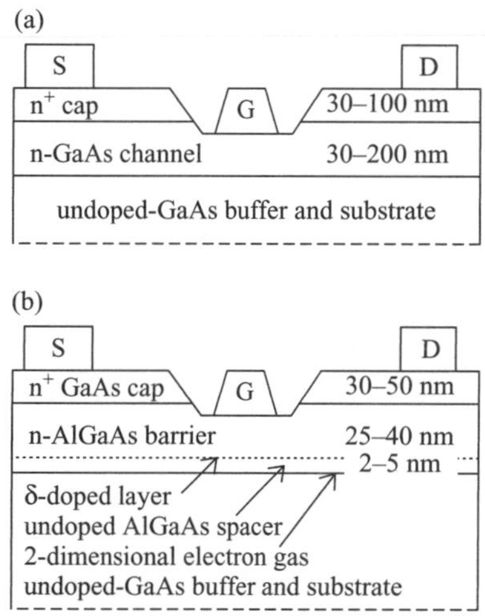

FIGURE 8.10 Illustration of typical structures for (a) MESFETs and (b) HEMTs.

Several types of HEMTs (FIGURE 8.10(b)) are in commercial use and many variations are subjects of continuing research. In the pseudomorphic HEMT (PHEMT) structure, a strained InGaAs layer is inserted between the AlGaAs/GaAs channel interface of FIGURE 8.10(b). The InGaAs alloy has better transport properties (higher mobility and saturated velocity) than GaAs, which lead to higher f_t and f_{max}, and better noise properties. Strained InGaAs will relax by forming misfit dislocations and render the PHEMT unusable if the In mole fraction or the total thickness exceed a critical value. InGaAs with 20% In and 15 nm thickness is commonly used. PHEMTs were initially developed for military and space applications due to the higher cost of MBE epitaxy in the early 1990s. They have since become prevalent in consumer applications due to cost reductions in the epitaxy, now that high-throughput MBE and MOCVD technologies have matured.

Metamorphic HEMT is a term used to describe HEMTs with a wide range of InGaAs compositions that are grown on GaAs substrates. Whereas PHEMTs operate with less than about 20% In, MHEMTs range from 30 to 60% In. The MHEMT approach consists of growing a special buffer directly on GaAs that is graded gradually in lattice constant from the GaAs value to the value for the desired InGaAs channel composition. Generally, the buffers are grown with stepped changes in composition as the growth progresses. As each layer of the buffer exceeds a critical layer thickness, the strain is relieved with misfit dislocations which coalesce into threading dislocations as the growth progresses. The attractive aspect of these step-graded buffers is the annihilation of dislocations through abrupt changes at the interfaces between buffers that change compositions. This improves the quality of the material, although not to the level of GaAs-on-GaAs growth. MHEMTs have been shown to produce noise performance consistent with that expected for the In composition in the InGaAs channel. At this time of writing, MHEMTs still face challenges in becoming qualified for power-amplifier applications.

MESFET structures are defined by the doping profile and the geometry of the gate-source and gate-drain region. The former is the subject of Section 8.3, while the latter was addressed in Section 7.3. The design of the MESFET channel is based on straightforward considerations of the depletion-region calculations derived from expected doping profiles from epitaxy or ion implantation. Textbook calculations or one-dimensional simulations give only a very basic estimate of a few parameters such as the threshold voltage. Better results are obtained from two-dimensional simulation where both the channel and the source and drain regions are considered. These methods are used mostly as a first-pass design tool for new structures or analysis of process deviations or improvements. Detailed analysis of devices for circuit design more commonly uses equivalent circuit models that analyse both DC and RF data.

HEMT structures are also defined through one- and two-dimensional simulations. The AlGaAs/GaAs and InGaP/GaAs heterostructure interfaces form triangular wells that will have quantised energy levels for electron occupation. InGaAs channels will have square wells. Calculation of the potential well and the position of the energy levels requires simultaneous solution of the Poisson and Schroedinger equations, a capability that is now available with commercial simulation packages. As in the case of MESFETs, one-dimensional results can give basic starting guidelines and two-dimensional simulations are used for more detailed structure and process analysis.

Many of the device performance metrics can be directly related to the design of the material structure and to the quality of the growth. Uniformity of the doping and the active layer thickness are critical quality factors. Contactless resistivity maps are a useful tool for assessing uniformity because the technique is fast, inexpensive and non-destructive. Hall mobility is another useful technique which is used to deconvolve doping and mobility from the sheet resistance.

Purity of the GaAs substrate and epilayers is extremely important. Impurities can change doping levels or affect free carrier levels through changes in the Fermi level far away from the device (at the substrate/epilayer interface, for example). During epitaxial growth, many possible sources of impurities exist including chamber or gas line leaks, outgassing from internal chamber parts, purity of source materials, etc. Again, Hall mobility is useful for the reasons outlined above, while SIMS analysis (Section 2.5.3) can identify impurities at the parts per million level. Unfortunately SIMS only has a $10^{16}\,cm^{-3}$ detection limit and FETs can be sensitive to lower impurity levels than that.

Traps are a major factor affecting performance and the growth process (for epitaxy) or substrate quality (for ion implantation) is critical. Optimisation of epitaxial growth of III–V materials is always a tradeoff between several competing factors. Higher temperatures favour thermodynamic equilibrium and adequate surface mobility of reacting precursors at the expense of high desorption of some atoms (In is the most problematic in this regard) and the requirement of high equilibrium overpressures of the group V element. Other growth parameters have tradeoffs as well. Traps are not always well characterised, but in terms of crystal structure, they are always imperfections. Hall mobility is an excellent characterisation tool for the growth process because both contaminants and crystal imperfections degrade mobility, while their minimisation leads to high mobility.

Interface abruptness is another critical issue for epitaxy. The ideal interface transitions abruptly from one compound semiconductor alloy to another over one monolayer. Two types of non-abruptness are common. In one type, the transition occurs gradually over several monolayers. In the second type, the non-abruptness comes from intermixing of the two layers. The former type of non-abruptness happens when the arrival of the different reactants at the growing surface is not switched rapidly enough. This situation was common in MOCVD in the early days of AlGaAs/GaAs and AlGaAs/InGaAs HEMTs, but has since been remedied. The latter type of non-abruptness is mainly a problem for As/P interfaces and does not seem to negatively affect (Al, Ga, In) arsenide alloys. The lack of interface abruptness is difficult to analyse from the device standpoint and it certainly makes experimental data difficult to compare to simulation. TEM analysis of the heterointerface is the most accurate way to characterise the abruptness of the interface. However, it is probably true that higher mobility results from better interfaces, where carrier scattering is reduced.

The true significance of any of these material issues lies in the final FET performance. The ongoing responsibility of the device engineer is to make detailed analysis of the device performance, try to correlate it with growth and fabrication processes, and feed back the results to growth and process engineers to make improvements.

This brief overview of materials and FET structures is intended to place the subject in context for the following detailed presentation on FET front-end processes. More detailed treatments can be found in [1–3].

8.2.4 Overview of field effect transistor fabrication

A recessed-gate FET process sequence is illustrated in FIGURE 8.11. The process may start with SiN_x passivation of a GaAs substrate (for ion implantation) or an epitaxially grown wafer (FIGURE 8.11(a)). Device suppliers usually require the substrate or epitaxial wafer qualification to the vendors of such wafers (or the in-house suppliers). The wafers as received are generally "epi-ready" with desirable surface properties. A first-step passivation is used to encapsulate the surface and prevent processing-related variability in the surface states. Proprietary surface cleans were once used to establish a high-quality surface, but have largely been

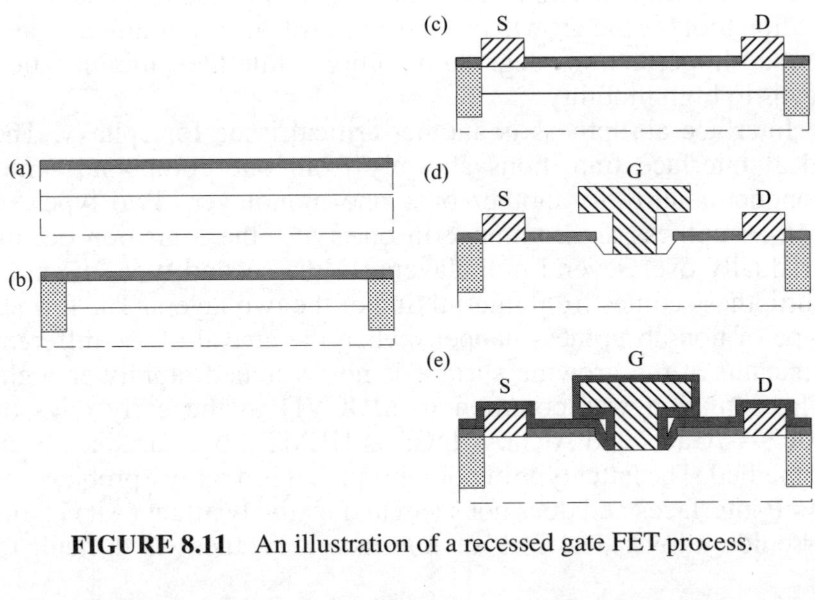

FIGURE 8.11 An illustration of a recessed gate FET process.

relegated entirely to wafer suppliers or have been greatly simplified from what was reported in the literature in the past. A surface-oxide removal step with or without subsequent oxygen treatment may be used prior to silicon nitride deposition (refer to Section 6.3.1 for details of this type of surface treatment). PECVD is the most common method of deposition of silicon nitride (Section 3.3.6). With the protective silicon nitride surface in place, all lithography steps use coatings on the silicon nitride layer and contamination of the GaAs surface is potentially eliminated. Processing of contacts to the FET are made through openings in the silicon nitride at the appropriate step.

At this point, the wafer is doped by ion implantation, if the active layers were not formed by epitaxy. After forming a set of alignment marks, the wafer is patterned, ion implanted and annealed after removal of the resist (Section 8.3 and FIGURE 8.11(a)). An optional implant-isolation step follows, where ion-induced damage reduces the substrate conductivity to very low levels (FIGURE 8.11(b)). This step may be omitted if the resistivity of the starting substrate is suitably high or a mesa etch may be used. Formation of the ohmic contacts is typically the next step (FIGURE 8.11(c)). A photoresist pattern is formed, followed by the removal of the silicon nitride in the open areas. A wet process for removing the silicon nitride would use a dilute HF mixture. A small amount of surfactant might be included since features roughly the size of a water droplet (near 50 μm) do not tend to wet reliably. The FET ohmic contact areas are much smaller and do not suffer from this problem. A dry process such as RIE may also be used. The ohmic contacts are then deposited, alloyed and tested (Section 6.3). The gate is patterned next. Again, the silicon nitride is removed prior to a gate recess etch, which can be either a wet or dry process (Section 7.3.1). Although the process sequence of FIGURE 8.11 is shown for a single gate recess, a double-recess process is used when higher breakdown voltage is required. The gate metal is then deposited (FIGURE 8.11(d)). At this point, an in-process DC test of the transistors is usually carried out to screen the wafers and flag process problems. Also at this point, another silicon-nitride passivation step may be carried out (FIGURE 8.11(e)), since the surface GaAs region between the edge of the gate and the edge of the recess can greatly affect FET performance. These process steps comprise the "active" processing steps of a recessed gate FET. The remaining frontside processing steps are used for the formation of passive elements and interconnects, and adding more conductive layers to ohmic or gate metals. Often a wafer thinning process including backside via formation and an electroplating to form backside ground plane follow the frontside processing.

Many variations of the recessed-gate FET process are possible. Most of these are unique processes that are required for specific gate metal structures or novel methods of reducing gate dimensions. Other types of recessed-gate processes include some of the following fabrication methods without the initial silicon nitride dielectric and etching mesas instead of using a planar process.

Critical steps in the fabrication of recessed-gate FETs are the gate lithography and the recessed-gate etch. Because the high-frequency performance of an FET is most strongly affected by the gate length, large investments are made to shrink gates in high-performance FETs. Traditionally, electron beam lithography has been used for gate-length reduction below 0.5 μm in military applications because the resolution of commercial electron beam writers is in the 20–50 nm range. However, electron beam lithography will impose high cost and low throughput, so cost-effective alternatives have always been desired. Today, most commercial fabrication of FETs relies on optical stepper lithography and phase-shift techniques. This technology receives continual attention from the commercial silicon equipment vendors and its capability has been improving gradually over the years. As of this time of writing, optical lithography is transitioning to sub-0.1 μm features with 190 nm light sources (deep UV) in the Si world, while older generation steppers used in the GaAs world are capable of 0.25 μm features with 365 nm light and phase-shift techniques. This trend continues to make optical lithography more attractive relative to e-beam lithography.

From FIGURE 8.10, it is clear that HEMT structures (for recessed-gate FETs) use highly doped cap layers. This situation is typical even for FETs made by ion implantation: it allows easy ohmic contact formation and provides low access resistance from the source ohmic contact to the channel region under the gate. In the recessed-gate process, the cap layer will be etched away so that the Schottky gate will contact a less heavily doped region. The depth of the etch, along with the doping and thickness of the layer under the cap, strongly affects the threshold voltage of the FET. No good measurement techniques exist to directly measure the depth of the etch and most etch methods lack the precision to maintain adequate manufacturing control, so other methods are used to determine the proper etch depth. The most common method is to monitor the "gateless" drain current during one or several interruptions of the recess etch. The gateless drain current can be correlated with the FET characteristics of a completed device to determine target current levels. The art associated with gate recessing as well as etch methods were the subjects of Section 7.3.1.

Another type of FET process is the self-aligned process. One illustration of self-aligned processing is shown in FIGURE 8.12.

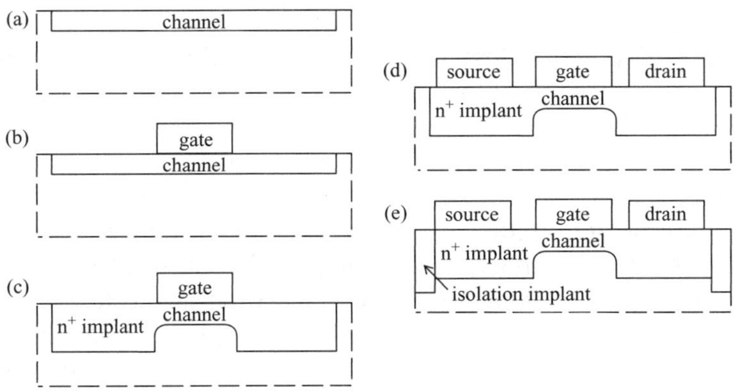

FIGURE 8.12 An illustration of a self-aligned gate FET process.

The first step would be the formation of alignment marks followed by ion implantation of the FET channel (if required, FIGURE 8.12(a)). This process may be repeated if desired for FETs with different threshold voltages on the same wafer. The next step is the formation of a refractory gate by sputter deposition (usually), patterning and reactive ion etching (FIGURE 8.12(b)). Various alternative metals such as WSi, WN and TiW are options at this step. After patterning the boundaries of the source and drain regions with photoresist and using the gate as an implant mask, the source and drain regions are formed by ion implantation (FIGURE 8.12(c)). After photoresist removal, the implants are activated with a high-temperature anneal, usually in the range of 800–900°C. Variations on the self-aligned gate process were the subject of Section 7.3.2. Next comes the patterning and deposition of the source and drain ohmic contacts and the ohmic alloying process (FIGURE 8.12(d) and Section 6.3.1). Following the ohmic contact is the implant isolation process (FIGURE 8.12(e)) and dielectric passivation (not shown). These steps comprise the basic active device processing sequence. As is the case for recessed-gate FETs, additional process steps for the formation of interconnects, passive elements and backside processing may be used as well.

8.3 DOPING FETS

FETs are doped either by ion implantation or during epitaxial growth. Only a few reports exist of doping by diffusion and those efforts were largely abandoned for FETs.

An ideal dopant offers the possibility of high concentration doping with low diffusivity and comes in a high-purity easy-to-use formulation. Unfortunately, no ideal dopant exists. GaAs n-type doping has an upper limit of the mid to high 10^{18} cm^{-3}, while lower doping limits below $10^{16}\,cm^{-3}$ are possible depending on purity and defect levels in GaAs. Many n-type dopants do diffuse appreciably, but the most common one, Si, does not. The p-type dopants offer higher doping levels ($10^{19}\,cm^{-3}$ levels), which are not needed for GaAs FETs, and they vary greatly in diffusivity.

Doping during epitaxy is described next. Silicon is commonly used and it is available as high-purity solid sources for MBE and silane sources for MOCVD. It diffuses little at the temperatures used in GaAs growth and processing and it can be incorporated over the range of 10^{16} to $5 \times 10^{18}\,cm^{-3}$. Silicon is an amphoteric dopant that can be n-type when it resides on the Ga sublattice and p-type on the As sublattice. It mostly, but not completely, resides on the Ga sublattice until it reaches the low $10^{18}\,cm^{-3}$ doping levels. Above the mid $10^{18}\,cm^{-3}$ doping level, it populates both types of lattice sites more or less equally. Tin (Sn) can reach slightly higher doping levels by epitaxial methods but it too is limited to the high $10^{18}\,cm^{-3}$ level. At the temperatures used in MBE or MOCVD growth (generally 650°C and below) no diffusion issues exist for these dopants. Even if high temperature annealing (800–900 °C) is used, as in self-aligned gate processing, epitaxial Si diffuses very little.

Ion implantation is an alternative process for doping semiconductors that is used extensively in both Si and GaAs processing. An ion implanter is a machine that accelerates atoms to very high energies (20–400 keV in common semiconductor processing) so that they penetrate a solid to a range of depths that are proportional to the ion's energy and whose concentration tails off as a Gaussian function for depths greater than the projected range. The ion implanter is a high-vacuum system that uses electron bombardment to ionise a suitable source of the intended atom (such as SiF_4 for Si). The ionised atom (ion) is separated from the other ions by a mass spectrometer before it is accelerated to the desired energy and directed at the sample. Ion separation by the mass spectrometer reduces unwanted contamination. Silicon is often implanted using the mass 29 isotope because mass 28 could have possible N_2 and CO contamination; both of these are common residual gases in the implanter.

Ion implantation doping requires high uniformity, as might be expected. During setup, the ion implanter can be "tweaked" to enhance uniformity by monitoring the current at a Faraday cup sensor while the instrument is adjusted. The optimised settings can be evaluated from uniformity maps that are generated by implanting into special wafer-shaped samples coated with a sensitive polymer. The implanted areas will darken and the transparency of the sample is measured optically, yielding a map of the implant uniformity.

The Gaussian profile at the end of an implanted ion's projected range helps keep the doping profile well defined and this is important for making shallow channels. If the ion is directed along an open path of high symmetry in the lattice, it can penetrate much deeper due to "ion channelling" along the unobstructed path. An ion's projected range depends on the ion encountering a random distribution of lattice atoms in the host crystal. GaAs is aligned away from the directions of major symmetry, or about 6–8° away from the ⟨001⟩ direction and rotated 30° in the azimuthal direction in order for the GaAs lattice to present an approximately random view to an incoming ion. Even so, a small number of implanted atoms will recoil from a collision with Ga or As and continue the trajectory along a direction with high symmetry. In this case this minority of atoms will propagate well beyond the projected range, providing a "tail" to the implanted profile, which may be several percent of the peak dose. This limits the sharpness of the doping profile and is one reason for the use of either epitaxy to define the channel or buried p-layers with ion-implanted channels.

The ion-implantation process causes considerable lattice damage to the GaAs crystal. Post-implant annealing is essential to "repair" most of the damage. It is important to use a sufficiently low flux of implanted ions that the GaAs does not become amorphous during the ion implantation. Thermal annealing in the temperature range of 800–900°C is used to allow the lattice atoms to regain their crystalline order while they incorporate the dopant into an active lattice position. Si is the preferred n-type dopant in ion implantation because it diffuses little even up to temperatures of 900°C and even in the presence of considerable lattice damage. Both tube-furnace and rapid-thermal anneals are used for implant activation, with the latter being more common because they offer greater control of the process. Practical considerations for the RTA are similar to those discussed in Section 6.3.1 with regard to equipment, sample placement and temperature measurement. However, the higher temperatures required for implant activation also pose extra challenges. For example, the GaAs wafer is subject to mechanical deformation if the heating or cooling process occurs too rapidly. In addition, preferential As loss can become severe and must be dealt with through dielectric encapsulation or with an As overpressure during the anneal. When GaAs wafers cool too rapidly, the outer parts of the GaAs wafer can cool more rapidly than the centre areas and the resulting stress will cause the formation of slip lines on the GaAs.

Arsenic loss during annealing occurs because the gas/solid-phase equilibrium condition results in As evaporation unless a fairly substantial As overpressure is used. One can remedy this situation by providing an As overpressure (commonly done for tube furnaces), by encapsulating the GaAs with a dielectric such as silicon nitride or silicon dioxide or by a special technique using a sample susceptor such as that described in Sections 6.3.1 and 7.3.2. The latter method involves impregnating the inside of the susceptor with a thin film of solid As so that, upon annealing, it provides the required gaseous As overpressure within the small space that exists between a GaAs wafer and the susceptor lid.

When dielectric encapsulants are used, the film must be mechanically robust at the required temperatures. PECVD silicon nitride and silicon dioxide films can incorporate up to 40% hydrogen during the film deposition. During implant activation, such high hydrogen levels will cause bubbling or cracking of the films. In addition, films that are in tensile stress with the GaAs will crack during a high-temperature anneal. The deposition process (PECVD, ECR or sputtering) must be optimised for any encapsulant film to withstand the annealing temperature without cracking, peeling or allowing significant interdiffusion with the GaAs. In addition, the film must be characterised to be certain the anneal yields reproducible and uniform activation of the implanted dopant. SiO_2 films will extract small amounts of Ga during the annealing process, but this can be an advantage for Si implantation. By extracting small amounts of Ga, the probability of the Si encountering a vacant Ga site is enhanced (because Si_{Ga} is a donor) and the activation efficiency is increased relative to that obtained with silicon nitride films. A well optimised silicon nitride film, on the other hand, is often preferred because encapsulation occurs without any significant outdiffusion in the O-free film.

Once the samples are implanted and activated, the material may be characterised. Many techniques are available and most are used when a process is first being established. SIMS (Section 2.5.3) is used to establish the depth profile of the implanted species as well as the existence of possible impurities. SIMS is usually performed as a service by companies that specialise in semiconductor analysis techniques. Capacitance-voltage measurements (Section 2.5.4) are used to establish the doping profile as a function of depth. Contactless resistivity measurements (Section 2.5.4) will yield sheet resistance maps of the wafer. A Hall measurement (Section 2.5.4) is often performed, principally because it is a good diagnostic tool to determine whether the mobility has been maximised for a given doping and process condition. A maximised mobility is usually an indication that the overall material quality is very high. Contactless resistivity measurements are performed on a regular basis in production mode. Other than contactless resistivity, the other techniques are usually destructive and are used mainly during initial process characterisation and infrequently thereafter.

Repeatable, highly uniform doping profiles and sheet resistivity are desired from the ion implantation and activation process. Well-optimised implantation and annealing processes are necessary but not sufficient conditions for realising the best reproducibility and uniformity. Achieving the desired final result also requires suitable high-quality, very uniform GaAs substrates. The early days of GaAs manufacturing were fraught with instances of high variability in the starting material which led to variability in the implant activation. GaAs manufacturers worked with material vendors to establish correlations of the implant activation process to crystal growth conditions and surface preparation. Many things were learned. For example, crystal vendors must control the stoichiometry of the melt, the purity and levels of EL2 and the uniformity of dislocations. In addition, process technologists established special, often proprietary, surface-preparation techniques for the GaAs wafers. Today, these correlations have been well established and it is generally the responsibility of the crystal vendor to qualify the GaAs wafers so that they meet the specifications of the customer.

8.4 ISOLATION OF FETS

FETs are three-terminal devices that must have good isolation between devices so that they will be immune to the operation of nearby devices. Isolation can be accomplished by three methods: mesa etching, implant isolation or simply using insulating regions of the semi-insulating substrate.

In mesa etching, the material between devices is removed by either dry or wet means after appropriate patterning. Wet methods are perhaps more desirable due to their ease of use and the lack of demanding feature size. The mesa dimension is usually a minimum of 6 μm, which does not present problems for the patterned etching of several hundred nanometres. The main considerations for wet-etch isolation are the cross-sectional profile (for ease of metal-step coverage), subtleties associated with the etching of differing epitaxial layers (InGaAs, in particular) and the surface roughness after the etch (mostly for cosmetic reasons, but sometimes for other special requirements).

Wet etching is often crystallographic (Chapter 4) and will yield different profiles along the [011] and [01$\bar{1}$] planes (aligned with the flat and perpendicular to it). Along the [011] planes a sloped profile of the GaAs will result for many wet etches (Chapter 4), while a dovetail profile will result along the [01$\bar{1}$] plane when a hard mask is used. Since the dovetail profiles will cause breaks in the metal coverage of interconnects and since orientation-dependent layout rules are undesirable, one needs to address this etch issue in the process. One approach is to etch shallow mesas and use thick metals. Another is to use soft (elastic) masks that will deform during the etch (due to surface tension effects), allowing greater liquid access and sloped profiles for a greater range of etch depths.

Another subtlety associated with the isolation of pseudomorphic HEMTs (PHEMT) is that the gate metal will cross from the edge of the mesa to the semi-insulating GaAs buffer. The gate metal should have low reverse leakage currents, which should be controlled by the metal/semiconductor interface of the given epitaxial structure. However, at the mesa edge, the gate may contact the InGaAs and be a source of gate leakage current since the InGaAs has a lower bandgap energy than GaAs. This parasitic gate leakage is more of an issue for InP substrates, where InGaAs with 49% In is commonly used, but nevertheless should be characterised for GaAs PHEMTs or MHEMTs as well. If it is found to be a significant source of leakage, the mesa etch can be modified to finish with a preferential InGaAs etch (Section 4.5.2 and FIGURE 4.11).

Ion implantation is a second method for isolation of FETs. Doped, conducting layers outside the active FET areas can be implanted and the resulting lattice damage will make these layers highly resistive, if the implant is done properly. The proper way of isolating by ion implantation is to create an optimum concentration of defect-associated traps with the implantation process. If there are too few defects, doped areas will exist that can join together and provide conductive paths. If there are too many defects, electrons can hop from trap to trap under the influence of an electric field and provide leakage paths. With just the right trap density, the conductivity of a doped layer can be reduced by 8 orders of magnitude or more from the peak drain current levels.

One must also take care to design the implant sequence with the proper projected range and with regard to the total thermal budget in fabricating the transistor. For example, the other annealing processes during the fabrication of the FET can reduce the trap density and lead to suboptimal isolation. Different implant doses (and optimisation) are required depending on whether the implant is done before or after the ohmic alloying process (refer to FIGURES 8.11 and 8.12). A wide variety of ions may be used to provide isolation by this damage mechanism including O, B, N and He. Generally, lighter atoms are preferred because they have a greater projected range at a given energy. Generally, H is not used because it shows considerable long-term mobility and can diffuse to regions far from the originally implanted region. It has the potential to passivate dopant atoms or cause hydrogen poisoning effects (Section 7.5).

Implant isolation is generally desired before the gate-metal process so that the parasitic gate-metal capacitance (that associated with the gate pad and not the active transistor) is minimised. However, in a self-aligned process (refer to FIGURE 8.12), an 800–900°C anneal will follow the gate-metal process. Lattice damage to GaAs from an isolation implant will generally be repaired at these temperatures. Generally speaking, implant isolation by a purely damage-based mechanism may work up to subsequent anneals at 600 to 650°C. An exception is a void formation process with a very-high-dose Al implantation into GaAs [5]. Such an implant is stable up to 900°C. However, the common procedure is to use selective-ion-implantation doping for self-aligned GaAs MESFETs to avoid a pre-gate isolation step. One could also use mesa isolation.

For self-aligned HFET structures using AlGaAs-doped layers, a thermally stable isolation process may be used prior to the gate process. Oxygen implantation will continue to isolate by the lattice-damage mechanism when the wafer is annealed at temperatures below 650°C. At higher doses, it will form a chemical complex with AlGaAs at temperatures of 700–900°C, which will neutralise the AlGaAs dopants.

Finally, selectively doped FETs may not need isolation if the resistivity of the substrate is high enough. If the need for an isolation step is eliminated for this reason, it is still necessary to control surface leakage currents and ensure process compatibility with subsequent high-temperature steps. In general, it is not easy to achieve this type of control and may be safer to incorporate an isolation step as an insurance policy.

8.5 SOURCE AND DRAIN OHMIC CONTACTS

An ideal FET has low channel resistance for high-performance applications. The resistance (ohmic contact and semiconductor resistance) associated with the transport of electrons from the source to the channel adds delay to the FET transit time and is considered a parasitic effect. It is very desirable to minimise the effect of ohmic contact resistance on the overall source resistance. Other important factors for GaAs ohmic contacts are reliability, thermal stability and shallow and minimally reacting interfaces. These factors as well as fabrication considerations have been thoroughly reviewed in Chapter 6.

GeAuNi ohmic contacts have been the mainstay for FET fabrication for more than two decades and nothing on the horizon appears to threaten that position. Contact resistance of less than 0.1 Ω-mm is routinely obtained. With today's high-frequency devices that have pushed performance above the 200 GHz range for GaAs PHEMTs (and approaching 400 GHz in InP HEMTs), it would be desirable to have improvement paths available for ohmic contacts, as well. However, the contact technology is rather mature and limited in part by peak doping levels achievable in GaAs and sheet resistance levels achievable in GaAs-based FETs. Unfortunately, no obvious improvements in electrical characteristics appear to be on the horizon for GaAs-based contacts.

The improvements in GaAs contacts over the last several decades have been in the areas of fundamental understanding of their reactions and mechanisms, thermal stability and the formation of shallower junctions (refer to Chapter 6 for more details). PdGe ohmic contacts were developed as an alternative contact for these reasons and were seriously considered as a replacement for GeNiAu for enhanced reliability. However, today most FET processes, both research and manufacturing, continue to use GeNiAu.

8.6 GATE METAL CONTACTS

As was seen in Sections 8.2.4 and 7.3, a wide variety of gate metal processes exist for GaAs-based FETs. Processing details were described in those sections. The choice of gate metal for an FET depends on the types of process used, the performance requirements, reliability issues, and manufacturability. All else being equal, a higher Schottky barrier height is preferred to reduce gate leakage and forward conduction. As described in Section 7.4, the Schottky barriers of commonly used contacts to GaAs vary within a fairly narrow range of approximately 0.2 eV. In spite of thermally activated reactions of Au with GaAs and the unsatisfactory diffusion barrier properties of Ti and Pt, Au-based contacts can be reliable against gate sinking (Section 7.5). These types of contacts raise concerns about hydrogen poisoning, especially for hermetic packages. Refractory metal contacts are preferred for self-aligned processes and provide the best diffusion barriers in low-temperature processes. Some of the best refractory contacts are $WSi_{0.45}$ and WSiN because these remain amorphous for better diffusion barrier properties at sufficiently high temperatures to prevent Ga and As outdiffusion.

8.7 PASSIVATION

The GaAs surface is one of the least understood areas of GaAs technology, but it remains an area of intense interest and active research. Moreover, the fundamental understanding that does exist is based on well-defined starting conditions, which often do not correlate well with the conditions present on partially-processed GaAs wafers. The electrical properties of a GaAs FET surface profoundly affect the device characteristics. The crystal structure of the surface, defects and chemical impurities can affect the Fermi level deep in the semiconductor. In addition, these factors can contribute to deleterious surface currents during device operation. For these reasons, GaAs FET device engineers have sought methods of surface passivation that can eliminate or at least stabilise these effects.

In Section 8.2.4 it was noted that passivation of GaAs FETs using silicon nitride deposition is sometimes performed as the first step of processing (FIGURE 8.11) in order to stabilise the surface in the as-received condition, at which time it is likely to be at its most reproducible state. Other process engineers have found ways of controllable processing that can return the GaAs surface to its most favourable, or at least its most reproducible, condition at various points in the process, usually by plasma oxidation treatments followed by oxide removal and native-oxide formation (Section 6.3.1). A more stable method of passivation is needed after the active device processing steps. Silicon nitride passivation was implemented in the 1970s as a "temporary" solution to the passivation needs of GaAs FETs. It is still in widespread use today!

The first effect of a good passivation is to physically protect the device from handling and environmental contaminants. SiN_x is a good choice for this purpose. While SiO_2 will take up water vapour and is permeable to mobile ions (even as thermally-oxidised single-crystal Si), SiN_x provides a good barrier to both moisture and ions. It is also a mechanically hard material capable of providing scratch protection. The SiN_x film can also have excellent insulating properties so that it serves well as an insulator between the electrical contacts.

It is important to protect the GaAs surface against oxidation (Section 8.8.3), since long-term degradation can occur if GaAs surfaces are allowed to oxidise. SiN_x provides this protection well as long as no oxygen contamination is present during the deposition. Sufficiently high purity source gases are available for this purpose. The trace gases in vacuum systems used for SiN_x deposition must be considered potential contaminants as well. Water vapour readily adsorbs on the walls of vacuum chambers that are vented to air for sample introduction. A loadlocked system will minimise this source of contamination. Alternatively, the gas-purging procedures as well as the vacuum integrity and operating pressures of the deposition system must be adequate to minimise contamination from residual sources of water vapour. The deposition conditions should be optimised to avoid energetic ions impinging on oxygen-containing parts (such as anodized Al) within the vacuum chamber since these may sputter and contaminate the sample.

The SiN_x film must also be electrically passivating. If this type of passivation is not performed, the GaAs FET is susceptible to high levels of RF dispersion and hot electron degradation (Section 8.8.2). Unfortunately, most of the available art in achieving electrical passivation is only available in proprietory settings rather than as publications of careful research studies. If a need arises for high-quality SiN_x passivation, the burden will rest on the individual process engineer to optimise his or her own process to arrive at the parameters for depositing a suitable film with the equipment available. A procedure will be outlined to do this.

First, one must characterise the deposition process and strive to develop a high-quality baseline film. This involves evaluation of the refractive index, stress, film composition and the film etch rate in diluted HF (or buffered-oxide etch, BOE). These properties can be evaluated by ellipsometry, a stress gauge (based on laser deflection, usually) and RBS for general composition or Fourier transform infra-red spectroscopy for hydrogen content. A low-stress dense film with a low BOE etch rate deposited at a temperature near 275°C is a good starting point.

Then, one should conduct carefully controlled experiments where one monitors the drain I-V characteristics on an FET before and after passivation using a candidate film. One should measure the DC performance and compare it with the pulsed I-V characteristics. For the pulsed I-V characteristics, one should begin by biasing the FET with a quiescent bias condition (the bias conditions that are imposed after each pulsed bias condition) of a gate voltage near V_{th} and a drain voltage near 5 V. The pulsed conditions may be adjusted based on what is observed. For each candidate film, one should compare the pulsed I-V with the DC I-V. Those films for which the DC I-V most closely matches the pulsed I-V, especially near the knee, are the best candidate films. Once passivated FETs with low dispersion (as observed in the pulsed I-V measurement) are obtained, the FETs are ready to be evaluated for their intended application.

Another important passivating-film property is the film stress. GaAs-based FETs are sensitive to the film stress through the piezoelectric effect. A stress in GaAs will change the electric field in the GaAs due to this effect. The electric field change is manifested as a change in V_{th}. At the very least, the SiN_x film should have a reproducible stress. It is even better for the film to have as little measurable stress as possible.

8.8 DEGRADATION OF FETS

Like almost everything, FETs can wear out, meaning they degrade gradually over time, and finally fail in the course of operation. They can also "burn out", or fail abruptly. Burn-out failures occur most often in the initial stages of operation, but not always. In this section, mechanisms of gradual FET degradation will be the main topic. The reader should be aware that many of these mechanisms also apply to burn-out. Manufacturers of FETs often apply "burn-in" stress tests to their entire population of deliverable parts. This procedure may involve a moderate temperature elevation and an electrical stress. The purpose of burn-in is twofold: to accelerate any normal changes in electrical characteristics and to eliminate as many early failures as possible. The burn-in procedure goes a long way towards compensating for our often inadequate knowledge and ability to control all failure mechanisms.

FETs can wear out through interactions (or complications) with their surrounding environment (the package, for example). We will ignore these problems for the purpose of this chapter and concentrate instead on degradations intrinsic to the FET. Chemical reactions can degrade the surfaces and interfaces of the FET. These reactions affect primarily the contacts of the FET (Section 8.5 and Chapter 7), exposed surfaces, and interfaces with the passivating film. The reactions can occur at ordinary temperatures over long periods of time, but they are more likely to be an issue with higher temperatures and other sources of environmental stress such as high humidity. The temperature of FETs can rise with the ambient, but can also rise with the power generated by electrically operating them.

FETs can also fail due to electrical changes in the semiconductor or its passivating films. Compound semiconductors can have many defects, most of which are electrically active. Changes in the charge states of these defects or "traps" can rearrange charge and cause devices to fail. Even where few defects existed initially, large electric fields can create new defects in the course of normal device operation.

It is the responsibility of materials, processing, device and design engineers, to understand how growth, processing and device structures affect the electrical operation of devices. The same applies for how they affect electrical degradation.

8.8.1 Definition and characterisation of hot electrons

The FET is always being engineered to yield higher performance. For example, higher power, higher gain and higher efficiency are always desired in power FETs. One approach to achieving the output power goal is to add gain stages and to add to the total gate width of the output stage of the final device. However, this approach has its limits. As the total current increases, the device impedance becomes extremely small (less than 1 Ω is not uncommon) and difficult to efficiently match to the load impedance (50 Ω, commonly). In order to increase power without reducing the device impedance, higher drain-voltage operation is desired. In addition, high-performance applications require short gate lengths, so peak electric field values can be very high, even for moderate drain voltages.

Channel electrons of an FET accelerate in an electric field, alternately gaining energy and losing it in scattering events with phonons, dopant impurities, defects and other scattering centres. On average, they gain kinetic energy in proportion to the magnitude of the electric field. For low electric fields (such as in the source region), the electron population has an average kinetic energy much lower than the thermal energy of the lattice and is considered to be in thermal equilibrium with it. As the electric field increases in the channel region beneath the gate and on the drain side of the gate, the electron kinetic energy greatly surpasses the thermal energy and the electrons are said to have become "hot". For an FET with a 0.25 μm gate with its drain biased at 5 V, the average electric field can approach 2×10^5 V/cm; the peak electric field can be even higher. Such fields are near the onset of avalanche (multiplication of electron hole pairs due to energetic electron collisions with atoms in the lattice) breakdown (4×10^5 V/cm) for GaAs.

Hot electrons can exceed the bounds of normal FET operation in several ways. They may have sufficient energy to escape the quantum-well confinement of HEMTs, which may have conduction band offsets in the range of 0.2–0.3 eV, or about 10 times the thermal energy. In that case, electrons can escape the (In)GaAs channel into the barrier layer. These electrons can be captured by deep level traps in the barrier layer, surface states, interface states of the Schottky gate or the passivation dielectric. Changes in the charge states of traps can change the threshold voltage of the device if they occur under the gate or at the Schottky interface. Although they do not necessarily cause threshold voltage shifts, they can cause changes in drain current, transconductance, and other electrical parameters when captured by deep levels in the drain or drain surface region.

A population of hot electrons reaches a point where a small but significant fraction possesses energy in excess of the bandgap. One of these electrons that scatters with the lattice can transfer enough energy to promote a valence electron to the conduction band in a process called impact ionisation. The hole that is left behind contributes to the drain current as it is pulled towards the source. However, some of the more energetic holes make their way to the gate, especially if it is biased more negative than the source. A negative gate current component thus accompanies impact ionisation. For FETs with low leakage, this source of gate current is the predominant one at high drain bias and represents a useful indicator of impact ionisation. An example of how gate current correlates with impact ionisation is given in

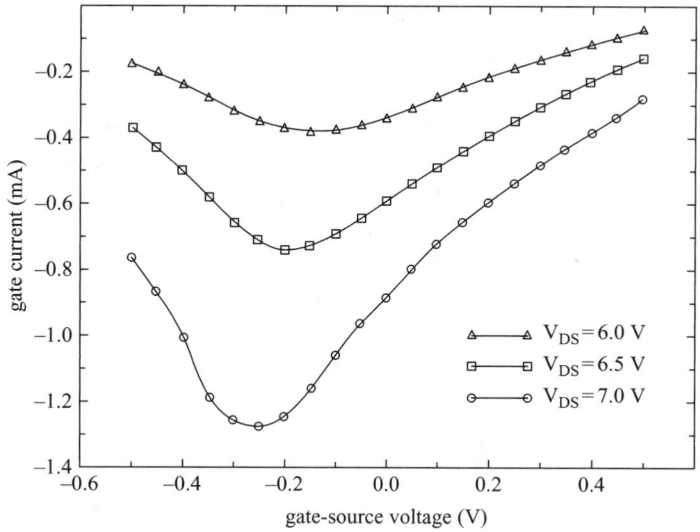

FIGURE 8.13 The bell-shaped form for gate current of a GaAs FET is a signature for impact ionisation. (*Semicond. Sci. Technol.* **13** (1998) p.1053–63, reprinted with permission.)

FIGURE 8.13 [6]. At drain voltages below 6 V, the gate current is comfortably below −0.4 mA. For drain voltages between 6 and 7 V, gate currents increase and peak at near −0.2 V. At very negative gate voltages, the gate current is low because the drain current is low (the channel is near pinchoff) and few channel electrons are available to participate in impact ionisation. At high gate voltage, the gate current also drops; the gate-drain voltage is reduced and impact ionisation drops dramatically. The bell shaped curve for gate current is a signature for impact ionisation. In particular, the I_G/I_D ratio is characteristic of impact ionisation and can be measured non-destructively (at low levels). I_G is proportional to the holes generated by impact ionisation and I_D is proportional to the number of electrons in the channel.

Of course, impact ionisation can also be observed by a sharp increase of drain current with drain voltage. There are two problems with this method of testing. One is that the device is easily damaged by such a test. The other is that the test is not necessarily definitive for impact ionisation; electron release from traps can result in a similar increase in the drain current. In the former type of test, one must make comparisons between several separate devices, introducing uncertainty since all devices may not be sufficiently "identical".

Another useful measure of impact ionisation is by band-to-band radiative recombination of cold channel electrons (near the source) with holes generated by impact ionisation. In a unipolar device such as the FET, impact ionisation is the most likely source of holes, so band-to-band electroluminescence is a sensitive complement to gate current measurements.

The maximum reverse bias that can be applied between the drain and gate of an FET defines the gate-to-drain breakdown voltage. It is important to distinguish between off-state and on-state breakdown [7]. In off-state breakdown, the gate-to-drain voltage is measured at a low drain current level with V_{GS} biased so that the channel is pinched off. In such a case, the channel current is low. On-state breakdown is defined as a locus of drain voltages and drain current pairs that converge to a given drain voltage at a given gate current.

8.8.2 Hot electron degradation

Hot electrons can considerably stress an FET and lead to degradation. Virtually any of the characteristics or performance parameters of a device can be affected. These include gain and power output in power amplifiers, noise or gain in low-noise amplifiers, and threshold voltage, drain current and transconductance in either digital or analogue FETs.

In Section 8.2.3, frequency dispersion and trap-related effects were introduced. Although these effects can have tangible consequences for high-frequency circuits that are proportional to the magnitude of the effect, they do not necessarily imply that degradation is inevitable. For example, power amplifiers with low to moderate gate lag may or may not be vulnerable to hot electron degradation under long-term bias stress. In other words, the initial magnitude of trapping effects is not a predictor of future degradation. It is, therefore, important to study and understand the similarities and differences between initial trap-related effects and those that contribute to hot electron degradation.

Power FETs experience some of the greatest stress from hot electrons because RF overdrive produces high gate-drain voltages. The hot electrons may become trapped in states located at the interface between the semiconductor and the passivation layer. They may also get trapped in the passivation dielectric. The accumulation of negative charge between the gate and drain locally enhances surface depletion. The most susceptible region in a recessed-gate FET is the space between the gate edge and the recess edge. In planar, self-aligned FETs, any part of the surface between the gate and drain is susceptible. The surface depletion causes an increase in the drain resistance, and may cause a reduction in drain current, transconductance, and RF parameters as well. These effects cause "power slump", a gradual and permanent reduction in output power. They may also cause "power drift", a partially recoverable form of power slump [8].

This degradation effect was first discovered in MESFETs as a result of DC and RF overdrive stress tests. The effect was found to be greater in HEMTs and PHEMTs. The cause is due to electron capture in the gate-drain access region. The drain-current reduction and drain resistance (R_D) increase was found to be proportional to the impact ionisation rate, i.e. the I_G/I_D ratio [9]. Newly created traps or deep levels of varying activation energies accompany the degradation [10].

Significant V_{th} shifts during hot electron stress of HEMTs have also been observed by many groups. The shifts may often be reversible, at least partially. In many cases, it has been postulated that negatively charged traps exist in the AlGaAs layer under the gate and holes created by impact ionisation can be captured by these traps to neutralise the negative charge. Such an effect causes a negative shift in V_{th}, since positive charge accumulated under the gate favours electron accumulation in the channel. Similar negative V_{th} shifts have been observed in high-temperature storage tests of some commercial PHEMTs at temperatures of 120°C to 240°C, indicating that thermal detrapping of electrons in the AlGaAs may be responsible. This thermally activated process has an activation energy of 0.21 eV and is reversible in tens of hours at room temperature [11]. In cases where the V_{th} shift is permanent, positive charge may be stored in the thin interfacial layer between the Schottky metal and the HEMT barrier layer.

Some cases of permanent positive V_{th} have been observed after hot electron stress. Formation of deep levels (1.2 eV activation energy) under the gate or on the drain side of the gate have been postulated [10]. Competing effects are possible as well.

A curious effect happens with hot electron stress with respect to breakdown. Most new traps created by the stress either degrade performance or cause large enough changes over a long enough time that a burn-in to stabilise them is not a viable option. However, a slight increase in the number of traps created during initial operation of an FET may increase its breakdown voltage without significantly changing other FET parameters. To achieve this, an FET is stressed with an appropriate impact ionisation current (kept constant during the stress) and the drain voltage is observed to rise gradually, resulting in "breakdown walkout". This effect may happen in seconds or minutes and the extent of the breakdown walkout is often several volts, as illustrated in FIGURE 8.14. The effect was first discovered for unpassivated AlGaAs/GaAs HEMTs [12], where it was attributed to oxidation of the exposed AlGaAs. The effect was much smaller for passivated HEMTs and MESFETs. Commercially-passivated PHEMTs are also susceptible, and negative charge at the surface of the drain access region is thought to be responsible [13]. The latter effect is clearly related

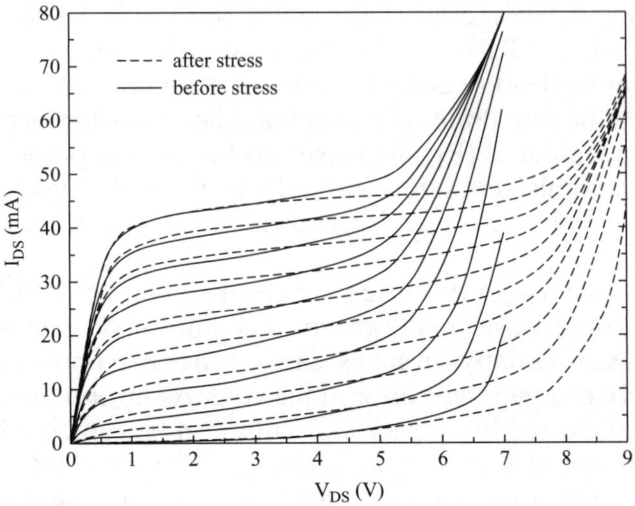

FIGURE 8.14 The breakdown voltage of a GaAs FET will "walk out" with drain bias stress during its initial operation. (*IEEE Trans. Electron Devices* vol.43, (1996) p.543 reprinted with permission.)

to hot electrons and is clearly not a degradation because it does not significantly change other FET parameters. Although not a degradation, breakdown walkout is included in this section because of similarities to the mechanisms of hot electron degradation.

This effect has a certain elegance to it. Device designers strive mightily to reduce peak electric fields in a device in order to maximise breakdown voltage. Yet, if they don't completely succeed, hot electrons naturally find these regions, are drawn to them, and create the negative surface charge that moderates the peak electric field. Slight moderation of the peak electric field need not cause degradation of other device parameters, but once the walkout effect is completed, hot electron stress should be reduced in order to reduce any ensuing gradual degradation.

In Section 8.2.2, short timescale, *reversible* trapping effects were described. These trapping effects manifest themselves on a timescale of sub-microseconds to hundreds of milliseconds. Pulsed I-V measurements (either gate or drain is pulsed) are a powerful tool for characterising these effects. Transient effects involving gate pulsing are known as "gate lag" and those involving drain pulsing are known as "drain lag". Because of the obvious similarity with degradation effects discussed in this section and the potential for confusion, we shall briefly contrast these effects in a way that emphasises generalised rather than complete descriptions.

In gate lag, traps are populated with negative gate biases, usually near but more negative than pinchoff. A lesser dependence on drain voltage exists. Traps may have various physical locations, but in well-optimised materials, the FET surface near the drain edge is the most susceptible. In such cases, the effect of gate lag will be to modulate the drain resistance of the device (and properties related to it, such as the drain current and transconductance). In other cases, the gate lag affects traps under the gate and changes the threshold voltage of the device. Most often, gate lag is correlated with the processing and geometry of the gate, although material growth conditions can also have an effect. Gate structure design and surface-passivation solutions to gate lag have been reported.

Generally, recessed-gate structures are less susceptible than planar structures and shorter gate-edge-to-recess-edge distances have lesser gate lag than longer ones. Structural designs that place the physical locations of surface traps in regions of lower peak electric fields are an effective strategy. Surface-passivation effects are less well understood. Many new approaches to surface passivation are described in Chapter 3 and new understanding of how to manipulate surface properties is gained every year. To date, none of these fundamental approaches to surface passivation has replaced the oldest method using amorphous, partially hydrogenated silicon nitride (Section 8.7). This "temporary" fix to transient surface effects has been with us for more than two decades and a basic understanding of the origin of its effectiveness is still lacking. Somehow the surface trap energetics are modified by the SiN_x/(Al)GaAs interface so that negatively charged traps either are not energetically favourable, can de-trap on a fast timescale or can conduct the charge away through slight surface or interface conduction.

In drain lag, traps in the substrate can be affected by changes in the drain bias of the FET. These traps can deplete the channel from the backside, as illustrated in FIGURE 8.15. As in gate lag, rapid changes in drain bias do not allow sufficient time for the traps to respond. One will generally observe a threshold voltage shift as the main effect due to drain lag. Drain lag effects can be more indicative of growth defects, or other types of substrate defects, than of process conditions.

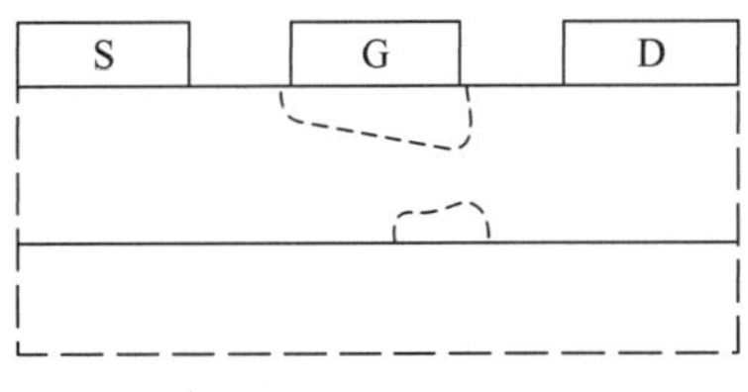

FIGURE 8.15 In "drain lag" traps are populated (or depopulated) by high drain bias. These traps can cause depletion changes at the back of the channel leading to V_{th} shifts.

Gate and drain lag exist in an FET because of traps that are created by the way it is grown or processed. Although low gate or drain lag is desirable for performance reasons, FETs with gate or drain lag can be stable in operation, and therefore reliable. By contrast, degradation of an FET can occur because of traps or other fixed charge centres that are usually *created* by hot electron stress *after* growth and processing. The susceptibility to such trap creation may be affected by growth or process conditions. Hot-electron stress tests are important to determine safe operating limits, to prevent degradation and to use as feedback for optimising device structures, growth and process conditions.

Although gate lag need not lead to degradation, a screening test based on gate lag is possible. Reversible gate lag will exist to a small degree in even the best devices. However, in some devices that were later shown to degrade, early changes in the magnitude of the gate lag were observed as well, while other devices that did not degrade showed stable gate lag [8].

From the preceding discussion, it is clear that the properties of SiN_x are critical to the *electrical* as well as the physical passivation of GaAs FETs. Little is actually known about its electrical properties, except that many interface states exist when MIS diodes are fabricated using SiN_x as the MIS dielectric. Section 3.3.5 contains a discussion of MIS diodes and interface states.

A wide variety of possible explanations for the effectiveness of SiN_x exist. SiN_x could play a role in rapidly detrapping deep levels by providing a conductive path along the SiN_x/GaAs interface. A Si-rich film could play a role in the interface conductivity. SiN_x could also modify the traps through interface reconstructions. The modified traps could then cease to trap mobile charges under the conditions of FET operation or could change their time constants for de-trapping such that slow transient effects in the FET are no longer possible. Whatever the cause, no detailed physical explanation has yet been conclusively verified.

8.8.3 Other types of degradation

The main other source of degradation is due to oxidation of the GaAs. One main purpose for SiN_x passivation is to act as an impermeable film to protect the GaAs device from environmental sources of contamination. When this protection is not in place, the GaAs surface can oxidise according to the following reactions:

$$2GaAs + 3O_2 = Ga_2O_3 + As_2O_3 \tag{8.5}$$

$$2GaAs + As_2O_3 = Ga_2O_3 + 2As \tag{8.6}$$

These reactions, described in greater detail in Section 10.2, create free As at the GaAs/SiN_x interface and change the interface state properties. As was described in Sections 8.8.1 and 8.8.2, the GaAs/SiN_x interface in the gate-to-drain region of the FET is especially sensitive to its electrical properties. As seen in Sections 8.3 and 8.7, one goes to great lengths to create high-quality GaAs surfaces prior to critical process steps such as dielectric passivation in order to promote stability and reproducibility in the FET.

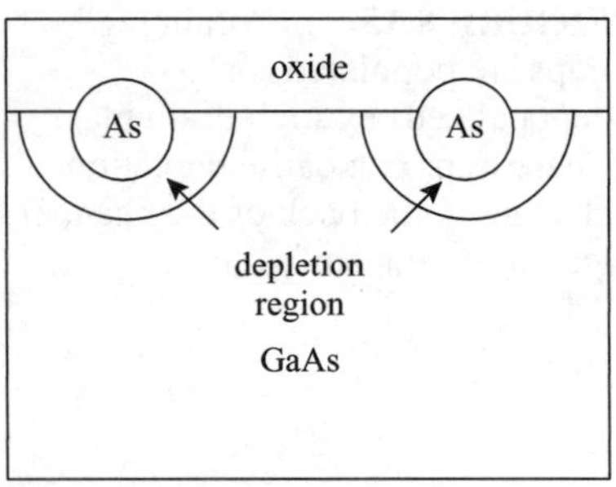

FIGURE 8.16 Continued oxidation of GaAs can lead to As precipitates in addition to free surface As. The precipitates act as local centres for depletion regions.

Arsenic is problematic both as isolated atoms and as precipitates. While As atoms create undesirable midgap states, the condensation of atoms into multiatomic precipitates produces many tiny depletion regions centred on the precipitates (FIGURE 8.16).

This phenomenon is the origin of the insulating character of thermally annealed low-temperature-grown GaAs, which is deliberately grown rich in As to enable formation of As precipitates upon annealing. Continuing production of As due to long-term slow oxidation of the GaAs surface may lead to the formation of local Schottky barriers and the formation of a non-uniform potential distribution at the interface.

For this type of degradation to occur, the FET passivation must be permeable to sources of oxygen (i.e. SiO_2) or contaminated with oxygen. High-density SiN_x films are generally impermeable to oxygen and water vapour but do incorporate oxygen in the films fairly easily at levels of a few percent. Whether the oxygen is securely immobilised in covalent bonds or available to oxidise the GaAs surface is open to question. However, it is entirely plausible that energetic electrons can gradually free up oxygen to oxidise the GaAs surface, which will lead to gradual degradation of the FET. For this reason, most manufacturers of GaAs devices strive to produce oxygen-free SiN_x films.

8.9 CONCLUSION

The FET is greatly affected by the surface and interface properties of GaAs and its contacts. These process-sensitive elements of the FET need to be understood and controlled by process development engineers. FET operation, testing and performance characterisation supply the basic information that materials, processing and device engineers need to understand the often complex materials and process interactions. Many FET process choices are available and some of the main ones have been reviewed in this chapter. High-quality contact choices are readily available for GaAs FETs, as long as one considers the thermal stability and hydrogen poisoning issues. Minefields must be navigated carefully for certain device and process design issues. Peak electric fields must be suppressed to the extent possible in order to provide adequate breakdown voltage and reliability consistent with performance requirements. One approach to achieving this end is to optimise the geometry of a recessed-gate process or the LDD implants for a self-aligned process. Finally, a high-quality SiN_x passivation process must be developed and implemented in the present absence of perfect understanding.

REFERENCES

[1] W. Lui [*Fundamentals of III-V Devices: HBTs, MESFETs, and HFETs/HEMTs* (John Wiley and Sons, New York, 1999)]

[2] M. Shur [*GaAs Devices and Circuits* (Plenum Press, New York, 1987)]

[3] S.M. Sze (Ed.) [*High-Speed Semiconductor Devices* (John Wiley and Sons, New York, 1990)]

[4] A.S. Grove [*Physics and Technology of Semiconductor Devices* (John Wiley and Sons, New York, 1967)]

[5] K.-Y. Ko, S. Chen, S.-T. Lee, G. Braunstein [*Appl. Phys. Lett. (USA)* vol.60 (1992) p.1223]

[6] R. Menozzi [*Semicond. Sci. Technol. (UK)* vol.13 (1998) p.1053]

[7] J.A. Del Alamo, M.H. Somerville [*IEEE J. Solid-State Circuits (USA)* vol.34 (1999) p.1204]

[8] J.C.M. Hwang [*Solid-State Electron. (UK)* vol.43 (1999) p.1325]

[9] G. Meneghesso, A. Paccagnella, Y. Hadab, C. Canali, E. Zanoni [*Appl. Phys. Lett. (USA)* vol.69 (1996) p.1411]

[10] R. Menozzi [*Semicond. Sci. Technol. (UK)* vol. 13 (1998) p.1053]

[11] G. Meneghesso, C. Canali, P. Cova, E. De Bortoli, E. Zanoni [*IEEE Electron Device Lett. (USA)* vol.17 (1996) p.232]

[12] P.C. Chao, M. Shur, M.Y. Kao, B.R. Lee [*IEEE Trans. Electron Devices (USA)* vol.39 (1992) p.738]

[13] R. Menozzi, P. Cova, C. Canali, F. Fantini [*IEEE Trans. Electron Devices (USA)* vol.43 (1996) p.543]

Chapter 9

Heterojunction bipolar transistors

9.1 CHAPTER SCOPE

This chapter will cover most of the basic active processing steps in GaAs heterojunction bipolar transistors. The active steps are those that are related to processing the GaAs and whose nuances require an understanding of the GaAs semiconductor surface and its interface to metal junctions for etching and contact formation. HBTs present many active processing subtleties with important implications for device design and operation. A clear understanding of how GaAs surface technology impacts HBT design and operation is a major objective in the presentation of this chapter's subject matter.

The basics of bipolar transistors and HBT operation and performance are given in Section 9.2. That section begins with a tutorial and continues to relate HBT operation to the main material and processing methods. It also highlights the main reliability issues and relates them to HBT structures, processing and materials. Sections 9.3, 9.4 and 9.5 present the details of the main active processing methods: mesa etching, ohmic metal formation and HBT passivation. In Section 9.6, alternative methods of HBT processing to enhance performance are summarised. Finally, specific HBT reliability and degradation issues are presented in Section 9.7.

9.2 HBT BASICS

The HBT is a type of bipolar transistor. Transistors are types of active or nonlinear devices in which the output response is modified in a nonlinear way from the input. The transistor was invented in 1947 at Bell Telephone Laboratories. Called a point-contact transistor and made from a piece of germanium with three wire contacts, it functioned more or less as a bipolar transistor. The invention of the transistor ushered in the modern age of microelectronics, which has so radically changed society in the latter half of the twentieth century.

9.2.1 Bipolar transistor tutorial

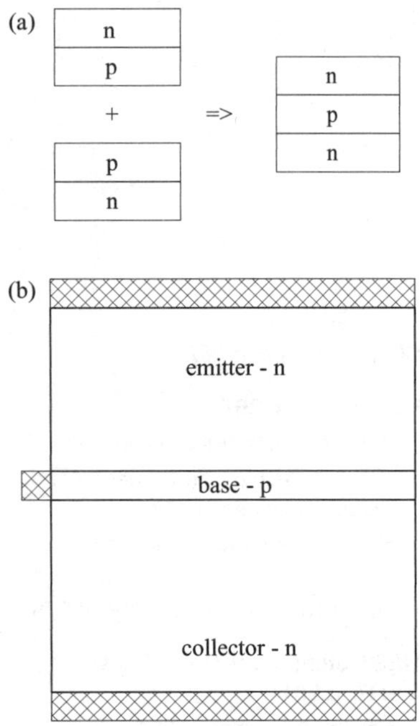

FIGURE 9.1 Schematic illustration of bipolar transistor.

This section presents a simplified tutorial on the operation and characterisation of bipolar transistors. Some knowledge of semiconductor physics is necessary, but the transistor action is described without any prior knowledge of transistors being assumed. A more complete treatment of HBT operation is given in [1,2].

The bipolar transistor is a semiconductor device that is the union of two pn junction diodes of opposite polarities in which one of the semiconductor regions is common to both diodes as illustrated in FIGURE 9.1(a). The most common arrangement is the npn transistor that is illustrated in FIGURE 9.1(b), where a p-type semiconductor is the common region between the two diode pn junctions. The most useful operation of a bipolar junction transistor (or BJT) is as a current amplifier. To achieve current amplification, one pn junction is operated in forward bias. Its n-region is termed the emitter and its p-region is the base. The second n-region is termed the collector for reasons that will become apparent. For the pn junction diode between the emitter and the base, the forward bias condition acts to inject electrons into the base where they can recombine with holes. The base contact needs to supply hole current to accommodate the recombination. In an isolated pn junction diode, the emitter and base currents are equal. In the BJT, the base can be made thin enough so that most of the electrons diffuse into the n-type collector before they recombine in the base. In the extreme of a very thin base, the base current is small compared with the emitter and collector currents, which are nearly equal. The current due to electrons arriving at the collector contact is amplified compared to the base current by a factor β, which describes the DC current gain:

$$\beta = I_c/I_b \tag{9.1}$$

The current gain is a function of the collector current density and can vary with the design of the transistor.

The current flow described in the previous paragraph can operate in reverse for the emitter-base pn junction. Holes from the base will be injected into the emitter for a forward biased junction. This hole current must come from the base contact, adding to the total base current. This component of the base current is undesirable since it reduces the current gain according to EQN (9.1). Accordingly, BJTs are designed with doping levels in the base that are substantially lower than those in the emitter in order to limit the reverse current. This approach addresses the issues of achieving adequate β in a BJT, but has other drawbacks. The low doping in the base results in a higher than desirable base resistance for many applications and limits the high-frequency performance. The HBT can be designed to eliminate this drawback of BJTs.

The key to HBT design is to use an emitter material with a wider bandgap than the base in order to provide a valence-band offset at the emitter-base junction, as illustrated in FIGURE 9.2. The valence-band offset at this junction provides a potential barrier to hole current from the base to the emitter and greatly reduces the reverse hole current. The net effect is that the β of the transistor can be high without resorting to low base doping. The high-frequency performance can be increased by increasing the base doping in order to reduce the base resistance. AlGaAs and InGaP emitters have been commonly used for high performance HBTs for almost two decades. Comparing HBTs to BJTs in a common material system, Si/SiGe HBTs have clearly demonstrated better performance than Si BJTs for comparable dimensions, as have GaAs HBTs over GaAs BJTs.

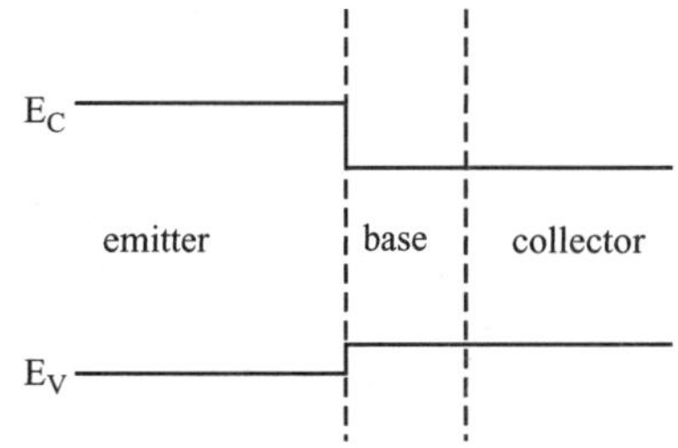

FIGURE 9.2 Conduction and valence bands for a heterojunction bipolar transistor, ignoring band bending.

The active layers of an HBT are usually vertically situated on a suitable (usually semi-insulating) substrate, which is schematically shown in FIGURE 9.3(a). The HBT can have either the emitter or collector as the top layer, with the base as the middle layer. The most common configuration is that with the emitter up. The cross-section of a finished device that uses a triple-mesa process is illustrated in FIGURE 9.3(b).

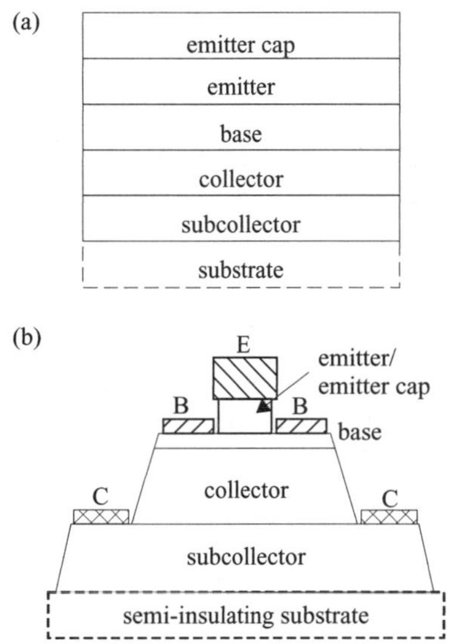

FIGURE 9.3 (a) Device layers for an emitter-up HBT.
(b) Schematic cross-section for a triple mesa HBT.

The transistor operation of an HBT is usually characterised by a few basic DC measurements as well as RF or microwave measurements. The most common DC measurements are the collector transfer characteristics, illustrated in FIGURE 9.4, and the Gummel plot, illustrated in FIGURE 9.5. The collector transfer characteristics are a measure of the dependence of the collector current on the collector voltage for different values of the base current. The emitter-base junction is forward biased to give the initial desired base current. While the collector voltage is at zero, the base-collector junction is also forward biased and a small negative current flows to the collector. As the collector bias is increased, some of the electrons from the emitter start to overcome the less forward biased base-collector junction and current to the collector starts to increase. The collector voltage at which collector current crosses zero is called the offset voltage of the HBT. It is less than 0.1 V for the HBT of FIGURE 9.4. The collector current continues to increase almost linearly with collector voltage until it begins to saturate near the knee voltage. At this point, the current becomes limited by the number of electrons injected into the collector, and to a lesser extent, other factors such as the resistance for the electrons to traverse the collector to the collector ohmic contact. The DC current gain is calculated as in EQN (9.1).

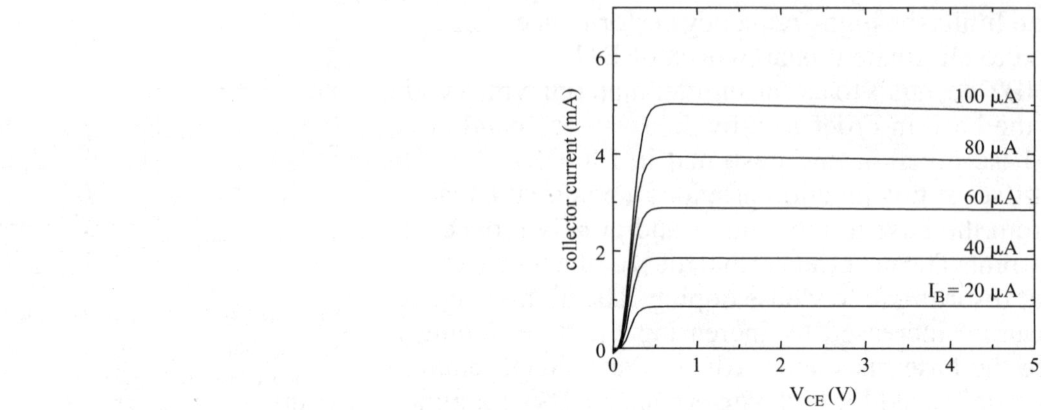

FIGURE 9.4 Common emitter I-V characteristic of an HBT.

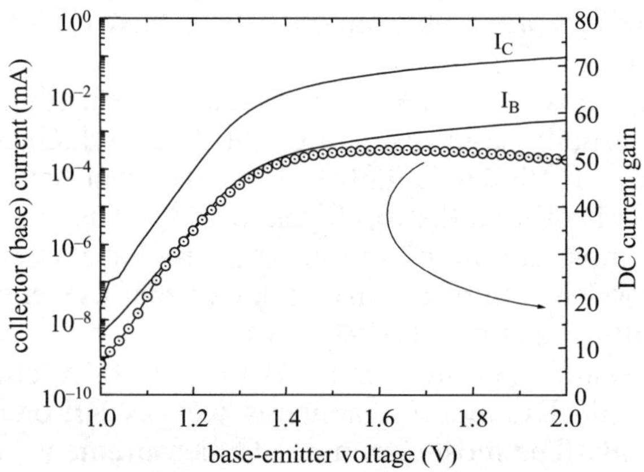

FIGURE 9.5 Gummel plot of an HBT.

The Gummel plot of FIGURE 9.5 shows the collector current and the base current as a function of base-emitter voltage for the collector voltage tied to the base voltage. Usually the current gain is plotted along with the collector and base currents. A plot of this type allows the device engineer to analyse a number of device characteristics such as the ideality factors of the collector and base currents extracted from the equation

$$I = I_o \exp(qV_a/nkT) \tag{9.2}$$

where I is either the base or collector current, I_o is a pre-exponential factor, q is the electron charge, V_a is the applied voltage to the emitter-base junction, k is the Boltzmann constant, T is the temperature and n is the ideality factor. Analysis of the ideality factor can give important insight into recombination in the intrinsic (the region under the emitter) and extrinsic (regions away from the emitter) base regions. It also gives characteristic signatures for abrupt and graded interfaces at the base-emitter junction. For example, one can separate out the contribution of different recombination centres such as base bulk recombination, emitter-base space-charge recombination (recombination in the depleted base-emitter junction) and surface recombination in the HBT by analysing the base current ideality factor [1].

The measurements of FIGURES 9.4 and 9.5 have been described as DC measurements, but may also be made with short current pulses (as short as 100 ns) to separate out thermal effects and trapping effects. Even in the HBT data for FIGURE 9.4, thermal effects are readily apparent in the increasing negative slope of collector current with collector voltage as the collector current (and self-heating) increases.

The high-frequency characteristics of bipolar transistors and of amplifiers built from them are described by S-parameter measurements. At high enough frequencies, direct measurements of voltages, currents and impedances are not accurate because the sinusoids describing them have wavelengths that are short compared to the length of the test interconnections. This situation makes the measurement impossible or, at best, highly dependent on the placement of the test probes. S-parameters (scattering parameters) are elements of a matrix used to transform voltages, currents and impedances at high frequencies that are easily measured from reflection and transmission coefficients of a device under test. Other device and circuit characteristics such as current gain and power gain are derived from S-parameter measurements. A plot of current gain versus frequency of a GaAs HBT is shown in FIGURE 9.6 as a means to derive its unity-current-gain cutoff frequency, f_t. Likewise, from S-parameter measurements the unilateral power gain (U) is plotted versus frequency to derive the f_{max} (unity U or unity maximum stable gain). The types of transformations necessary to extract f_t and f_{max} were described in Section 8.2.1. A particularly useful equation relates f_t, f_{max} and the base resistance of a bipolar transistor:

$$f_{max} = (f_t/2\pi R_B C_{BC})^{1/2} \tag{9.3}$$

where R_B is the base resistance and C_{BC} is the base-collector capacitance. Most device researchers focus on high-frequency measurements such as f_t and f_{max}, but other high-frequency measurements such as amplifier gain and noise figure are important as well.

As for the case of FETs, DC, pulsed I-V and S-parameter measurements are used to develop equivalent circuit models for HBTs that can be used in circuit simulators to design high-frequency amplifiers and other circuits.

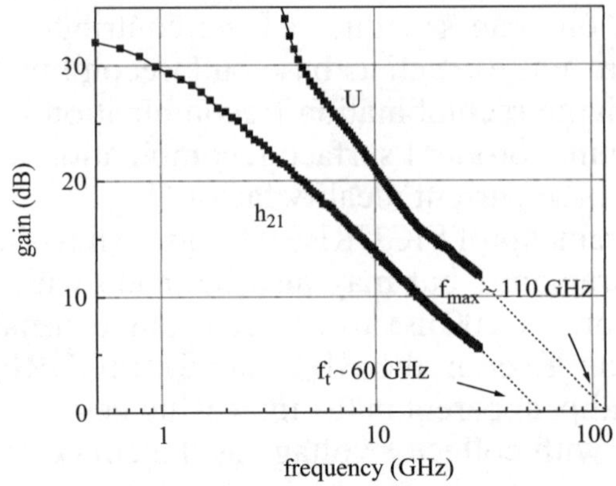

FIGURE 9.6 Plot of current gain, h_{21}, as a function of frequency. The extrapolated unity gain value gives the f_t of the HBT.

9.2.2 Other GaAs HBT performance and reliability issues

HBTs are a specific case of bipolar transistors where the emitter has a wider bandgap than the base. The specifics of the device structure will be introduced in Section 9.2.3. This subsection is a logical continuation of Section 9.2.1 which will transition from tutorial aspects of bipolar transistors to address some general performance and reliability issues that affect high-performance GaAs HBTs. Specific material and process issues that affect HBT performance will be addressed in Sections 9.2.3, 9.5 and 9.6.

Thermal effects are extremely important considerations for high-performance HBTs. Parameters such as f_t and f_{max} are often found to peak at current densities near $10^5\ A\ cm^{-2}$. It is often observed that the DC current density also peaks near the same current density. A typical power amplifier application might use 10 to 30 emitter fingers with dimensions of $2 \times 30\ \mu m^2$ in a typical cell. Assuming that drain voltage is 3 V, as for cell phone handsets, each emitter finger dissipates up to 180 mW at the peak of the RF signal. Such high power densities will result in large increases in the junction temperature during operation of the power amplifier. The close proximity of 10 or more emitter fingers would add to the thermal load. Calculation of the junction temperature depends on how the power is thermally dissipated. The following equation establishes the relation between the temperature rise and the power dissipated for an HBT sample in thermal contact to a perfect heat sink, i.e. one that does not experience a temperature rise in response to a thermal load:

$$\Delta T = R_{th} I_C V_{CE} \tag{9.4}$$

ΔT is the junction temperature rise from the heatsink, R_{th} is the thermal resistance of the semiconductor and I_C and V_{CE} are the collector current and collector-emitter voltage averaged over the RF frequency. The thermal resistance R_{th} is dependent on the geometry and the thermal conductivities of the materials in contact with the HBT.

GaAs is a poor thermal conductor compared to Si and to some of the metals commonly used as interconnects, packaging materials and heatsinks. Junction temperatures for GaAs HBT power amplifiers can commonly attain and even exceed 150°C. High junction temperatures have important implications for device, processing and reliability engineers. First, electron transport degrades with increasing temperature. The thermal energy of common lattice vibrations (phonons) increases with temperature and causes more electron scattering. Both electron mobility and saturated velocity decrease with increasing temperature. This negatively affects the transit time of HBTs and, therefore, their high-frequency performance. Moreover, performance is temperature dependent, and due to self-heating, HBTs will operate at different junction temperatures with different input-power levels. These thermal processes affect the linearity of any semiconductor device. For applications in which linearity is important (i.e. communications), thermal effects must be managed carefully and HBTs may be used at current densities well below what their performance maximum would dictate. In other applications, linearity is less important (radars) and HBTs are operated at higher current densities and higher junction temperatures.

High junction temperatures also have the potential for accelerating the aging of some of the material constituents of the HBT. Chapters 6 and 7 discussed how GaAs reactions with many common contacts such as Au, Ni, Pt and other contact constituents occur within minutes or seconds at temperatures between 300 and 400°C. It is not unrealistic to expect that some of these reactions can occur at temperatures of 150–200°C if the time is long enough. In HBT investigations, both thermal and current-density acceleration are considered important factors that can affect reliability. The classic reliability equation for the median (or mean) time to failure (MTTF) was introduced in Chapter 2 and is given by

$$\mathrm{MTTF} = A_0 \exp(E_a/kT) \tag{9.5}$$

where E_a is the activation energy, k is the Boltzmann constant, A_0 is a constant and T is the absolute temperature. The median time to failure is determined by testing a given set of samples until half of them have failed. Reliability experiments are performed at temperatures high enough to accelerate the failures so that the MTTF is measurable within hundreds or thousands of hours. Extrapolation to the expected junction temperature is then made with EQN (9.5) if the activation energy is known. Such studies are commonly performed for HBTs and some of the most significant findings will be summarised in Section 9.8.

Performance degradation of GaAs HBTs past 150°C is usually unacceptable and provides a common upper limit for HBT junction temperatures. Thermal management solutions are sought to keep below this upper junction temperature limit. Some common techniques in practice are the thinning of the GaAs substrate, sometimes to as thin as 50 μm, and the use of a thermal shunt to the emitter of the HBT. The thermal shunt is a thick electroplated metal which contacts the emitter directly and therefore is situated directly over the heat source.

As a practical matter, there are various theoretical and experimental techniques to determine the R_{th} of EQN (9.4) for a GaAs HBT and manage its thermal properties. Theoretical approaches can include using reasonable thermal simplifications to build equivalent-circuit thermal models and using available material parameters to optimise them. Theoretical approaches may also use three-dimensional thermal solvers to optimise a thermal approach. Experimental approaches require a determination of the junction temperature and then measuring it as the power density of the HBT is changed. Direct junction temperature measurement techniques are often unsatisfactory because they lack the resolution and possibly also the accuracy to determine the hot spot of an HBT. For example, an infrared image will smear the emitter image over several microns from its true location and give an average temperature over that area which is less than the actual peak temperature. Nevertheless, with skilled practitioners, they can still be useful. Indirect methods are often electrical in nature. One common method is to relate the temperature of the emitter-base pn junction to its bandgap and to measure it in an actual device by monitoring the emitter-base turn-on voltage and comparing to carefully calibrated structures.

HBT performance peaks at high current density, often near 10^5 A/cm^2. As will be elaborated on in Section 8.7, reliability of HBTs has been difficult to achieve for current densities greater than 2×10^4 A/cm^2. As is the case for FETs, both bias and current stress have the potential for degrading HBTs. Device engineers are forced to make practical tradeoffs such as these between performance and reliability considerations.

Other performance and reliability issues relate to materials choices made for the HBT structure (Section 9.2.3). Specific reliability issues involving base doping (carbon and beryllium), emitter choice (AlGaAs or InGaP) and processing choices are discussed in Section 9.7.

9.2.3 HBT device structure and material issues

Both AlGaAs and InGaP are suitable emitter materials for GaAs-based HBTs. AlGaAs was used earlier because most MBE systems lacked a phosphorus source in the early days and because MOCVD development of HBTs lagged MBE. AlGaAs is not an ideal emitter for an HBT because about two thirds of the change in bandgap is manifested in the conduction band offset at compositions of interest (below 45%). The conduction band spike can cause current blocking in an HBT (by reducing the number of electrons injected into the base), while a valence band offset to GaAs is more desirable. The effect of the conduction band profile on current blocking is illustrated in FIGURE 9.7. Contrasting with FIGURE 9.2, it is clear that accounting for band bending due to the depletion region at the emitter-base junction leads to a conduction band spike at the E-B junction. Depending on the combination of the conduction band offset and the emitter doping level, electrons may traverse the spike by thermionic or thermionic field emission. Modelling in one- or two-dimensions can be useful in assessing the effect. AlGaAs emitters often use compositional grading gradually changing the Ga to Al ratio during growth to eliminate the spike.

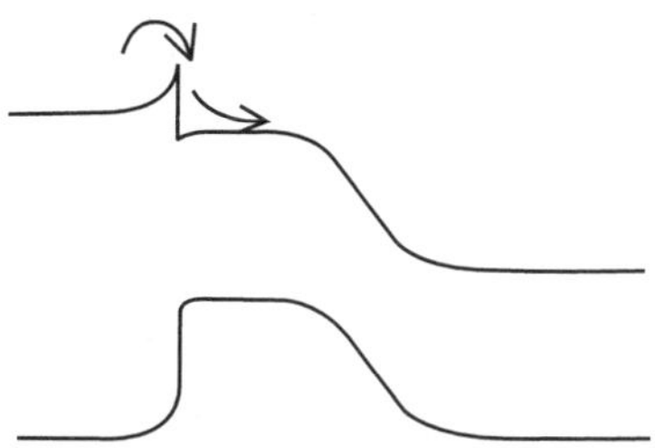

FIGURE 9.7 Conduction and valence bands of a npn HBT with band bending.

The bandgap of lattice-matched InGaP is near 1.9 eV. Both its value and the conduction-band offset (ΔE_C) vary with the degree of ordering in the InGaP, with ΔE_C ranging from 0.03 to 0.39 V. The lower value is ideal for an Npn HBT, while the higher value is not. The ordered state with a now uniform distribution of In and Ga on the group III lattice site can be grown by MOCVD at relatively high temperatures. The larger valence band offset, ΔE_V, of ordered InGaP is better suited to blocking the base-emitter hole current than that for AlGaAs. However, AlGaAs has two practical advantages. First, it is nearly lattice-matched to GaAs and slight variations in composition do not introduce strain into the structure. Second, there is no mixed group V chemistry to deal with for AlGaAs; P and As intermixing may affect the quality of the InGaP/GaAs interface. These advantages are slight and InGaP emitters have gained favour in recent years to reduce back injection and for processing and reliability reasons.

A typical structure is outlined in TABLE 9.1. The layers of the emitter contain the intrinsic emitter layer (emitter-base junction), the extrinsic emitter, which is a highly doped GaAs layer to transition to the contacts, and optionally a cap layer, which is often InGaAs for a degeneratively doped layer for non-alloyed ohmic

TABLE 9.1 A typical epitaxial structure for GaAs HBTs.

Description	Layer	Composition	Thickness (nm)	Doping
Emitter cap	InGaAs	50%	30	$1\text{–}2 \times 10^{19}$
Ext emitter	GaAs		100	2×10^{18}
Emitter	(InGaP or AlGaAs)	49% In 20–35% Al	50–70	5×10^{17}
Base	GaAs		50–100	$2\text{–}4 \times 10^{19}$
Collector	GaAs		300–500	$1\text{–}2 \times 10^{16}$
Subcollector	GaAs		500–1000	2×10^{18}
Substrate	SI GaAs			

contacts. InGaP or AlGaAs are the common choices for the emitter layer and they are generally doped in the mid 10^{17} cm^{-3}. Neither the doping, the dopant used (usually Si), nor the thickness is critical, though they may have to be optimised in conjunction with the rest of the structure. AlGaAs emitters are usually compositionally graded to smooth the transition to the base (in order to eliminate current blocking), but InGaP emitters are not, both because it is more difficult and because it is less desirable with its smaller ΔE_C. GaAs and InGaAs are the common choices for the extrinsic emitter and the cap layers. Grading of the composition between the binary and ternary alloys (GaAs/InGaAs, GaAs/AlGaAs) may be employed to smooth the barriers to electron transport.

The base layer and the base/emitter interface are often the key distinguishing features for the epitaxial growth of HBTs. High base conductivity and fast transit times of the minority carrier electrons across the base are required for high-frequency applications. These requirements are at odds with their separate optimisation. Base conductivity is increased with a thicker base layer and higher base doping, while the minority-carrier transit time is inversely proportional to the base thickness. Minority-carrier recombination is increased with the higher base doping. These factors suggest that an optimised solution will have thin base layers that are highly doped. Fabrication and current gain specifications impose practical limits that are within the ranges shown in TABLE 9.1. Clearly, the quality of the epitaxial growth can affect the limits of these ranges.

Many growers assess HBT material quality across different laboratories by a metric defined as the current gain divided by the base sheet resistance (in Ω/square) to account for differing base doping levels and thicknesses. This ratio is measured for a large-area HBT ($100 \times 100\ \mu m^2$ emitter, for example) fabricated with a triple-mesa process and measured at a current density near 800 A/cm^2. A value near 0.4 is considered excellent [3].

Another quality factor arises from the need to keep sharp composition and doping profiles. Be, Zn and C are some choices for doping the base, with Be being the historical dopant of choice for MBE. Be diffusion becomes extremely problematic at common epitaxial growth temperatures at concentrations significantly greater than the mid $10^{18}\,cm^{-3}$ and will smear the base-emitter pn junction. Consequently, C doping was developed as an alternative, first for MOCVD and later for MBE. Special procedures were developed to limit Be diffusion in MBE and it has been used in production by some major manufacturers. Often an undoped GaAs spacer layer borders both the emitter and the doped base layer. The thickness of this layer is of the order of 5 nm and it will accommodate small amounts of diffusion. More significantly, most MBE growers drop the temperature of the growth below 500°C for growing the base spacer and emitter layers to control Be diffusion.

C doping does not suffer from any diffusion problems. The main issues with C are H incorporation during growth and build-up of strain as the doping increases beyond $10^{20}\,cm^{-3}$. Since fast minority carrier recombination limits the practical doping level to $4 \times 10^{19}\,cm^{-3}$ in all but the thinnest bases, strain issues are usually secondary. H incorporation passivates the C acceptor, reducing the base doping. If the C-H passivation complex is stable with time, its effect can be calibrated out of the expected results. However, under high current density, energetic electrons can impact the complex and release free H. The C acceptors are depassivated and the hole concentration then increases, which results in a reduced current gain. In order to minimise this problem, the growth conditions can incorporate a high-temperature process, either at the end of the growth or during an interruption at some point in the emitter growth, which serves to reduce the amount of incorporated H.

The growth of the collector involves a tradeoff between breakdown voltage and collector resistance. As seen from TABLE 9.1, the collector doping is kept low to allow high breakdown, while the subcollector doping is high to reduce the collector resistance. The collector depletion width, w, is given by the expression

$$w = (2\varepsilon\varepsilon_0(V_{BC} + V_{bi})/qN_{Dc})^{1/2} \tag{9.6}$$

where ε is the dielectric constant of the collector (GaAs: $\varepsilon = 13.1$), ε_0 is the permittivity of free space (8.85×10^{-14} F/cm), V_{BC} is the applied base-collector voltage, V_{bi} is the built-in voltage of the base-collector pn junction, q is the electron charge and N_{Dc} is the donor concentration of the collector. The collector doping is chosen such that its dopant-dependent breakdown is sufficiently high for the application of interest. Often, the collector breakdown is designed to be equal to the intrinsic doping-level-dependent breakdown or it can be designed such that a certain collector-base bias results in complete depletion of the collector layer prior to intrinsic breakdown. In the latter case, breakdown occurs in the subcollector at slightly higher bias, a situation termed punch-through. The collector resistance is minimised by making the collector thin, consistent with the requirements for breakdown voltage, and by doping the subcollector highly. The structure of TABLE 9.1 will result in breakdown voltage in the range of 10–30 V.

GaAs is the most commonly used collector and subcollector material. Highly doped InGaAs is not a reasonable choice for a subcollector material because a thick layer is needed and the lattice strain in such a layer will have to be accommodated by lattice relaxation with the formation of misfit dislocations, which will propagate upward as threading dislocations. However, it is not uncommon to use thin etch-stop layers of InGaP near the collector/subcollector interface or near the subcollector/substrate interface for control of mesa etching or to allow selective undercut (Section 9.6).

The vast majority of research and device development work on HBT design is done by experimenting with the structure and growth of HBTs. Some variations on the typical structure will be briefly mentioned.

An electric field in the base can greatly increase the transit time across the base (compared to diffusion) and improve high-frequency performance. Two approaches to create a quasi-electric field are the grading of the composition in the base (more Al near the emitter or more In near the collector) and the grading of the doping (higher p-doping near the collector). Both of these approaches create a gradient in the conduction band relative to the Fermi level and aid in accelerating minority carriers through the base.

Another design approach is the use of a wide-bandgap collector for higher breakdown voltage with a thin collector. InGaP is a natural choice, but the base/collector interface must be optimised to avoid current blocking.

Various approaches for engineering emitters have been tried. One is to use a composite emitter. A thin InGaP interface with the base is used for a fully depleted ledge (Section 9.5.1) with an AlGaAs emitter above it. This type of structure is electrically similar to an AlGaAs emitter and the main advantage is for a selective wet etch that stops at the ledge. Another engineered emitter approach is placing the pn junction in the GaAs (heterostructure emitter).

Other design approaches that combine process and growth methods will be described in Section 9.6.

9.2.4 Overview of HBT fabrication

In both the GaAs and the Si world, the bipolar transistor may be expressed as a multitude of device designs with many permutations of processes or materials. Field effect transistors certainly have a number of permutations in some of the processes and materials, but they tend to converge in function and implementation much more than do bipolar transistors. It then becomes a real challenge to present enough material of utility to a process engineer without sinking hopelessly in the details. The choice of subject matter in this chapter is weighted towards those devices and processes that produce the most performance or reliability advantages. Many other possibilities will be left as subjects for advanced study.

We will mainly discuss the fabrication of Npn (the convention is to capitalise wide bandgap layers) emitter-up HBTs. Much research on collector-up HBTs exists in the literature because of the possibility of achieving high f_{max} by reducing C_{BC} as in EQN (9.3). However, the collector-up structure has suffered its share of problems such as low current gain due to injection from the emitter into the extrinsic base and the difficulty of precisely etching down to thin base layers of the same material (GaAs) as the collector. It is not at the present a widely used technology.

The basic Npn process is a triple-mesa process. A schematic illustration of the triple-mesa process is shown in FIGURE 9.8. First an emitter metal is deposited (FIGURE 9.8(a)). Common choices for the emitter metal are GeAuNi, TiPtAu and refractory metals such as WSi. The choice of the emitter metal depends on the strategy chosen for forming the emitter ohmic contact and the emitter mesa etch. These issues will be discussed further in Sections 9.3.1 and 9.4.1. In many cases, the emitter metal serves as a mask for the emitter mesa etch (FIGURE 9.8(b)), which may be a wet etch or a dry etch, depending on the feature size and the dimensional control required. The base metal alignment and evaporation is next (FIGURE 9.8(c)). Following the base ohmic contact, the base mesa is patterned and etched (FIGURE 9.8(d)). Again, both wet and dry etches may be used. The base mesa is etched to isolate the base layer and expose the subcollector layer. Collector ohmic contact and collector mesa processes follow the base mesa (FIGURES 9.8(e) and (f)). The collector mesa etch will isolate the subcollector region from the semi-insulating substrate. Finally, the active device processing is completed by dielectric passivation. A completed HBT integrated circuit process will also include further thin film metal and dielectric layers for interconnects, transmission lines and passive elements such as resistors, capacitors and inductors. These interconnects and passive elements are deposited directly on the semi-insulating substrate or

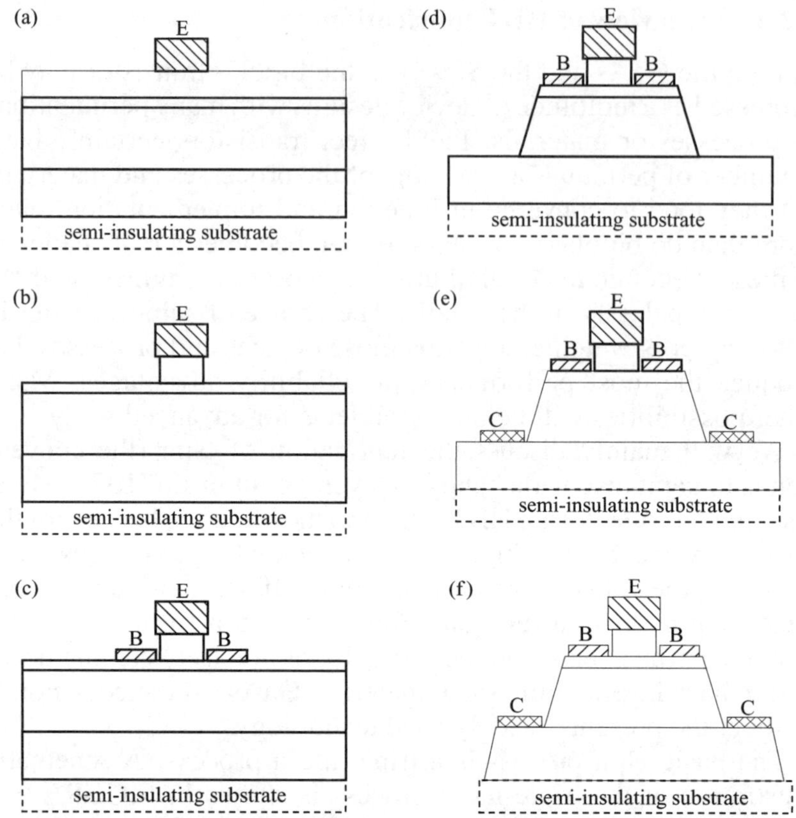

FIGURE 9.8 The steps of a triple mesa HBT process: (a) emitter metal deposition, (b) emitter mesa etch, (c) base metal deposition, (d) base mesa etch, (e) collector metal deposition and (f) collector mesa etch.

on a suitable insulating dielectric. These non-active steps will not be further considered in this chapter.

Device engineers will have to choose between self-aligned and non-self-aligned emitter schemes. The self-aligned process uses the emitter metal as a shadow mask for the base metal evaporation to place the base ohmic contact in close proximity to the intrinsic emitter and therefore lower the base resistance. The self-aligned process is also efficient in minimising the total area needed for the base mesa and thus minimising C_{BC}. These factors can lead to higher f_{max} as described in EQN (9.3). The self-aligned process also leads to several drawbacks. First, it places strict demands on the processing. During the emitter etch, a controlled amount of undercut of the emitter metal is required to prevent shorting of the base and emitter without severely undercutting the emitter metal. For short emitter widths, the consequences of excess undercut can be the loss of mechanical support of the emitter metal. Second, the self-aligned process reduces the current gain, β, of the HBT because of the proximity of the base contact to the intrinsic emitter region. The base/contact interface recombination velocity is even greater than the recombination velocity of the free GaAs surface. Since the HBT will be required to have a certain minimum β, the relative merits of self-alignment must be evaluated against other aspects of the device design that impact the current gain. For example, higher doping in the base layer will also improve R_B and reduce β. The device engineer may then have to choose between a self-alignment scheme or increasing the base doping to push the limit of the current gain specification.

Non-self-aligned processes can have quite good realignment tolerances by using state-of-the-art optical lithography steppers. Based on realignment tolerance of 0.15–0.2 μm, contact spacing from the edge of the emitter metal can be as little as 0.25 μm. Such good alignment tolerance allows one to control the base metal to emitter spacing and avoid the need for a self-aligned process.

Many of the challenges of the active HBT fabrication steps involve understanding the interplay between design and process. It is not enough to know the etch chemistry for the layer that needs to be etched. HBTs have at least one heterointerface and typically two or three. Layers near the interface often do not etch the way a bulk layer does. Implementation of etches for specific device designs requires understanding of the surface technology issues. The next section explores specific etches in GaAs-based HBTs.

9.3 MESA ETCHING FOR GaAs-BASED HBTS

9.3.1 Emitter mesa etch

The key requirement for emitter etches is that they remove emitter material to expose the base layer with minimal damage and without excessive undercut. In the case of a self-aligned process, a small amount of undercut of the emitter metal is needed so that the contact shadows the emitter sidewall during the base metal evaporation to prevent shorting to the emitter. Wet etches are suitable for larger emitter widths, while dry or combination wet and dry etches may be more suitable for smaller width emitters. The cross-over between wide and narrow emitters depends on the total thickness of the emitter epitaxial layers, but is generally near the 1 μm range. Since the emitter is the smallest dimension in the HBT, GaAs HBT dimensions have relaxed geometries compared with GaAs FETs and Si devices. Emitter etch chemistry, discussed in the next two sections, depends on the materials in the emitter stack. As noted in the previous section, the most common emitter materials are AlGaAs and InGaP.

If the emitter metal is used as an etch mask for the emitter mesa, it must be compatible with the etch chemistry used. If a wet etch is used, the emitter metal must be impervious to the chemicals used. If a dry etch is used, the metal must not erode too significantly during the emitter etch. Care must be taken to avoid sputtering Au (or other metals), which may cause micromasking and roughening of the semiconductor (Section 5.10.3 and FIGURE 5.16). Finally, the emitter metal patterning process must be compatible with the emitter etch.

9.3.1.1 AlGaAs emitters

Wet etching AlGaAs emitters in GaAs-based HBTs is mainly a non-selective process in which the etch is controlled by carefully monitoring the time and the etch rate. As was described in Chapter 4, several wet chemistries exist for etching GaAs selectively to AlGaAs and AlGaAs selectively to GaAs, but significant problems prevent most implementations of any of these in HBT emitter etches. HF can etch AlGaAs, but is only effective for compositions over 45% Al and most HBTs use an Al composition between 20 and 35%. Heated HCl can etch AlGaAs compositions down to about 30%, but they also attack the Ni of GeAuNi metals. KI/I_2 can etch AlGaAs, but it also attacks Au.

A good non-selective etch for AlGaAs is $H_3PO_4 : H_2O_2H_2O$ with a 1 : 4 : 45 ratio. This chemistry etches AlGaAs and GaAs at nearly equal rates of approximately 5 nm/s. Other etch chemistries have been described in Section 4.3. The main concern of a non-selective wet etch is to avoid cutting excessively into the base layer or its resistance will increase, causing high-frequency performance degradation. Even a 10 nm removal of the base layer can cause noticeable degradation of the HBT performance. The emitter layers are typically 200–300 nm in total thickness and step height measurements are usually made over another 200–300 nm metal layer. A 10 nm tolerance for the etch represents only 2–3% uncertainty of the total measured thickness of the epitaxial growth and metal deposition. The tolerances for growth or deposition of these layers plus the uncertainty of the metrology tool make a 10 nm total tolerance in etch depth challenging, given that the etch rate must also be tightly controlled. Clearly great attention to the fabrication details is required. Some suggestions for controlling wet etches are given in Section 4.4.

An electrical method to determine whether the etch is nearing completion involves surface probing of the breakdown voltage. An etch time is selected based on the calibrated etch rates and is purposely short of the target etch depth by approximately 20%. Profilometry is used in combination with the surface probing method. Based on the measured etch depth a new etch rate is calculated and a new time is chosen to target the etch depth within 5% of the nominal value. The new etch depth is now within the uncertainty of the profilometry and the uniformity of the epitaxial layers. The surface breakdown is then measured by directly probing the semiconductor. When the probes touch down on AlGaAs, a surface breakdown significantly greater than 1 V is measured. However, when the probes contact the heavily doped base layer, a breakdown under 1 V is measured. By using the surface probing technique, one can limit the over-etch to within 10 nm and can also better calibrate the profilometry tools so that direct probing is not necessary on every wafer.

Selective dry etching down to the AlGaAs layer can be a great aid in making the emitter etch more uniform. The emitter can be controllably undercut for self-aligned processes. When one uses a selective dry etch the thickness of the AlGaAs is known from the growth, and control of a subsequent wet etch with a slight over-etch into the base is easier than if the entire structure is non-selectively wet etched. Dry etches generally cause some electronic damage to (Al)GaAs (Section 5.10.2). The HBT will manifest the damage in two ways. The emitter-base junction will show leakage at low forward bias (0.2–0.7 V), in which region the E-B diode ideality factors become $\gg 1$. Comparison to a wet-etched E-B diode will allow the damage to be quantified. One can also examine a base TLM (Section 6.3.2) to determine dry etch damage. Using either of these test methods, one may adjust the parameters of the etch process (to reduce the DC bias), use a wet etch or clean to remove the shallow damaged region, or use a post-etch anneal to remove the damage.

Sometimes a wet emitter etch is extremely fast. It has been observed that certain emitter metals enhance the etch rate to anomalously high levels. This effect can happen due to an electrochemical effect that was discussed in Chapter 4. The etch mechanism of compound semiconductors involves an oxidation step and an oxide removal step. The oxidation step happens with the diffusion to the surface of holes to complete the electrochemical reaction. It is this step that can be influenced by the choice of emitter metal. The surface of the AlGaAs being etched can act as one electrode (the oxidation electrode) in an electrochemical reaction and the emitter metal can act as the other electrode (the reduction electrode). In an electrochemical reaction, the reaction rate follows the following equation:

$$R = k \exp(E_a/kT) \tag{9.7}$$

where k is a constant and E_a is the electrochemical potential between the electrodes, k is the Boltzmann constant and T is the temperature. The choice of metal determines the electrochemical potential in this reaction. It has been found that a GeAuNi electrode enhances the lateral etch rate of the AlGaAs by more than an order of magnitude compared to the vertical etch rate and causes severe undercutting of the emitter metal. Consequently, GeAuNi metal may not be used as exposed emitter metal contacts with a wet emitter etch.

Selective etches to a combined AlGaAs/InGaP emitter can also be used (see Section 9.3.1.2). Electrically, these emitters behave more like AlGaAs emitters (Section 9.7) and the InGaP is placed in the structure as an etch stop and for ledge passivation of the extrinsic base (Section 9.5.1).

9.3.1.2 InGaP emitters

The InGaP alloy can have conduction and valence-band offsets that are preferable to AlGaAs, when grown as an ordered alloy (Section 9.2.3). It also has more flexible wet etch chemistries than AlGaAs. The HCl-based wet etches of InGaP are selective to GaAs and AlGaAs as long as they lack an oxidiser such as H_2O_2 in the mixture. HCl-based mixtures can also etch InGaP with minimal undercut under some circumstances. Suitable wet etch chemistries are also available for selective etches of InGaAs and GaAs layers on top of the InGaP.

A selective wet etch process for InGaP emitters can use any of several etch chemistries. Both InGaAs and GaAs can be selectively etched relative to InGaP with etches such as $H_3PO_4 : H_2O_2 : H_2O$ and citric acid : $H_2O_2 : H_2O$. Overetching of the InGaAs and GaAs layers to undercut the emitter metal can be used if overhang for a self-aligned liftoff is desired. Then HCl is used to etch the InGaP. The etch rate of HCl is very fast for a thin InGaP layer and concentrated HCl can also cause the loss of photoresist adhesion. $HCl : H_2O$ and $HCl : H_3PO_4$ are two ways of diluting the HCl. The former mixture can be troublesome because it is found to leave scum on the surface of the base layer for high-dilution mixtures. But, a 3 : 1 dilution in water is sometimes used. The latter etch with H_3PO_4 dilution does not suffer from the scum problem. A 7 : 1 H_3PO_4 : HCl etch mixture etches InGaP at approximately 2–3 nm/s and is suitable for emitter etching. In contrast to GaAs or InGaAs, creating an undercut of the InGaP is tricky. In spite of the high selectivity, undercuts are not easy to obtain due to the highly variable orientation dependence of the lateral InGaP etch [4]. The [011] planes etch up to 100 times more slowly than 55 degrees towards [0$\bar{1}$1]. Not only are undercuts extremely hard to control, but unpredictable etching at emitter corners can be troublesome.

Even with the flexibility offered by selective etches, a combination dry/wet etch can be used to better control feature sizes for emitters at or below the 1 μm level. The constraints regarding possible electrical damage for the InGaP etch are similar to those for GaAs/AlGaAs. Since the upper epitaxial layers (extrinsic emitter and emitter cap) are typically the same, a dry etch process can be used for these.

Finally, it should be noted that the termination of InGaP etch processes affects the quality of the base ohmic contact and etch initiation of the base mesa etch. These issues will be further discussed in Sections 9.4.2 and 9.3.2.

9.3.2 Base mesa etch

The base mesa etch is often fairly straightforward. Many of the same considerations for the emitter etch also apply to the base mesa etch. The dimensions are usually fairly tight for high-performance devices in order to minimise the base mesa area and maximise f_{max} as in EQN (9.3). For this reason, the photoresist overhang of the outside edge of the base metal is often small or non-existent. Often the base mesa is etched to undercut the base metal and photoresist to reduce C_{BC}. This must be done with concern for the structural stability of the completed feature so that manufacturability will not be compromised.

Most HBTs use GaAs as the collector and subcollector with a similar structure to that in TABLE 9.1. In this case, the base mesa etch will be non-selective. Most of the GaAs-based etches described in Chapter 4 can be selected and $H_3PO_4 : H_2O_2 : H_2O$ is often a good choice. Using a well-calibrated etch with care to maintain temperature control and consistent agitation is usually sufficient to etch down to the subcollector. It is also possible to use a thin InGaP layer or another material as an etch stop layer in the subcollector buried near the transition to the collector without materially affecting the electron transport in the subcollector.

Etch initiation issues are among the biggest problems when etching multilayered structures. Often multilayer selective etches are found to roughen and stop etching when iterated several times. Some of the surface conditions that cause these problems and possible solutions have been described in Chapters 4 and 5. In the case of the base mesa etch, these issues are present when a selective etch has been used for the InGaP emitter etch. A non-selective $H_3PO_4 : H_2O_2 : H_2O$ etch will roughen because some areas etch immediately and other areas begin etching after a delay due to a non-uniform surface. Possible causes of this surface issue are As, P intermixing at the InGaP/GaAs interface, either during growth or during the completion of the emitter etch. In either case, it has been found that treatment of the surface with concentrated HCl prior to the $H_3PO_4 : H_2O_2 : H_2O$ etch can eliminate the surface roughness in the base mesa etch and makes reproducible etches possible.

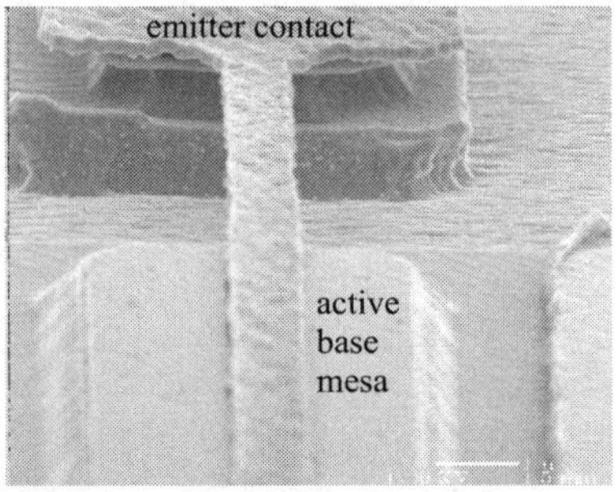

FIGURE 9.9 An SEM image of an emitter "airbridge" structure.

Other base-mesa-etch issues can depend on specific processing variations. One variation of interest is the "airbridge" emitter process illustrated in FIGURE 9.9. In this process, the emitter contact is placed away from the base mesa in cases when the emitter is very narrow and will not be contacted directly. The base-collector capacitance will suffer an unnecessary increase if the base mesa is left continuous under the entire length of the emitter metal. In order to prevent this situation, the emitter metal extrinsic to both the HBT mesa and the emitter contact is undercut during the base mesa etch. The extrinsic emitter metal is narrowed to ensure that the depth of the base mesa etch is sufficient to laterally undercut the extrinsic emitter metal.

9.3.3 Collector mesa etch

The collector mesa etch does not present any particular issues not already covered in Sections 9.3.1 and 9.3.2. In most cases, the collector mesa etch is a non-selective GaAs-based etch using any of the GaAs wet etches described in Chapter 4, but often the $H_3PO_4 : H_2O_2 : H_2O$ etch. For a double heterojunction HBT (DHBT) with an InGaP collector, one will use an HCl-based etch. However, since most DHBTs will use a grading scheme to avoid current blocking effects, a non-selective HCl etch with an oxidiser (such as H_2O_2) will be used at least for the transition from base to collector.

9.4 OHMIC CONTACTS FOR GaAs-BASED HBTS

Ohmic contacts to GaAs-based HBTs are mostly straightforward due to the typical high doping levels present, but do present a few special issues due to the uniqueness of the HBT process.

9.4.1 Emitter metal ohmic contacts

The emitter structure incorporates a semiconductor bandgap and doping level optimised for its interface to the base. The doping level is moderate. Extrinsic to the emitter area are additional semiconductor layers whose purpose is to reduce the emitter resistance and facilitate the transition to the emitter ohmic contact. As seen from a sample structure in TABLE 9.1, the extrinsic emitter and emitter cap layers are highly doped to facilitate easy ohmic contact formation. The main requirements for the n-type ohmic contacts to emitters are low contact resistance, low stress and not reacting to depths that would short the base and the emitter. Stress in emitter metals may cause defect creation during temperature and bias stress (Section 9.7).

The GeAuNi ohmic contact can meet these requirements and has a long history of reliability. However, unless completely protected by resist, it is not suitable for use with wet or combination wet and dry emitter etches because of severe lateral undercutting due to an enhanced electrochemical etch. Consequently, many HBT processes focus on the use of non-alloyed emitter contacts. The most straightforward way of achieving non-alloyed ohmic contacts involves the use of a surface InGaAs layer, as described in Section 6.4.4. The incorporation of In into GaAs increases the lattice constant and produces compressive strain. Thin layers of InGaAs or layers with minimal In incorporation accommodate the strain without relaxing by point defect creation. However, HBT emitter cap structures are usually much thicker than the critical layer thickness above which relaxation must occur. For the desired In concentration of 30–50%, the cap layer will relax by point defect creation. It is necessary that point defects generated in this layer stay confined to this layer and do not propagate to the base or the emitter-base space charge region. Growth conditions to achieve suitable high quality relaxed InGaAs cap layers have been achieved in many laboratories. Indium composition in the 50% range allows n-type doping greater than 10^{19} cm^{-3} to be achieved, which facilitates the formation of non-alloyed ohmic contacts.

The GeAuNi metallurgy has been described in detail in Chapter 6.

WSi is a good choice for the emitter metal in combination with highly doped InGaAs cap layers. Low ohmic contact resistance has been achieved with WSi contacts to n$^+$ InGaAs. WSi is patterned by reactive ion etching, which is well suited to patterning micron-sized or submicron structures. WSi also acts as a good etch mask for wet chemical etching of emitter structures without contributing to enhanced lateral electrochemical etching. WSi is typically deposited by magnetron sputtering and a wide range of W/Si ratios, determined by the sputter target composition and deposition conditions such as pressure and RF power, will be suitable for this application.

TiPtAu is another common choice for emitter metal deposition. It also does not suffer from enhanced electrochemical etching.

9.4.2 Base metal ohmic contacts

The base of an HBT is a critical layer that is extremely thin, of the order of 50–100 nm. Its conductivity is critically important to

Refer to Section 6.5 for a summary of p-type ohmic contacts to GaAs.

its high-frequency performance. Consequently, the ohmic contact resistivity of the base metal must be low, and the contact must be extremely shallow, not only as a result of the alloying process, but also during long term, high current density operation.

AuBe or AuZn ohmic contacts are among the most widely used ohmic contacts for GaAs-based optoelectronic applications. However, these contacts have the potential to penetrate too deeply for use as the base ohmic contact of an HBT unless modified in some way such as limiting the total available Au and taking steps to insure uniform interfacial reactions. A number of other options are available for producing shallow, low-resistivity contacts including alloyed and non-alloyed TiPt contacts. Because base doping above 10^{19} cm^{-3} is usually desired in high-performance HBTs, properly implemented non-alloyed contacts can have contact resistivity that is quite good. A non-alloyed ohmic contact is not intentionally reacted. However, recall from Section 9.2.2 that junction temperatures of power HBTs are expected to reach 150°C during routine operation. Such high temperatures must be considered for their alloy reaction potential and possible contact degradation. When considering the total temperature history during both processing and the expected life of the HBT, the "non-alloyed" ohmic contact may in fact not be so.

From Chapter 6, it is clear that refractory metals would make true non-alloyed ohmic contacts over the total temperature history of the HBT. Though sometimes used, these refractory metals, mostly W-based, are usually sputter deposited because of the extremely high radiant heat generated from electron beam evaporation. Sputter deposition can be problematic with the liftoff process used for base metal contacts because they tend to coat the resist sidewalls (the emitter sidewall must be protected as well). Nevertheless, an example of a WSi-based base ohmic contact is given in Section 9.6.

All of these metal recipes make good non-alloyed contacts. After deposition and measurements of TLM structures, good non-alloyed contacts can display contact resistivity less than $10^{-6}\ \Omega\,\mathrm{cm}^2$. From the previous discussion it is clear that these contacts must still have good contact resistivity and be stable during long-term operation at temperatures of 150°C. Another consideration is that the base ohmic contact process often comes before the collector ohmic contact process. Certainly, a process can be designed to reverse the order, but it is not necessary. The collector metal also contacts n$^+$-GaAs in the majority of cases and needs to be alloyed because of the doping limitations of n-GaAs. Therefore, barring a process order reversal between collector and base ohmic contacts, it is reasonable to design for alloying the base contact at a temperature at least equal to that needed for the collector contact. That temperature would be near 400°C for a GeAuNi contact, but could be in the range of 175–200°C when using a PdGeAu ohmic contact for the collector (see Section 6.4.2). Either temperature offers reasonable prospects for establishing reliable ohmic contacts that will withstand normal operating conditions for the HBT. The greatest concern then turns on the design of the contact for a shallow reaction depth that is stable over time and standard operating conditions.

The solid-phase-reaction ohmic contacts described in Chapter 6 offer one class of choices, which are occasionally reported in the HBT literature for base ohmic contacts.

Another class of choices are the limited-Au ohmic contacts, and of particular interest would be any of the shallow-reacted ohmic contacts which can be proven to be reliable under continuous operation at 150°C. Some of the more common ohmic contacts reported for base contacts to GaAs HBTs are TiPtAu, TiPt, PtTiPt and PtTiPtAu. The latter two contacts have among the lowest contact resistivities for contacts annealed at 400°C. The first Pt thickness for the PtTiPtAu contact is the most critical of the three constituents, while the Ti and Au can vary somewhat [5]. Its purpose is to react completely to a shallow depth with the GaAs base as insurance against any interface issues, while the Ti limits further reaction up to temperatures near 400°C. Specific contact resistance below 10^{-6} Ω-cm^2 has been achieved for Pt thicknesses greater than 5 nm, while thicknesses of 10 nm or less offered the greatest stability of contact resistance after high temperature treatment.

The GaAs/Pt and GaAs/Ti reactions were described in Section 7.2.2.

A workable recipe for the PtTiPtAu p-type ohmic contact uses 15 nm Pt, 10 nm Ti, 15 nm Pt and 60 nm Au and is alloyed at 400°C.

9.4.3 Collector metal ohmic contacts

The subcollector of an HBT is not a critical layer in the structure. It is generally thicker than 500 nm and is highly doped in order to provide high conductivity to the extrinsic collector layer. Conventional GeAuNi ohmic contacts or any other n-type contacts shown to be reliable in FET applications would work well for this ohmic contact.

9.5 PASSIVATION OF GaAs-BASED HBTS

The base of the HBT is a heavily p-doped layer that is both a transport layer for the input signal in a common-emitter HBT amplifier and a conduit for minority-carrier transport from the emitter to the collector. The transport from the emitter to the collector occurs largely due to diffusion, which is an otherwise non-directional process that is driven by a concentration gradient. Diffusion is an inefficient process that can send the minority carriers far from their desired destination, such as laterally to the surface of the extrinsic base region. As was discussed in Section 9.2.1, recombination of minority carriers in the base will reduce the number of electrons reaching the collector and will, therefore, limit the DC current gain. The surface of the extrinsic base region is one common area where recombination can readily occur. Bipolar transistors with a large emitter-periphery-to-area ratio are susceptible to surface recombination effects. Emitters of the order of 5 μm wide or less in GaAs HBTs are especially susceptible to these effects. The term passivation refers to methods designed to modify a semiconductor surface to reduce surface effects such as recombination in HBTs. Passivation of HBTs can refer to a semiconductor-based passivation, a dielectric passivation or a treatment that modifies the surface in a way that is said to passivate the surface. Each of these subjects will be treated in the following subsections.

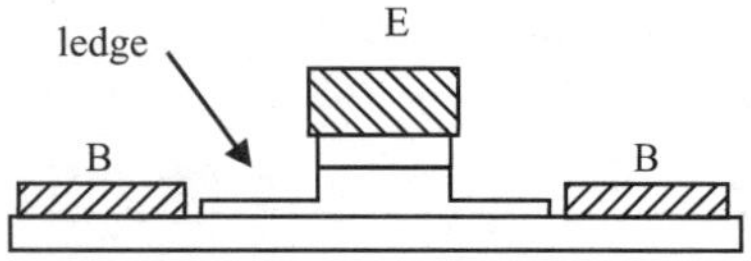

FIGURE 9.10 A schematic cross-section illustrating ledge passivation of an HBT.

9.5.1 Ledge passivation

An emitter ledge of an HBT is illustrated in FIGURE 9.10. It is a thin region of a wide bandgap semiconductor that surrounds the emitter in the extrinsic base region of the HBT. The GaAs surface and its high surface-recombination velocity of approximately 10^6 cm/s is replaced by a GaAs/ledge interface with a much lower surface-recombination velocity of approximately 10^3 cm/s. By using a wide bandgap semiconductor for the ledge, the diffusion of electrons to the surface of the extrinsic base region is inhibited by the band bending of the depleted ledge (and to a lesser extent by the potential barrier of the conduction-band offset between the base and the wide bandgap semiconductor). Generally speaking, the farther the ledge extends from the emitter edge, the greater the reduction in surface recombination. However an optimum ledge dimension must be determined, since the extrinsic base resistance will also be increased by placing the base metal further away from the emitter. This optimisation depends very much on the intended application of the HBT.

The basic process for constructing the semiconductor ledge will illustrate the objectives and act as a basis for contrasting alternative approaches. The emitter wet or dry etch is terminated prior to reaching the base material. The remaining emitter material in the extrinsic base region is kept thin so that it will be fully depleted by the combination of the surface charge and the junction potential. Practically speaking it may be anywhere from 20 to 50 nm thick, depending on the emitter doping level. A photoresist process is used to mask the ledge and complete the emitter removal process. For such thin layers, wet etches are usually preferred. Non-selective etches work fine because of the small amount of material being etched, although selective etches are also used if available. Another photoresist masking process is used to align the base metal to the emitter ledge after ledge removal. With high-resolution steppers, this realignment tolerance is very manageable. However, other methods are also available to self-align the base metal to the ledge, if desired.

One ledge self-alignment method [5] involves depositing the base ohmic contact directly on the ledge after aligning it to the emitter edge, as in FIGURE 9.11. It will contact the base after a shallow alloy reaction. This method requires careful optimisation of the ledge thickness and ohmic metal composition and sufficiently good control over the processes to repeatably achieve these thicknesses. Too shallow an alloy reaction will result in poor contact resistance due to a p-metal/wide bandgap n-type semiconductor interface. Too deep an alloy reaction will run the risk of local areas of spiking that will short the base and the collector. These alloy reactions are avoided if the ledge material is etched away prior to metal deposition.

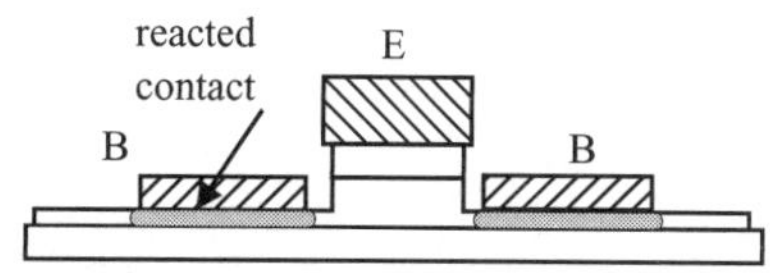

FIGURE 9.11 A schematic cross-section of a base contact reacting through the ledge of an HBT.

One may also apply a sidewall process to shield the extrinsic emitter. This process is illustrated in FIGURE 9.12. After terminating the emitter etch prior to reaching the base, a dielectric such as SiN is applied over the sample. A highly anisotropic etch-back, such as by RIE, is used to remove the dielectric over the planar areas of the sample, i.e. over the emitter and over the unshadowed extrinsic base region. Lack of a lateral dielectric etch leaves a protective coating over the emitter sidewall and the shadowed part of the extrinsic base. After the sidewall is in place, the emitter etch is continued to remove the remaining emitter material in regions not protected by the dielectric. This process is followed by the base metal deposition.

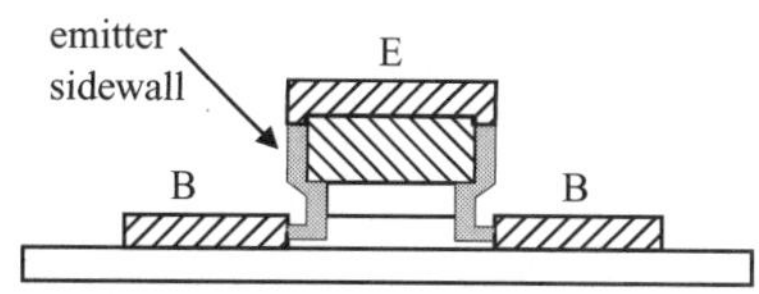

FIGURE 9.12 A schematic cross-section illustrating a self-aligned sidewall ledge passivation of an HBT.

The two basic types of ledges are AlGaAs and InGaP; the choice is dependent on the design of the HBT emitter. Both materials have lower surface recombination velocities than GaAs and are good passivation choices. In cases where InGaP is used as a ledge, it is also possible to use AlGaAs as part of a combination emitter. In this way a selective ledge etch can be used and a reproducible ledge thickness is more readily attainable. Because of the space-charge properties of AlGaAs (higher space-charge recombination than InGaP), the electrical characteristics of a combination emitter HBT resemble those of an AlGaAs emitter in many ways, but without having to grade the AlGaAs/GaAs emitter-base junction.

9.5.2 Dielectric passivation

As the area of an HBT inevitably shrinks in the quest for higher performance, the surface area of the exposed emitter sidewall becomes a significant fraction of the emitter itself. Any surface is associated with electronic states, which in turn are associated with depletion regions that can be considered a parasitic reduction in the emitter dimensions. A dielectric covering of the emitter sidewalls can, in principle, provide a similar surface modification to that of the emitter ledges described in the preceding section, but none to date are actually promoted for this purpose. Nevertheless, dielectric passivation provides an extremely important function. The exposed emitter edges are extremely susceptible to environmental exposure and damage. Thus, the current gain of an HBT may change, sometimes dramatically, due to unpassivated environmental exposure. The dielectric passivation provides a method of fixing the surface condition of the emitter edges and preventing further changes due to environmental contamination.

Silicon nitride is one common dielectric used for this purpose. When deposited by plasma-enhanced chemical vapour deposition (PECVD), it is mostly an amorphous combination of Si-Si, Si-N and Si-H bonded complexes. Because oxygen is a common contaminant in PECVD, it is easily incorporated in amounts of several percent. Ideally, silicon nitride is an amorphous or crystalline solid of a single chemical entity, Si_3N_4, which chemically forms from silane and ammonia precursors at temperatures near 700°C. This temperature is too high for GaAs processing because of the rapid formation and desorption of As-oxides. Fortunately, PECVD and related deposition methods, which can be used at temperatures of 300°C and lower, provide chemically similar dielectrics, in terms of the chemical properties that can stabilise dangling bonds at the (Al)GaAs/dielectric interface. Nevertheless, specific process issues associated with this dielectric can adversely impact HBTs if they are not understood and addressed.

Through careful control of the emitter-edge surface preparation during the emitter etch and subsequent surface cleans, an acceptable condition is attained and stabilised by a suitable dielectric deposition. The goal is to achieve a stable and acceptably high current gain after dielectric passivation. The stability comes about if the interface bonds, reconstructions and other relevant effects do not change significantly over long periods of time and if the dielectric prevents the diffusion of environmental contaminants that could cause these changes. Silicon nitride has long been an excellent choice for this type of passivation even though the scientific understanding of what occurs at the interface is mostly lacking. There is most certainly a need for process optimisation in which current gain and high-frequency performance, reliability and other considerations are monitored as a function of process conditions. Often process conditions suitable for high-quality dielectrics in other applications (dielectric interconnects, capacitors) are useful starting points.

One major process issue arises because hydrogen can passivate both n-type and p-type dopants in HBTs. PECVD silicon nitride typically contains of the order of 25% hydrogen. That this amount of hydrogen can be available and not cause HBT degradation problems is remarkable, but no lifetests to date have suggested that it is a serious problem. However, it is important to recognise the possible methods of degradation and test the HBTs with various PECVD deposition conditions. Little of this type of research has been reported in the literature. It has so far largely been left to process engineers at each institution to optimise their own processes. Some symptoms of hydrogen passivation have been documented from reliability studies on devices with H-implant isolation [6,7].

When p-type dopants are passivated in the base, the base sheet resistance will increase due to reduced hole concentration and the current gain will also increase (since recombination increases with higher doping level). Likewise, the subcollector sheet resistance will increase if n-type donors are passivated with H. Since some of these effects occur with ion damage, which can also be present in PECVD, RIE and high-density plasmas, it is necessary to do carefully controlled studies.

An SiN passivation step will usually change the DC current gain of a small-area HBT by a small amount (β changes by less than 15%, typically), although before and after measurements are not usually made. After the process steps of FIGURE 9.8, large-area HBTs are usually measured. Small-area HBTs are usually not measured until after another interconnect metallisation step. Large-area HBTs are affected much less than small-area HBTs (typically β changes by a few percent) by SiN passivation because of their large area-to-periphery ratio. Ledge-passivated HBTs are also less affected by SiN passivation, since much of the critical area where SiN interface modification can take place is already protected. Nevertheless, subtle effects are present with SiN passivation, even in these cases. Piezoelectric effects induced by SiN stress are one type of passivation effect to be considered. Such an effect is illustrated in FIGURE 9.13 where the DC current gain is both emitter-length and orientation dependent for a small-area HBT without a ledge. Stress-induced electric fields cause differing amounts of electron drift to the extrinsic base region and affect surface recombination and DC current gain. The effect was of larger magnitude for PECVD deposited SiN compared to ECR SiN for comparable stress [8], indicating that ECR films are more benign to the exposed HBT surfaces.

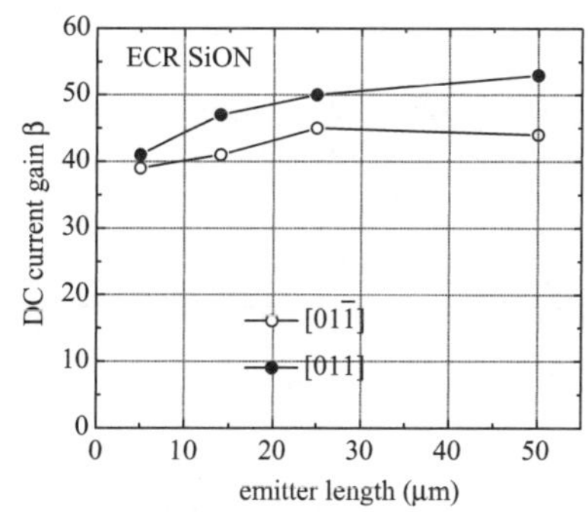

FIGURE 9.13 An example of orientation and dielectric overlayer effects in GaAs HBTs.

Dielectrics deposited with high-density plasmas (ECR and ICP) often contain much less hydrogen, of the order of 10–20%. However, we emphasise that mobile hydrogen is the concern, and process conditions may produce acceptable films by either deposition method.

9.5.3 Sulphur passivation

In the late 1980s, it was found that sulphur treatments of a GaAs surface could drastically modify the GaAs surface and change its electrical properties. The details of sulphur passivation are presented in Chapter 3. It was quickly verified by HBT engineers at different laboratories that this treatment reduced the surface recombination of GaAs in the extrinsic base region and led to increases in current gain for small emitter dimensions. Many different methods of sulphur treatment were studied while seeking the best approach. However, it was also quickly found that the sulphur-modified GaAs surface was not stable and the current gain degraded in a matter of hours in air without further treatments. This led to a further search for a method of stabilising the sulphur passivation. Although many promising results were achieved, none were found satisfactory for stabilising the current gain increase in HBTs. Two basic problems were never overcome. First, the sulphur-modified surface was very sensitive to plasma exposure and was never satisfactorily combined with dielectric passivation. Second, HBTs operate at a junction temperature near 150°C. Most of the chemical treatments to the sulphur-modified surface were not stable at that temperature.

9.6 VARIATIONS ON HBT PROCESSING

Many dozens of variations on HBT processes are possible. Some of the more interesting and useful ones have been selected for presentation in this section to give a flavour of how processing is used to improve devices or tailor them for certain needs. As will be seen, many examples exist where performance was enhanced significantly.

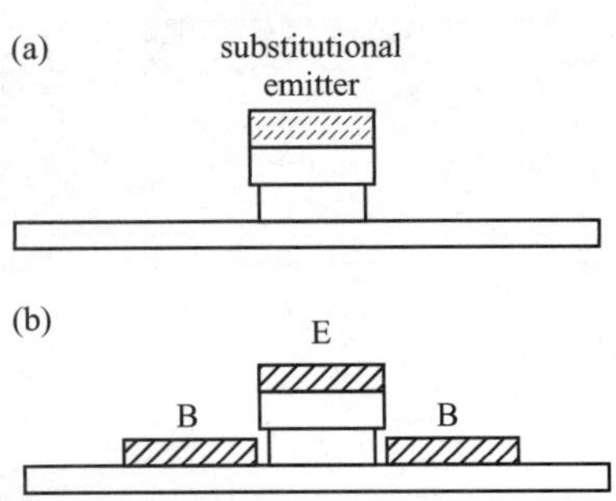

FIGURE 9.14 A substitutional emitter HBT process.

In Section 9.2.4, the basic triple-mesa process was described. The standard method of fabricating a self-aligned emitter is to use the emitter metal as an etch mask and as a shadow mask for self-aligned deposition of the base metal. An alternative self-aligned process involves defining the emitter using a dielectric or photoresist mask. Then the emitter is defined by wet or dry etching, preferably with undercut for metal shadowing (FIGURE 9.14(a)). After the emitter etch, the dielectric (photoresist) mask is removed and base and emitter metals are simultaneously evaporated (FIGURE 9.14(b)). This process is feasible if an InGaAs cap layer is used for a non-alloyed ohmic contact to the emitter.

A double-mesa process is sometimes used instead of a triple-mesa one. A basic drawback of the triple-mesa process is the introduction of significant topology from the various mesa steps. As seen in the cross-section of FIGURE 9.3 and the dimensions of the structure shown in TABLE 9.1, steps of approximately 2 μm are possible. The high-performance devices have minimal dimensions for the mesa and ohmic metal widths. The step height difference between different ohmic contact metals causes the photoresist to spin with different thicknesses over the contacts. High-resolution optical lithography steppers with near-UV sources have limited depth of field, which makes it difficult to pattern high-resolution features with different resist thickness and different depths. Contact lithography tools will lack contact between the mask and the resist at the lower elevation features (such as the collector contact), which degrades resolution of the exposed feature. These processing problems have reasonable solutions for GaAs HBTs, but with a price of extra processing complexity. A double-mesa process would use an implant for isolation of the subcollector. The implanted species might be chosen from He, B, O, or F, depending on the depth required and the implanter available. The isolation mechanism would be through a damage-creation process as described in Section 8.4. The double-mesa process is a reasonable way to minimise topography and has been widely used (FIGURE 9.15).

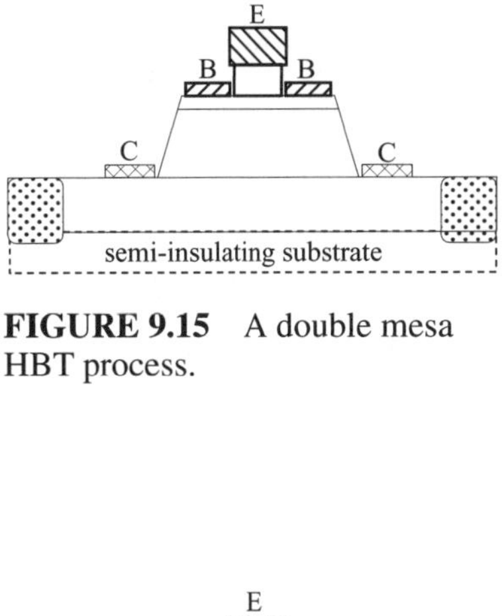

FIGURE 9.15 A double mesa HBT process.

Another possibility is to use ion implantation for isolation of all layers for a quasi-planar process [6]. However, ion implantation is not very effective for creating high-resistivity regions in highly doped InGaAs. At thc vcry least, InGaAs should be removed across the whole wafer during the emitter etch if it is used for an emitter cap layer. A cross-section of a quasi-planar HBT is shown in FIGURE 9.16. In this type of HBT process, the etching and metal deposition for base and collector contacts generally occur within the same lithography step.

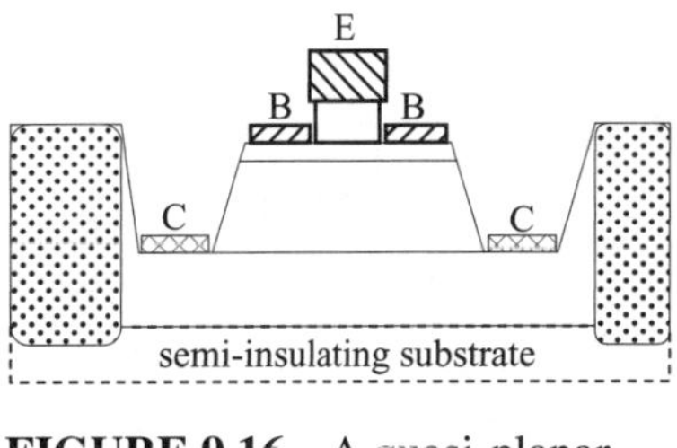

FIGURE 9.16 A quasi-planar ion implant isolated HBT process.

Hydrogen is an obvious element for the implant isolation process because the thickness of the collector and subcollector to be isolated can be close to 1.0 μm. Thc more common ion implanters have an upper energy of 200 keV and only some of the light elements have a projected range in GaAs of greater than 1 μm. Atomic hydrogen implants were a popular early choice for HBT isolation, providing good isolation properties. Unfortunately, it was found that the H was extremely mobile and easily moved across the device to passivate donors and acceptors within the HBT. The H-passivation complexes caused large changes in device operation that were not acceptable to device engineers. Consequently, B or

O have been used instead of H for implant isolation, depending on the projected range required and the implanter available.

Tradeoffs regarding base doping levels were discussed in Section 9.2.2. High doping levels will improve high-frequency analogue performance by reducing the base resistance, but they also reduce the current gain by promoting recombination centres. Another tradeoff is the desire for extremely thin base layers (to improve transit times), while still maintaining the ability to etch down to this layer and achieve reliable ohmic contacts that do not completely react through the layer. One approach to improve on these tradeoffs is doping the extrinsic base as illustrated in the cross-section of FIGURE 9.17. A dielectric such as SiN/SiO_2 covers the patterned emitter metal. SiO_2 can be patterned over WSi during the WSi etch, for example. After the emitter etch, a dielectric sidewall is formed (to passivate and protect the emitter) by a deposition and anisotropic etchback. Then Zn diffusion can be used to increase the doping for the intrinsic base. An open-tube diffusion using a $ZnAs_2$ source is one possibility for this step. A temperature of 550°C for 1 min results in a diffusion depth of 80 nm and a sheet resistance of 385 Ω/square [9]. The rest of the HBT fabrication can proceed in any of the ways described previously. Although the extrinsic base resistance is reduced by this Zn diffusion, the base-collector capacitance may be increased if the process is not optimised properly. The base and collector doping levels typically differ by more than three orders of magnitude, and it may be difficult to control the tail of the diffusion into the collector. If the Zn tail converts part of the collector into p-type GaAs, extrinsic C_{BC} will go up as the collector distance is reduced. Another concern is that Zn diffusion will degrade the reliability.

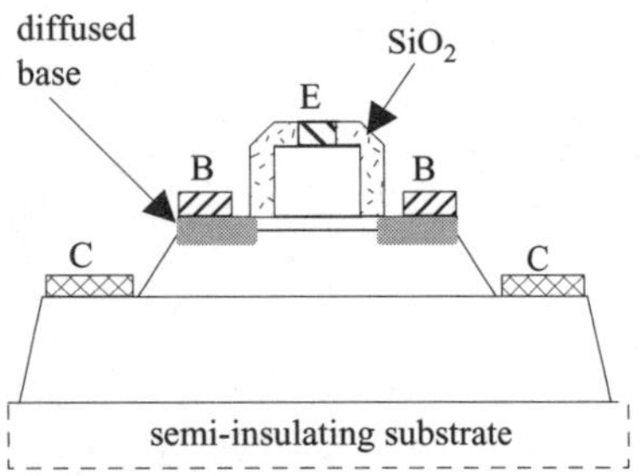

FIGURE 9.17 A schematic cross-section of an HBT with an extrinsic diffused base.

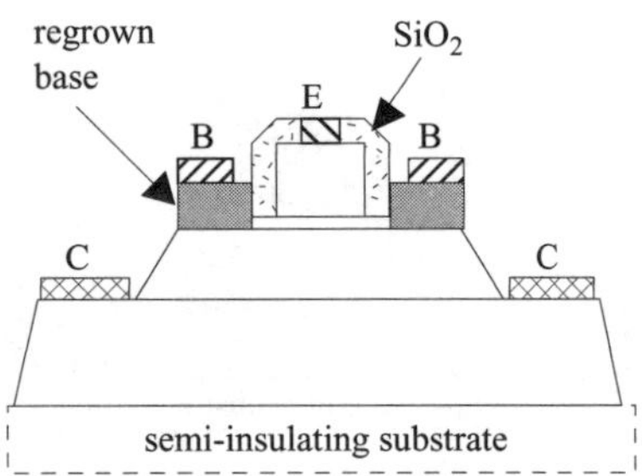

FIGURE 9.18 A schematic cross-section of an HBT with an extrinsic regrown base.

Another method with the same goal of reducing the extrinsic base resistance is to re-grow the extrinsic base [10]. Using a similar emitter protection scheme as that for Zn diffusion (FIGURE 9.18), the extrinsic base region is extended by selective regrowth of p^+ GaAs using MOCVD. The emitter is protected with SiO_2. Then the extrinsic base is selectively regrown on the GaAs. Fabrication of the rest of the HBT occurs according to the methods of FIGURE 9.8, for example. Although effective in principle, regrowth methods for GaAs HBTs have not found the same acceptance as for laser diodes due to the more cost-sensitive nature of HBT applications.

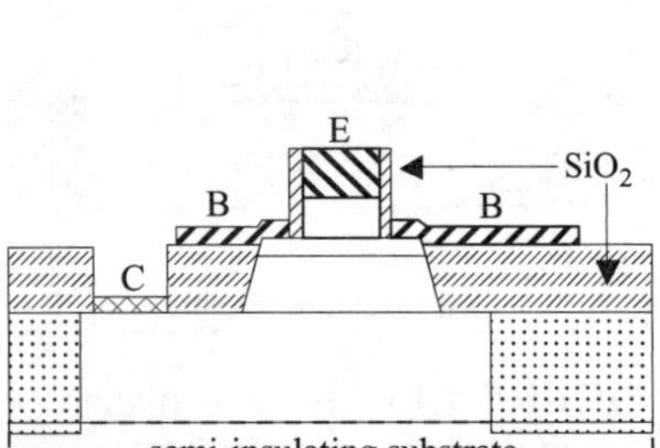

FIGURE 9.19 A schematic cross-section of an HBT with a buried SiO_2 dielectric.

As has been mentioned, methods to reduce C_{BC} can take many different forms. The method described next reduces the extrinsic base area to a bare minimum and is illustrated in cross-section in FIGURE 9.19. After emitter metal definition and emitter etch, a thick dielectric sidewall (SiO_2 works well) is formed by deposition and dielectric etchback. The thickness of the sidewall will be proportional to the original thickness of the dielectric. The emitter and sidewall will be the mask for a self-aligned base-mesa etch. After another lithography step and collector isolation by ion implantation, the outside of the base-collector junction is protected with dielectric by planarisation and etchback. An SiO_2 layer is deposited and the device is planarised with a spin-coated film such as photoresist. The entire structure is then etched back until the base is exposed. A thin sidewall is then deposited (by dielectric deposition and anisotropic etchback) to protect the emitter. The base metal is formed next. An etchable metal like WSi is deposited over the entire wafer and etched after protecting the base area with photoresist. The base ohmic contact is non-alloyed and extends little more than a transfer length from the emitter edge. At the same time, it is supported and isolated from the collector by a dielectric. Formation of the active elements of the HBT is completed after etching a window through the dielectric down to the subcollector and depositing the collector ohmic contact. In order to minimise plasma damage, an InGaP emitter can be left in place until its removal just prior to the base metal deposition. In addition to isolating the base from the collector, the buried SiO_2 has a lower dielectric constant than GaAs and helps minimise extrinsic C_{BC}. This process was shown to achieve f_t near 100 GHz and f_{max} near 200 GHz for InGaP/GaAs HBTs [11]. Its primary disadvantage is in the complexity and precision required during processing.

A selective buried subcollector has been used to enhance HBT performance by reducing C_{BC} as in EQN (9.3). A subcollector is grown by MOCVD or MBE and patterned by chemical etching. A wet etch process followed by oxide removal is preferable for subsequent regrowth. Using a sulphuric-acid-based etch yields a subcollector mesa in the shape of a trapezoid. The rest of the HBT structure can then be regrown by MOCVD. The HBT is then fabricated with the desired process. A schematic cross-section of the structure is illustrated in FIGURE 9.20(a). The base-collector capacitance is defined by overlap between the base mesa and the subcollector. As seen in the top view of FIGURE 9.20(b), the base mesa and the subcollector mesa extend in opposite directions and the overlap can be minimal. The base and collector mesa extend more to the left side of the emitter to allow room for the base contact, while the subcollector mesa extends more to the right side of the emitter for the same reason. The process is implemented by growing the subcollector first, patterning it by mesa etching, and then regrowing the rest of the HBT structure. When the HBT is processed, the step of one side of the subcollector lies very close to the emitter, and care must be taken that the regrowth does not create extra defects in this region as they will lead to

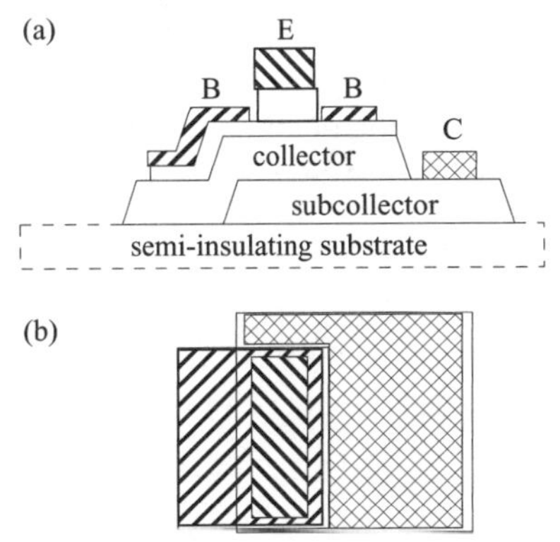

FIGURE 9.20 A schematic cross-section of an HBT with a selectively grown subcollector for C_{BC} reduction.

excessive recombination. An f_{max} of 140 GHz was reported using this process [12].

Another approach for making a selective buried subcollector uses ion implantation instead of etching to define the subcollector. This approach is completely planar and avoids regrowth on a faceted mesa edge close to the emitter. However, it is difficult to achieve the high conductivity necessary for the subcollector by ion implantation in GaAs. Epitaxial growth is much more effective for maximising the doping level and thickness compared to ion implantation. Ion implanted, buried subcollector HBTs with increased collector resistance will negate the benefit of a reduced C_{BC}.

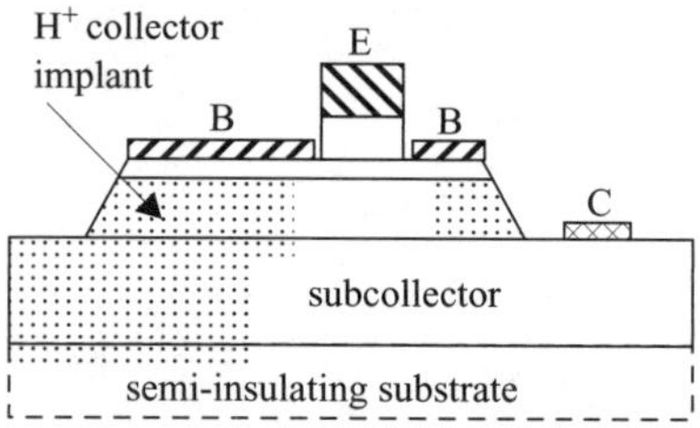

FIGURE 9.21 A schematic cross-section of an HBT with deep implants for C_{BC} reduction.

Ion implantation has been used in another way to reduce the base-collector capacitance. A deep H-ion implantation is used to damage the conducting collector and subcollector region below the extrinsic base, as illustrated in FIGURE 9.21. As in the case for the regrowth method, one side of the subcollector is left intact along with a narrow base metal and mesa. The other side of the subcollector is implanted with only a small separation from the emitter edge to reduce extrinsic C_{BC}. The H-implant only minimally damages the extrinsic base because the surface concentration of the damage profile is several times less than the peak damage at the projected range for the H ion. Another shallower, lower-dose implant can be used to damage the low-doped collector on the side of the emitter without the subcollector implant. As for the selective buried subcollector process, the collector contact is made only on one side of the emitter. With this type of process the subcollector resistance can be expected to increase in excess of 1 MΩ, while the base resistance increases from 175 to 265 Ω/square [13]. A separation of 1.2 μm of the implant from the emitter was chosen to minimise the damage to the intrinsic base region. An $f_{max} > 200$ GHz was achieved with this process [13]. Although these results are impressive, hydrogen passivation of donors and acceptors is a serious reliability concern. Use of another ion instead of H will, unfortunately, lead to greater damage to the base that may be unacceptable.

Another method for minimising C_{BC} utilises an undercut collector, a technique borrowed from InP HBTs. Emitter metal, mesa and base metal are formed as in the standard triple-mesa process. After base-mesa lithography, the collector undercut is performed. The key to this process is the design of etch-stop layers in the collector and subcollector. InGaP or AlGaAs layers can be used as the etch-stop layers. For the InGaP layers wet etching (e.g. $H_3PO_4 : H_2O_2 : H_2O$) selectivity can be used to achieve the desired collector undercut profile. To undercut the AlGaAs layers, RIE can be used. Active HBT fabrication is completed after collector ohmic contact and collector mesa formation. A cross-section of an HBT with collector undercut is illustrated in FIGURE 9.22. Though this structure looks very unstable, the mechanical strength of microstructures is often surprising. Hero results can be reported with air as the dielectric for the gap between the extrinsic base and the subcollector. An f_{max} of 170 GHz has been reported for this type of structure [14]. As a performance enhancement technique, the collector undercut method with an air gap is superior to the previously described C_{BC} reduction techniques. The previously described techniques used GaAs, with its high dielectric constant, as the insulator beneath the extrinsic base and will have greater parasitic capacitance.

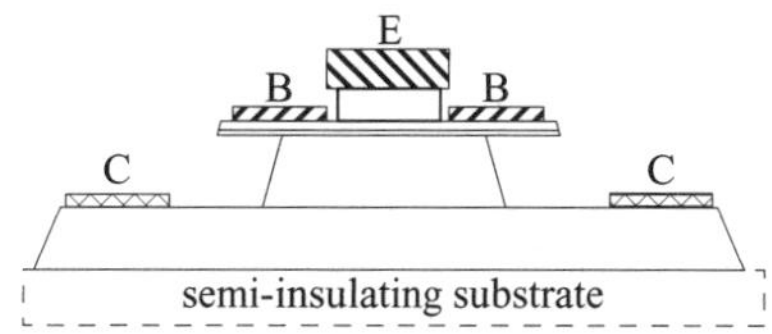

FIGURE 9.22 A schematic illustration of an HBT with a collector undercut etch.

9.7 RELIABILITY OF HBTS

HBT performance peaks at high current density, often at greater than 10^5 A/cm^2. However, reliability of HBTs has been difficult to achieve for current densities greater than 2×10^4 A/cm^2. In applications requiring high linearity, perhaps backing off on power levels to the lower current density levels may be an acceptable solution. However, many other applications would be better served with higher current density devices. The need for greater performance from reliable parts has spurred research into the degradation mechanisms and searches for improvements. Much has been learned and progress continues to be made. This section will summarise some of the important findings and current understanding of HBT reliability. This information should be an important consideration influencing many aspects of HBT processing.

Like FETs, both temperature and electric fields are important stressors. Thermal stress is especially difficult in HBTs because of self-heating effects in small emitters. Electrical stress is difficult because of the varied and subtle ways that the effects manifest themselves.

HBTs can fail by gradual degradation or by sudden failure. A gradual degradation is typified by a 10–20% change in an electrical parameter such as current gain. Depending on the application, other electrical parameter changes such as V_{BE}, I_C or output power, may provide the criteria for failure. Abrupt failure may be preceded by little degradation or by gradual degradation.

As for FETs, thermal stress will cause HBTs to fail through metal contact degradation. Also like the FET case, the thinnest active layers are the most susceptible to degradation. Activation energies are high for contact degradation mechanisms (1.3–2.0 eV). A number of ohmic contacts that degrade with this activation energy contribute to high reliability even for thin layers, if properly implemented. Some of these were described in Section 9.4. They can generally be optimised to last hundreds of hours at 300°C, which extrapolates to millions or more hours at 150°C.

The most common mode of HBT degradation is gradual β decrease, with device failure commonly defined as a percentage degradation from the initial value (10–20%, commonly). Though many specific types of failure are reported, two mechanisms receive the most attention: H passivation (and depassivation) effects and recombination-enhanced-defect-reaction (REDR). These will be discussed in order.

Hydrogen is known to passivate donor and acceptor impurities in many semiconductors, including GaAs. In HBTs, MOCVD growth is known to incorporate H in the base layer when it is C-doped. H incorporation at about 20% of the C level is not unusual. The H incorporates as C-H complexes, which act as passivated acceptors. The C-H complexes can dissociate thermally, which is probably what happens to 80% of them at MOCVD growth temperatures (near 600°C). They will also dissociate in the presence of electrons that diffuse from the base according to the reaction:

$$\mathrm{CH} = \mathrm{C}^- + \mathrm{H}^+ \tag{9.8}$$

The C^- and H^+ can be electrostatically bound and quasi-stable in the absence of electrons. Creation of these types of complexes in the first moments of HBT operation causes an initial drop in current which has been explained by H^+ acting as a recombination centre as follows [15,16]:

$$\mathrm{H}^+ + \mathrm{e} = \mathrm{H}^0 \tag{9.9}$$

$$\mathrm{H}^0 + \mathrm{e} = \mathrm{H}^- \tag{9.10}$$

H^0 or H^- either do not act as recombination centres or diffuse away from the base region. The electrical effect is that HBTs with C-H complexes in the base show a decrease in β during initial operation. As time passes the number of C-H complexes slowly decreases as the reaction in EQN (9.8) proceeds, leading to an increase in p-type doping in the base of the HBT. The current gain of the HBT decreases further in response to increasing C-doping. The former short-term effect can be dealt with through a burn-in process, while the latter long-term effect is a reliability issue. The fact that both thermal energy and electrons dissociate C-H complexes implies that both electron density and junction temperature are important factors in characterising the kinetics and reliability implications of H-incorporation in HBTs. Although most reliability studies emphasise current gain, H-depassivation also lowers V_{BE} (the base-emitter turn-on voltage).

This reliability issue is dealt with either by removing the source of hydrogen from the growth chemistry of the HBT or by removing the hydrogen after growth. By using MBE instead of MOCVD for the HBT growth, one incorporates much less hydrogen. For MOCVD growths, one resorts to post-growth annealing at temperatures near 650°C to remove the hydrogen. The timing (i.e. if an interrupted growth is used) and length of the anneal must be carefully optimised to avoid degradation of other aspects of the HBT structure.

After H-passivation effects, the REDR mechanism receives the most attention. As the HBT is operated, a fraction of the electrons injected into the base recombine non-radiatively. For example, electrons can tunnel from the emitter into a midgap trap level; once in this trap, the electron can recombine non-radiatively with holes. A non-radiative recombination event involves releasing energy to phonons, which carries a finite probability of rearranging the lattice and creating another defect. The process has positive feedback such that the generation of a point defect associated with a non-radiative recombination centre leads to the generation of more point defects associated with recombination centres. These point defects can join together to form more extended defects and more extended recombination centres which leads to the gradual degradation of the device. The best means of avoiding REDR is to start with high-quality HBTs with low levels of initial recombination. Well-optimised InGaP ledge passivation may be a good means of achieving high-quality, low recombination HBTs.

Evidence from HBT reliability studies [17] indicates that HBT degradation includes many similarities to LED degradation [18]. Activation energies associated with midgap trap formation are in the range of 0.4–0.7 eV. Such low activation energies are problematic because long times may be required to reveal evidence of degradation at reasonable temperatures. Attempts to accelerate the process at high temperatures is likely to activate another mechanism with a different (usually higher) activation energy and mask the degradation due to REDR alone. The evidence seems to suggest that high-quality HBTs that are free from degradation with higher activation energies will ultimately show degradation by REDR.

To the extent that the physics of REDR is accepted, it has important implications for the choice of emitter material for reliability considerations. REDR can occur in the emitter and emitter-base space-charge region if holes are injected into the emitter. AlGaAs with a valence-band discontinuity of 0.13 eV for an abrupt emitter and up to 0.25 eV for a graded emitter (25% Al), is a decent barrier to hole injection. However, ordered InGaP has a valence band discontinuity as high as 0.43 eV, leading to far less recombination in the high-bandgap emitter. The larger bandgap of the emitter leads to more energy associated with REDR and can negatively impact reliability. Also, most AlGaAs is initially more defective than InGaP making the incidence of REDR more likely. Today, GaAs HBT reliability is an evolving story. However, the evidence indicates that more progress is being made at this time in raising the maximum reliable current density using InGaP emitters with ledge passivation. Verification that the REDR mechanism is the main limiting factor may suggest that the base doping levels need to be considered and possibly lowered to provide the lower initial recombination levels.

Another example of gradual degradation is dopant diffusion. Beryllium diffusion into the emitter will extend the base-emitter junction from the heterointerface into the wide bandgap material and cause V_{BE} shifts. Beryllium was the predominant early preference for MBE doping of the base, but has largely been supplanted by C-doping to avoid this problematic diffusion of the base dopant. Nevertheless, Be-doped HBTs were designed to be reliable at current densities of 25 kA/cm^2 by optimising the structure and growth conditions, and by being conservative in rating the operating current density [19]. Zinc doping in MOCVD has similar diffusion concerns and has been supplanted by carbon doping.

Although we have focused thus far on gradual degradation of HBT electrical parameters, some devices fail suddenly, as in the case for LEDs and LDs. Failure modes include dislocation propagation, dislocation loops and dark line defects. These defects can be observed by electroluminescence and, with considerable patience, TEM. The dislocations may have propagated from the substrate or be the result of mechanical stress in the device. Low-dislocation substrates are one means of minimising the former type of propagation. Mechanical damage can result from stresses arising during sawing, scribing or bonding the chip. They can also result from process steps such as high-stress metals or other steps that may induce strain. Studies of process-induced defects are little more than exploratory case studies at this time. Relatively little is known about the true origin of these types of defects.

After long-term bias-temperature stress, carbon precipitates are sometimes observed in GaAs HBTs with a C-doped base. For example, both AlGaAs/GaAs and InGaP/GaAs HBTs with 4×10^{19} cm^{-3} C-doping of the base were biased at 6×10^4 A/cm^2 at junction temperature (T_j) of 150–260°C [20]. After an initial burn-in, a stage of gradual β degradation resulted, followed by an abrupt failure. InGaP/GaAs HBTs displayed both slower gradual degradation and longer times until abrupt failure than AlGaAs/GaAs HBTs. TEM analysis with EDX was performed on some of the failed samples. Both C-precipitates and occasional migration of metallic impurities were imaged by TEM/EDX for the InGaP/GaAs HBTs. The C-precipitates were 10–15 nm in diameter.

Another reported degradation mechanism of GaAs-based HBTs is a microtwin-like defect [21]. It has been suggested that these defects are similar to typical micro defects in hydrogenated Si, but it is not presently known if they are H-related in GaAs.

Reliability studies of GaAs-based HBTs have provided useful information that can guide certain technology choices. They have provided useful guides to operating current-density ranges. Best practice for the choice of ohmic contacts, base dopant, emitter material, and ledge-passivation schemes is also indicated from reliability studies. Information is readily available for minimising H-passivation effects. Companies involved in high-volume production run reliability evaluations on multiple lots every month and unquestionably will have access to information about many process choices that are not publicly available. Most reliability issues, however, are poorly understood, including the origin of various types of extended defects, the best methods of dielectric passivation and the best ways of avoiding process-induced stress. In addition, most device engineers are not satisfied with the limitations on HBT current density and would agree that much remains to be done.

9.8 CONCLUSIONS

GaAs HBTs are greatly affected by the unintentional introduction of defects in certain key parts of their structure. These process-sensitive parts of the HBT structure need to be understood and controlled by process engineers. HBT operation, testing and performance characterisation supply the basic information that materials, processing and device engineers need to solve and understand the often complex material and process interactions. Reliability characterisation is an equally important guide for HBT engineers to the extent that such information is available. Many HBT process choices are available and some of the main ones were reviewed in this chapter. As in most facets of engineering, knowledge and information are a starting point for a job that needs to be done better than is now possible.

REFERENCES

[1] W. Liu [*Handbook of III-V Heterojunction Bipolar Transistors* (John Wiley and Sons, New York, 1998)]

[2] W. Lui [*Fundamentals of III-V Devices: HBTs, MESFETs, and HFETs/HEMTs* (John Wiley and Sons, New York, 1999)]
[3] N. Pan et al. [*IEEE Electron Device Lett. (USA)* vol.19 (1998) p.115]
[4] M.J. Cich, J.A. Johnson, G.M. Peake, O.B. Spahn [*Appl. Phys. Lett. (USA)* vol.82 (2003) p.651]
[5] E. Neubauer, M. Mai, J. Wurfl, W. Osterle [*Semicond. Sci. Technol. (UK)* vol.15 (2000) p.818]
[6] F. Ren, S.J. Pearton, W.S. Hobson, T.R. Fullowan, J. Lothian, A.W. Yanof [*Appl. Phys. Lett. (USA)* vol.56 (1990) p.860]
[7] S.J. Pearton [*Int. J. Mod. Phys. B (Singapore)* vol.30 (1994) p.1247]
[8] A.G. Baca, C. Monier, P.C. Chang, R.D. Briggs, M.G. Armendariz, S.J. Pearton [*Solid-State Electron. (UK)* vol.46 (2002) p.797]
[9] T. Nittono, K. Nagata, O. Nakajima, T. Ishibashi [*IEEE Electron Device Lett. (USA)* vol.10 (1989) p.506]
[10] H. Shimawaki, Y. Amamiya, N. Furuhata, K. Honjo [*IEEE Trans. Electron Devices (USA)* vol.42 (1995) p.1735]
[11] T. Oka, K. Hirata, K. Ouchi, H. Uchiyama, K. Mochizuki, T. Nakamura [*IEEE Trans. Electron Devices (USA)* vol.45 (1998) p.2276]
[12] Y.-F. Yang, C.-C. Hsu, E.S. Yang, H.-J. Ou [*IEEE Electron Device Lett. (USA)* vol.17 (1996) p.531]
[13] M.-C. Ho, R.A. Johnson, W.J. Ho, M.F. Chang, P.M. Asbeck [*IEEE Electron Device Lett. (USA)* vol.16 (1995) p.512]
[14] W.L. Chen, H.F. Chau, M. Tutt, M.C. Ho, T.S. Kim, T. Henderson [*IEEE Electron Device Lett. (USA)* vol.18 (1997) p.355]
[15] F. Ren, C.R. Abernathy, S.N.G. Chu, J.R. Lothian, S.J. Pearton [*Solid-State Electron. (UK)* vol.38 (1995) p.1137]
[16] J.Y. Chi, K. Lu [*IEEE Electron Device Lett. (USA)* vol.19 (1998) p.408]
[17] T. Henderson [*Microelectron. Reliab. (UK)* vol.39 (1999) p.1033–42
[18] O. Ueda [*J. Electrochem. Soc. (USA)* vol.135 (1988) p.11C]
[19] D.C. Streit et al. [*J. Vac. Sci. Technol. B (USA)* vol.14 (1996) p.2216]
[20] O. Ueda [*Microelectron. Reliab. (UK)* vol.39 (1999) p.1839]
[21] H. Suguhara, J. Nakano, T. Nittono, K. Ogawa [*GaAs IC Symp. Technol. Digest* (IEEE, Piscataway, NJ, USA, 1993) p.115]

Chapter 10

Wet oxidation for optoelectronic and MIS GaAs devices

10.1 CHAPTER SCOPE

Most of the post-growth fabrication issues of importance in making electronic devices are equally applicable in making optoelectronic devices. Many of the performance characteristics are grown in by the specific design of mirror stacks, quantum-well numbers and thicknesses, dopant profile distributions, etc. Post-growth processing, with the exception of wet oxidation, is generally the same as that of electronic devices. The previous chapters in this book dealing with cleaning, passivation, etching and contact formation address the main requirements for fabricating optoelectronic devices. The details of the technique of regrowth, which involves etching a structure into a wafer and introducing it again into a growth chamber for deposition of additional semiconductor layers on the textured substrate, are beyond the scope of this book. To fabricate top-surface gratings for a DFB (distributed feedback) laser, proper mask alignment relative to the crystal axes and crystallographic etching can be used to form the grating, as discussed in Section 4.3.2. Crystallographic etching can also be used to form smooth, vertical mirrors with proper alignment of the mask relative to a wafer's crystal planes.

Because the principal additional fabrication technique for optoelectronic devices is the wet oxidation of buried Al-containing layers to make current apertures in VCSELs (vertical cavity surface emitting lasers), this chapter will focus on that process. Both fundamental issues and practical considerations for using a wet oxidation process are presented here. The chapter concludes with a discussion of attempts to make electronic devices, such as GaAs-on-insulator (GOI) MESFETs and metal-insulator-semiconductor (MIS) devices, using wet oxidation.

10.2 MECHANISM OF WET OXIDATION PROCESSES

The oxidation of Si has been the topic of exhaustive research because of its importance in forming the Si/SiO_2 interface, the basis of CMOS technology. By comparison, the oxidation of GaAs and AlAs is not as well understood. This section addresses some of the fundamental chemical and physical properties that profoundly affect the application of wet oxidation technology in GaAs devices.

10.2.1 Chemistry of wet and dry oxidation of AlGaAs

Dry oxidation of AlAs (and similarly of GaAs) can be described relatively simply by

$$2\mathrm{AlAs} + 3\mathrm{O}_2 = \mathrm{Al}_2\mathrm{O}_3 + \mathrm{As}_2\mathrm{O}_3 \tag{10.1}$$

which presumably started with Al-O and As-O surface species. In contrast, wet oxidation of AlAs or AlGaAs is much more complicated.

Water can dissociatively adsorb on AlAs to give Al-O, Al-OH and As-H. Some of the intermediate species in the wet oxidation process and their evolution with oxidation time have been identified using Raman spectroscopy [1] and the following suite of reactions [2] are consistent with the observed products of As and As_2O_3 and the surface species formed by water adsorption. (Molecular analogues, such as AsH_3 and AlO(OH), are used here to represent surface species, such as As-H and Al-OH, so the thermodynamic calculations are approximations to the actual situation.)

Reaction	ΔG (kJ/mol at 425°C)	
$2\mathrm{AlAs} + 3\mathrm{H_2O_{(g)}} = \mathrm{Al_2O_3} + 2\mathrm{AsH_3}$	−451	(10.2)
$2\mathrm{AlAs} + 4\mathrm{H_2O_{(g)}} = 2\mathrm{AlO(OH)} + 2\mathrm{AsH_3}$	−404	(10.3)
$2\mathrm{AsH_3} = 2\mathrm{As} + 3\mathrm{H_2}$	−153	(10.4)
$2\mathrm{AsH_3} + 3\mathrm{H_2O} = \mathrm{As_2O_{3(l)}} + 6\mathrm{H_2}$	−22	(10.5)
$\mathrm{As_2O_{3(l)}} + 3\mathrm{H_2} = 2\mathrm{As} + 3\mathrm{H_2O_{(g)}}$	−131	(10.6)
$2\mathrm{AlAs} + 3\mathrm{H_2O_{(g)}} = \mathrm{Al_2O_3} + 2\mathrm{As} + 3\mathrm{H_2}$	−604	(10.7)
$2\mathrm{AlAs} + 4\mathrm{H_2O_{(g)}} = 2\mathrm{AlO(OH)} + 2\mathrm{As} + 3\mathrm{H_2}$	−557	(10.8)

The Gibbs free energy for these reactions depends on the temperature; at 425°C, all have a negative change in free energy, ΔG, which means that the reaction should proceed spontaneously if not kinetically constrained. A general caveat should always be remembered when discussing reactions in terms of their thermodynamics. While thermodynamic calculations predict what reactions will have negative changes in Gibbs free energy (ΔG) and should, therefore, proceed spontaneously, the kinetics of the reactions often determine what actually happens. If there is a high activation energy (large $E^* =$ large kinetic barrier) for a reaction, it may proceed only very slowly, if at all, despite having a large $-\Delta G$.

The H^+ from water serves as the oxidising agent for the low-oxidation-state Al in AlAs. As_2O_3 is observed by Raman spectroscopy as an intermediate in the reaction, but the elemental H resulting from the reduction of the H^+ of water in the formation of Al_2O_3 or AlO(OH) can react readily with As_2O_3 to form As. An analogous set of equations can be formulated for GaAs, with AlGaAs being a composite of the two. Changes in the relative rates of formation and loss of these species can explain most of the observations regarding composition dependence, time dependence, layer shrinkage upon oxidation, changes in electronic properties in adjacent layers and the inability to form a good GaAs/oxide interface for MIS devices. This will be discussed further in Section 10.3.

The formation of elemental As is the most fundamental difference between wet and dry oxidation. It can readily volatilise as elemental As or react with H to form AsH_3, a gas under these reaction conditions. Significant quantities of AsH_3 are not detected in standard wet oxidation processes, so formation of volatile elemental As is apparently strongly favoured. Elemental analysis of completely oxidised AlAs or AlGaAs shows that very little As remains after oxidation. This loss of As favours formation of a relatively porous oxide material that facilitates transport of reactants and products. This porosity enables a linear dependence on time as the oxidation proceeds over relatively long distances, as will be discussed in more detail in Section 10.3.

Wet oxidation of AlGaAs can be suppressed by addition of even small amounts of oxygen to the gas. This may be due to the preferential consumption of the reduced H-species by oxygen, preventing formation of significant amounts of As. In the presence of oxygen, the process proceeds like a dry oxidation even when plenty of water vapour is present.

10.2.2 Electronic consequences of oxidation processes

While the generation of elemental As is highly desirable to permit rapid, deep oxidation of an AlGaAs layer, the presence of free As atoms is problematic for the electronic properties and crystallographic perfection of the semiconductor adjacent to an oxidised layer. Some of the possible reactions with adjacent GaAs are illustrated in FIGURE 10.1. Arsenic atoms can react with the surface Ga atoms, causing them to move forward onto a new lattice site and leaving a Ga vacancy (V_{Ga}) behind. Column III vacancies

$$As^o \rightleftharpoons As_i$$

$$As^o \rightleftharpoons GaAs + V_{Ga}$$

$$As_i + V_{Ga} \rightleftharpoons As_{Ga}$$

$$As_i + Ga_{Ga} \rightleftharpoons As_{Ga} + Ga_i$$

FIGURE 10.1 Defects generated by free As atoms.

play an important role in enhancing interdiffusion, which can degrade heterojunction interfaces. An As atom can diffuse into the semiconductor as an interstitial As (As_i). The As_i can move deeper into the semiconductor quite rapidly and undergo reaction quite distant from the initial point of generation. The interstitial can form anti-site defects (As_{Ga}) by binding at vacant Ga sites or by displacing a Ga from its site, forming a Ga interstitial (Ga_i). These defects quench luminescence. Arsenic precipitates resulting from the diffusion of interstitials have been observed at remote interfaces (tens of nm away from the oxide) and even at the GaAs surface of a wet-oxidised sample.

Problems due to defect-caused luminescence quenching and intermixing at heterointerfaces in optoelectronic devices will be discussed in Section 10.5.2. The importance of elemental As at the oxide/semiconductor interface is especially problematic when trying to achieve a very low interface-state density, as will be discussed in Section 10.6.

10.3 RATES AND PROFILE EVOLUTION

The wet oxidation of AlGaAs can be understood in terms of the classic model of Deal and Grove. In their model, the time dependence of oxidation can be described as the sum of two terms: $d^2/k_{diff} + d/k_{rxn} = t$ where d is the oxidation depth and t is the time. The linear term dominates when the oxidation is reaction-rate limited and the parabolic term is dominant under diffusion-limited conditions. In dry oxidation of AlGaAs, the parabolic, diffusion-limited term is dominant and rates slow rapidly as the oxidation front penetrates into an AlGaAs layer. However, it is most desirable to operate in the linear regime when trying to oxidise long distances (tens and even hundreds of nm) very controllably and in a relatively short time. Wet oxidation conditions can be found that produce primarily linear, reaction-rate limited behaviour for high Al-content AlGaAs. The specific conditions favouring linear oxidation depend on both wafer temperature and AlGaAs mole fraction and factors that favour linear rates are discussed below.

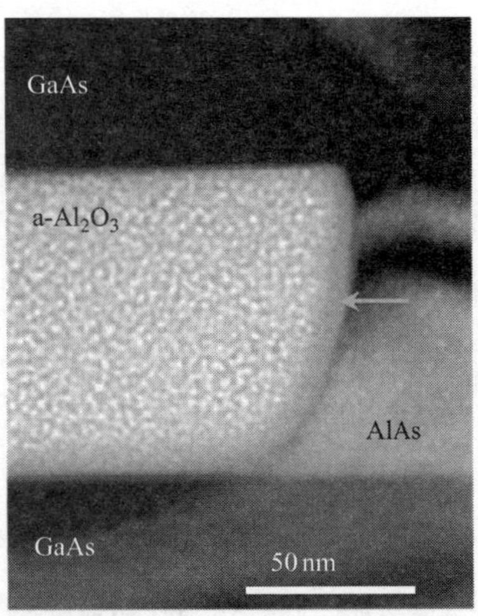

FIGURE 10.2 Dense amorphous interfacial layer at oxidation front (TEM by R.D. Twesten).

The time dependence of the oxidation rate will be determined by the relationship between the rate of formation of As_2O_3 and its conversion to As (EQNS 10.5 and 10.6 in Section 10.2.1) and the subsequent rate of loss of As from the oxidised layer to leave behind a relatively porous matrix of AlO_xH_y species, such as Al_2O_3 and AlO(OH). TEM studies have revealed the presence of a thin (a few nm), dense amorphous region at the oxidation front of a layer oxidised at 440°C (FIGURE 10.2). Behind this dense region lies a less dense region of amorphous AlO_xH_y that extends

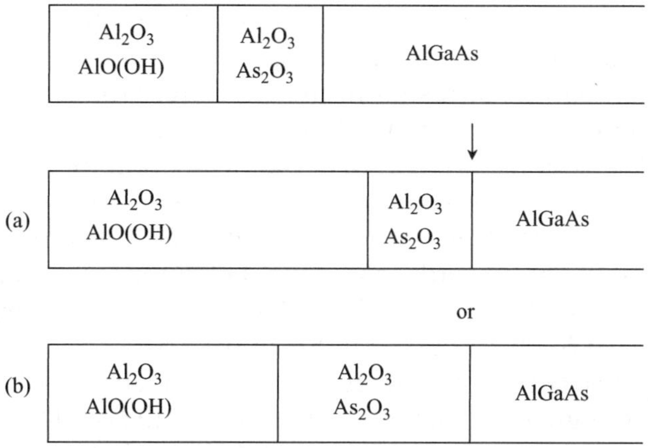

FIGURE 10.3 Evolution of oxide during wet oxidation of a layer: (a) linear: fixed thickness of interfacial layer, (b) parabolic: increasing thickness of interfacial layer [2].

back to the exposed mesa edge. The evolution of the thickness of this dense interfacial zone as the oxidation penetrates deeper into the layer will determine whether linear, parabolic or intermediate time dependence is obtained [2], as illustrated in FIGURE 10.3.

There will be a dense layer at the oxidation front composed mainly of AlO_xH_y and As_2O_3. When the reduction of As_2O_3 to As is sufficiently fast to balance the rate of formation of As_2O_3, this interfacial zone does not grow appreciably in thickness as the oxidation front moves deeper into the layer (FIGURE 10.3(a)).While this balance is maintained, a linear time dependence will appear because the diffusional contribution to the reaction rate does not increase significantly over the time required to oxidise the device. If reaction conditions are changed to preferentially increase the formation of As_2O_3 relative to As formation and loss, a steady increase in the thickness of the interfacial zone will occur (FIGURE 10.3(b)) and the diffusion-limited parabolic time dependence will become dominant. Raman spectroscopy has shown a greater amount of As_2O_3 to be present during wet oxidation at higher temperatures, consistent with the transition from linear to parabolic behaviour as oxidation temperature increases.

For extended reaction times at a given temperature, any deviation from linearity will become increasingly obvious. For conditions where the thickness is increasing only slowly, the initial stages of the oxidation will closely approximate a linear time dependence. This is usually observed with $Al_{0.98}Ga_{0.02}As$ under typical processing conditions, e.g. 30 min at 440°C for $Al_{0.98}Ga_{0.02}As$. Conditions producing faster oxidation tend to tilt the balance in the direction of parabolic dependence.

For oxidation rates of AlGaAs below 0.2 μm/min, the process appears reaction-rate limited (linear) while initial rates above 1.3 μm/min appear diffusion limited at longer times. Thermodynamic calculations of the temperature dependence of the important As-based reactions predict that the formation of As_2O_3 will become increasingly favourable versus the reduction to As as temperature increases. This is consistent with the generally observed shift from linear towards parabolic behaviour with increasing temperature for both AlAs (FIGURE 10.5) and $Al_{0.98}Ga_{0.02}As$ (FIGURE 10.6), as discussed in Section 10.3.1.

It is important to remember that wet oxidation of GaAs also occurs although at a much lower rate. While the oxidation front is advancing rapidly into the horizontal AlGaAs layer, the oxidised layer is also slowly growing vertically into the bounding GaAs or AlGaAs layers with lower mole fractions. Oxidation will occur along the entire length of the oxidised channel, and the vertical oxidation front will have progressed farther near the mesa edge, where it has been exposed longer to the oxidising atmosphere. If bounding layers are very thin, it is possible to oxidise completely through them during long oxidation times.

Several factors strongly affect the rate of wet oxidation: Al mole fraction, layer thickness and digital versus continuous-grade alloying. These plus the oxidation of materials other than AlGaAs will be discussed in more detail.

10.3.1 Al-mole-fraction effects

The addition of even small amounts of Ga to AlAs has a profound effect on the wet oxidation rate [3], as illustrated in FIGURE 10.4.

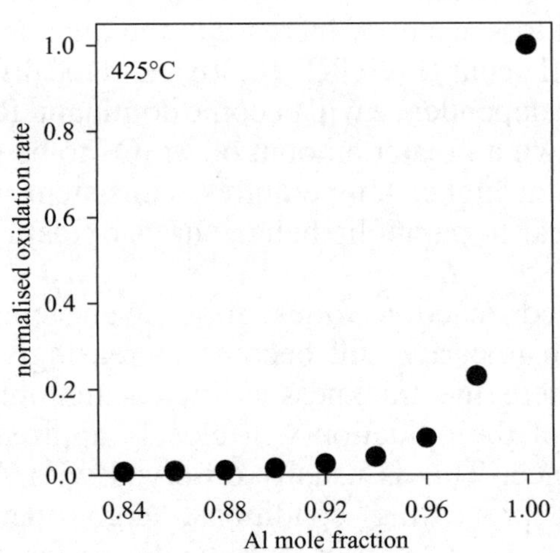

FIGURE 10.4 Wet oxidation rate dependence on Al mole fraction.

While $Al_{0.98}Ga_{0.02}As$ has only a 2% difference in Al content versus AlAs, a 435% difference in the wet oxidation rate is observed. Fortunately, it is commonplace to have sufficient control of Al mole fraction during growth that deviation in oxidation rate due to compositional non-uniformity is generally not a problem. The 98% composition is commonly used because it oxidises rapidly enough to be convenient for making oxide apertures in VCSELs while forming an oxide having greater mechanical robustness than that obtained with pure AlAs. Another practical advantage is the retention of nearly linear time dependence of oxidation depth (distance oxidised inward from the exposed mesa edge) for the depths and processing conditions commonly employed, as will be discussed below.

The wet oxidation of AlAs clearly displays the transition from reaction-rate limited (linear versus time) to diffusion-limited rates (parabolic) as the temperature is increased from 356 to 518°C, as illustrated in FIGURE 10.5 [4]. The dotted lines indicate the expected behaviour for a completely linear ($d = kt$) and completely parabolic ($d^2 = kt$) oxidation. Since relatively minor changes in temperature can have a pronounced effect on the time dependence of the oxidised depth, AlAs is less attractive for device applications where precise control of oxidised depths is essential. Useful

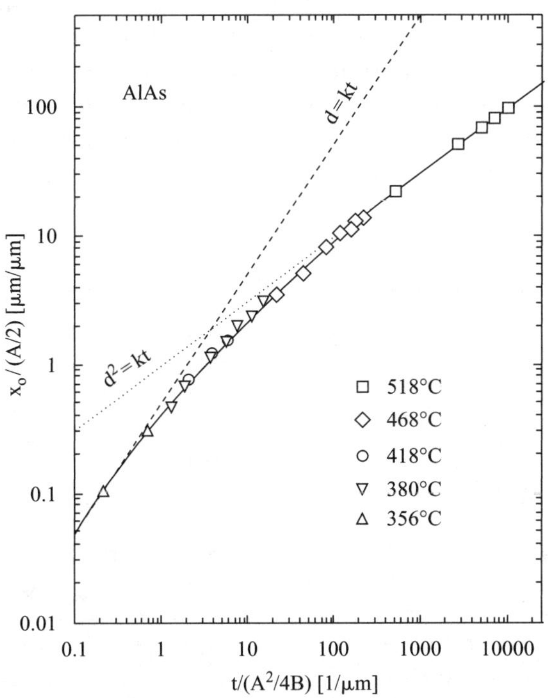

FIGURE 10.5 Temperature dependence of wet oxidation of AlAs from Ochiai et al. [4].

oxidation rates are achievable with 98% Al compositions while remaining within a virtually linear growth region.

The transition from reaction-limited (linear) to diffusion-limited (parabolic) behaviour is often difficult to detect for relatively short oxidation depths. A comparison of the oxidation behaviour of 94% Al and 98% Al layers at 400 and 440°C with 80°C water and an N_2 flow rate of 3.0 slm in a 4″ diameter three-zone furnace (FIGURE 10.6) reveals that the slower-oxidising material remains in the linear regime at both temperatures, while a shift from linear towards parabolic is seen for the faster-reaction 98% Al material at the higher temperature [2]. However, for shorter reaction times (shallower oxidation depths), the parabolic dependence is well approximated by a straight line and the assumption of linear time dependence would be adequate for selecting oxidation times for depths less than 15 μm. The specific times and temperatures at which the linear-to-parabolic transition will be observed will depend on the water vapour supply, which is controlled by the bubbler temperature and gas flow rate, and can vary depending on specific operating conditions.

From these studies, it is clear that faster oxidation rates, whether due to higher Al mole fraction or higher oxidation temperatures, will shift the balance from linear towards parabolic time dependences. It appears in general that for oxidation rates of AlGaAs below 0.2 μm/min, the process appears reaction-rate limited while initial rates above 1.3 μm/min appear diffusion limited at longer times. The transition region lies between these rates. The slower reaction rate obtained with higher Ga mole fractions enables the loss of As, forming the porous oxide, to keep pace with the advancing dense oxidation front. This allows such materials to remain

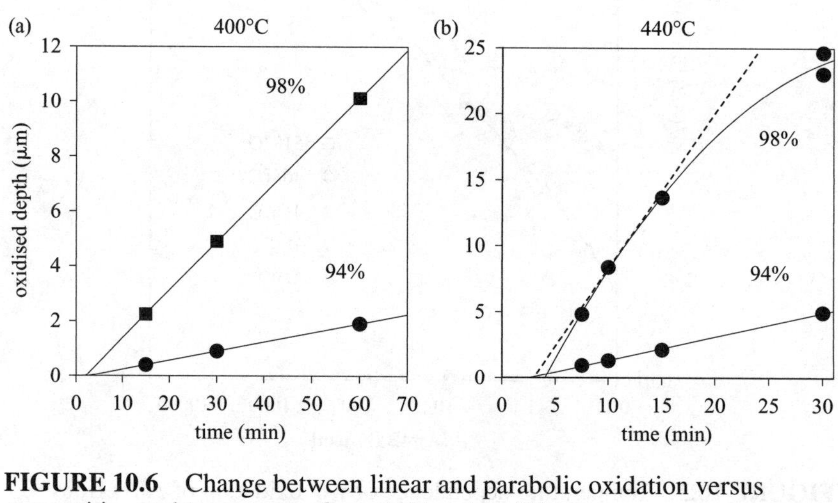

FIGURE 10.6 Change between linear and parabolic oxidation versus composition and temperature from Ashby et al. [2].

in the linear regime at higher temperatures and for longer times than their higher-Al analogues.

It is clear in FIGURE 10.6 that some time may pass between insertion into the furnace at $t = 0$ and the onset of oxidation. In these cases, $t = 0$ was defined as the time at which the wafer-loaded boat was returned to the central zone of a three-zone tube furnace through which water vapour was flowing entrained in N_2. The lag in the apparent onset of oxidation may be attributed to the much slower rates while the wafer and boat were reheating to the furnace reaction temperature. Thermal equilibration times will vary between different reactors and may be significant or negligible. This will be discussed further in Section 10.4.

10.3.2 Layer thickness effects

Above a certain layer thickness (about 100 nm), the oxidation rate for a particular AlGaAs composition is not appreciably influenced by the layer thickness. However, for thicknesses below 80 nm, a very strong dependence on layer thickness appears, as illustrated in FIGURES 10.7 and 10.8. This effect is probably related to the ease of diffusive transport of either reactants or oxidation products from the advancing oxidation front. A similar retardation of etching reactions is seen when trying to selectively etch a thin layer, as in selective channel recessing, discussed in Section 4.5.

The rate of diffusion-controlled reactions can be limited by the diffusion of either reactants or products. A study of the time

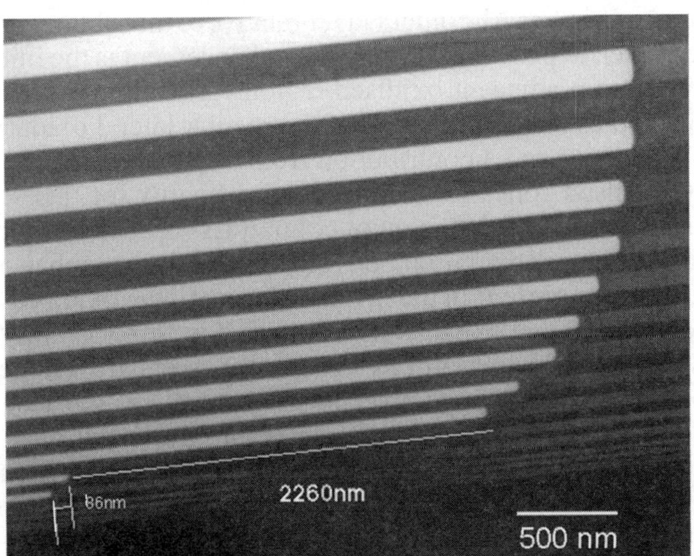

FIGURE 10.7 TEM of oxidation front region showing layer thickness effect.

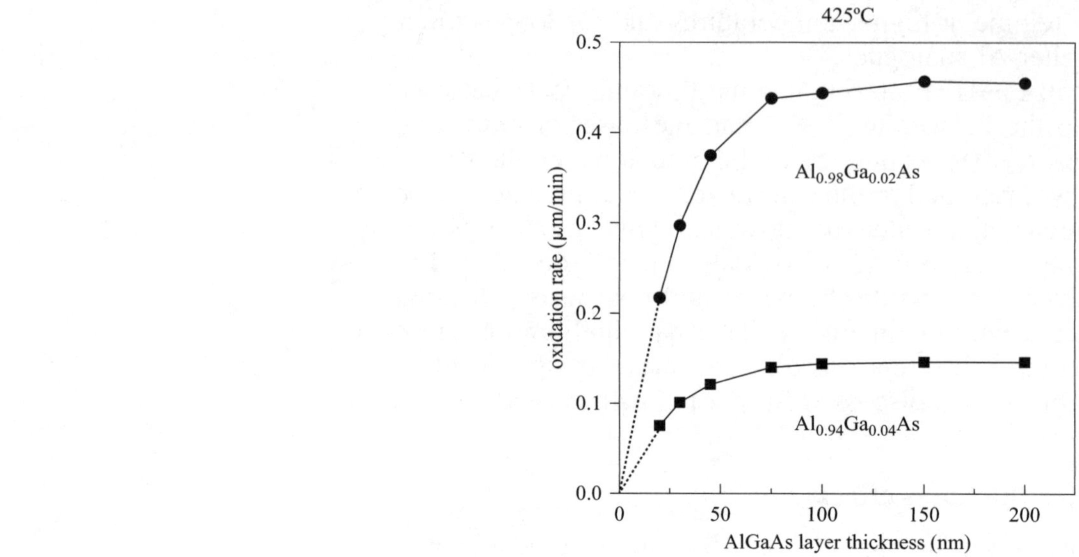

FIGURE 10.8 Effect of layer thickness on oxidation rate.

dependence as a function of layer thickness could distinguish between the two in this case, since a transition from linear to parabolic behaviour would signal the formation of a progressively thicker interfacial front if the formation and removal of products failed to keep pace in the thinner channels.

10.3.3 Proximity enhancement effect

The close proximity of a rapidly oxidising layer can accelerate the oxidation of a lower Al-content layer relative to that obtained for an isolated layer. A possible cause of the enhancement is the injection of defects from adjacent oxidised regions. The enhancement can be modelled as the combination of the regular lateral oxidation of an isolated layer and a contribution from a species diffusing from nearby oxidised zones [5]. The activation energy for this species is consistent with the diffusion of column III vacancies, which are known to enhance oxidation and whose diffusion is enhanced by strain, which is present in the layers adjacent to the oxidised layer. For example, lattice expansion is observed in thin (55 nm) GaAs layers in oxidised superlattices so adjacent layers are strained.

From a practical perspective, this enhancement must be considered when profile control is important. Compositional steps are used to fabricate tapered-oxide current apertures for optimum VCSEL performance (Section 10.5) so the observation of profiles that show appreciably deeper oxidation than expected from compositional effects alone is important. The enhancement resulting from adjacent regions with faster oxidation rates actually simplifies

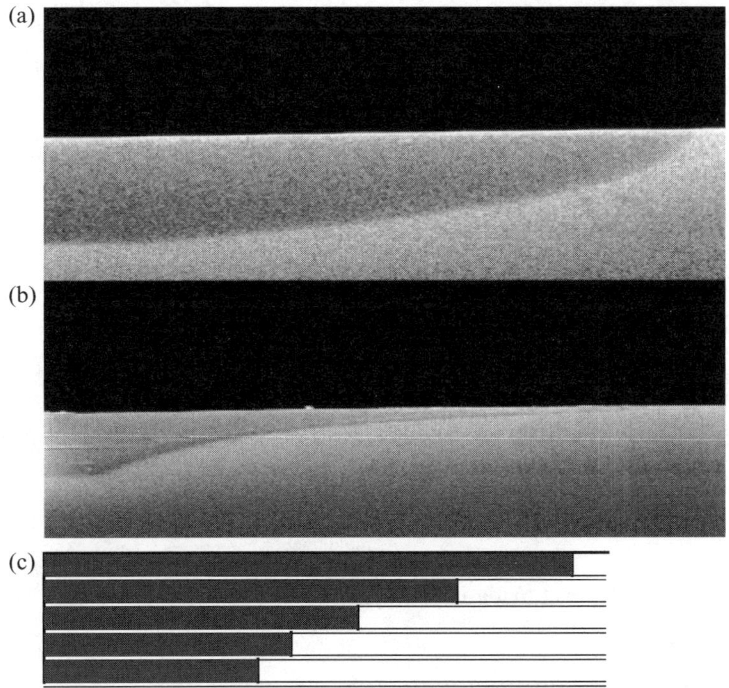

FIGURE 10.9 Enhancement effects from adjacent oxidised layers: (a) continuous compositional grade from 98 to 90% Al in AlGaAs, (b) step grade of 0.002% Al from 98 to 90% Al with 5 nm GaAs barriers between AlGaAs layers and (c) schematic of oxidised step-graded structure.

the grown-in structure for a sharp taper. A sharp, smooth taper can be obtained with only three layers and only two compositions, as shown in FIGURE 10.14.

Enhancement is most pronounced when a continuous gradation of Al percentage is employed in the structure (FIGURE 10.9(a)). Where the compositional grading was expected to produce a convex oxidation front replicating the Al-composition profile (as in FIGURE 10.9(b)), a concave oxidation profile is obtained due to enhancement from adjacent higher-Al-content regions.

The enhancement can be mitigated by insertion of thin (5 nm) layers of GaAs between layers of a step-graded structure that covers the same composition range (FIGURES 10.9(b) and 10.9(c)). The GaAs layers serve to impede the oxidation-enhancing diffusion from the oxidised layers and the final oxidation profile is closer to that expected based on composition alone.

10.3.4 Wet oxidation of other materials

While most wet oxidation work has focused on AlGaAs materials, other III–V semiconductors with high Al content also have the potential to undergo rapid wet oxidation.

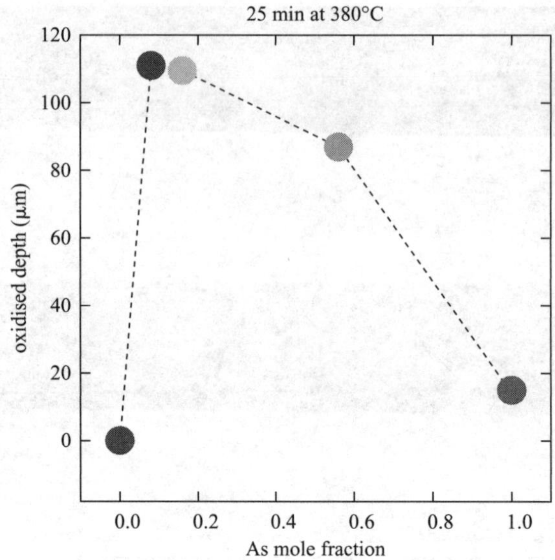

FIGURE 10.10 Wet oxidation of AlAsSb as a function of As mole fraction.

10.3.4.1 Sb-containing materials

Another III–V material family that might be amenable to wet oxidation is the AlAsSb family of compounds. Among the thermodynamically favoured products are As, which can mostly escape as in the AlGaAs case, and Sb, which will remain behind. Oxidation of these materials proceeds at temperatures lower than those required for AlGaAs (325–380°C.)

In the absence of As, wet oxidation produces a non-porous film and the oxidation proceeds at a slow, diffusion-controlled rate. The introduction of a small amount of As is sufficient to accelerate the reaction by increasing the porosity (FIGURE 10.10). The fall-off towards higher As mole fractions reflects the slower oxidation rates for AlAs at these relatively low temperatures.

The oxidation of AlAsSb lattice matched to InP has been studied [6]. The structure shown in FIGURE 10.11 consisted of a 109 nm cap of $In_{0.53}Ga_{0.47}As$, 288 nm layer of $AlAs_{0.56}Sb_{0.44}$, 109 nm layer of $In_{0.53}Ga_{0.47}As$ and an InP substrate. While surface oxidation of AlSb, GaSb and InSb produces a layer of Sb which moves ahead of the advancing oxidation front, in lateral oxidation this Sb layer segregates to one of the oxide/semiconductor lateral interfaces, typically the top one as shown in FIGURE 10.11, where deformation can accommodate the volumetric changes caused by oxidation. There is significant deformation of the surface layer as the front advances, as can be seen in FIGURE 10.11(a), and a dip where the fronts meet if the layer is completely oxidised (FIGURE 10.11(b)).

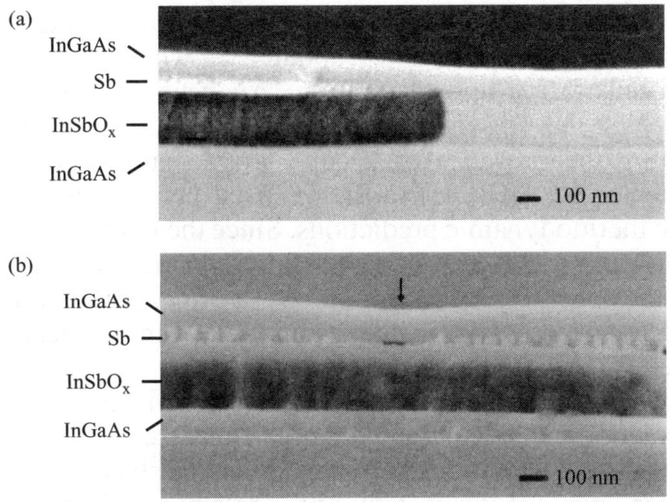

FIGURE 10.11 Sb segregation at interfaces in InAsSb wet oxidation from Blum et al. [6].

10.3.4.2 P-based materials

Wet oxidation of P-based materials has also been the subject of study. These materials differ appreciably in their principal chemistry from As-based ones. While post-oxidation materials analysis shows that most of the As is lost from the oxide, most of the P is retained. Thermodynamic calculations suggest that the principal products of this oxidation are very different for AlAs and AlP. While the formation of As, Al_2O_3 and AlO(OH) are most favoured for AlAs, the change in Gibbs free energy is twice as great for the formation of $AlPO_4$ than for P and Al_2O_3 or AlO(OH). Rates for phosphide materials are slower, and temperatures in excess of 500°C are employed.

The oxidation of $In_{0.5}Al_{0.5}P$ has been compared with that of $Al_{0.5}Ga_{0.5}As$. While InAlP oxidises more slowly than AlGaAs, it has a lower activation energy. With oxidation at 500°C, In is seen to build up at the oxide/semiconductor interface. At 550°C, the In is distributed uniformly throughout the film and a parabolic time dependence prevails. At 650°C, In is depleted at the interface [7].

An interesting and potentially useful property of GaP is the ability of a very thin layer (two monolayers) to serve as a diffusion barrier to the movement of As away from the oxidised layer and into adjacent semiconductor layers. The importance of this will be discussed in Section 10.6.

10.3.5 Miscellaneous observations

The observed change in layer volume upon oxidation of AlAs and AlGaAs is not as expected if calculated based upon theoretical molar volumes of the starting semiconductor and a dense, crystalline oxide. AlAs shrinks by about 13% while $Al_{0.98}Ga_{0.02}As$ shrinks only 6.7% upon oxidation. Analysis of oxidised films has revealed the presence of appreciable quantities of hydrogen, suggesting the presence of a considerable amount of hydrous species such as AlO(OH) in the oxidised layer. This is consistent with thermodynamic predictions. Since the relative amounts of AlO(OH) and Al_2O_3 are not well defined and may change depending on the specific process conditions, a priori prediction of film volume is not strictly possible at this time. Similar behaviour is expected for AlGaAs materials.

A serious mechanical problem can exist when oxidation is taken to completion, i.e. when the oxidation fronts from opposing mesa walls coalesce. The oxide/semiconductor interface is weak, and the upper layers of the structure will delaminate from the oxide unless a portion of unoxidised structure is present to hold the device together. This is naturally the case for oxide-aperture VCSELs, but use of wet oxidation for GOI (GaAs-on-insulator) devices will face serious problems unless provision is made for holding the device together. This problem is most pronounced for AlAs and is less problematic with AlGaAs alloys.

Changes in composition mean changes in dimension, producing stresses, especially in overlying layers. These can deform if they are not too physically constrained by a thick semiconductor overlayer. As seen in FIGURE 10.2, the oxidation-front profile tends to be slightly deeper at the top interface than at the bottom interface. This may be related to stress, although the origin of the effect is not definitively established and the effect is rather small.

It is generally not possible to oxidise a certain distance into a layer, pull the wafer from the furnace and then reinsert the wafer to oxidise further for a controlled final depth. Wet oxidation under these conditions is not a simple time-additive process. There appears to be a change in the chemical or structural nature of the oxide-semiconductor reaction front that discourages restarting the wet oxidation process, perhaps by inhibiting the free access of water to AlAs or AlGaAs for the generation of the H that is essential for the rapid removal of As from the oxidising layer.

When inserting a wafer into a furnace with wet gas flow, there may be a delay in the onset of rapid wet oxidation (FIGURE 10.6). This is more of a problem for depth control of shallow oxidations, since the apparent incubation time can be a substantial fraction of the total time in the furnace. For such processes, it may be especially important to load into dry gas, thermally equilibrate to the final reaction temperature and then introduce the wet gas flow. Since the heat capacities of dry nitrogen and water-saturated nitrogen are different, it may be necessary to use different flow rates of dry and wet nitrogen to maintain a constant furnace temperature when switching gases. This will require calibration for a specific reactor.

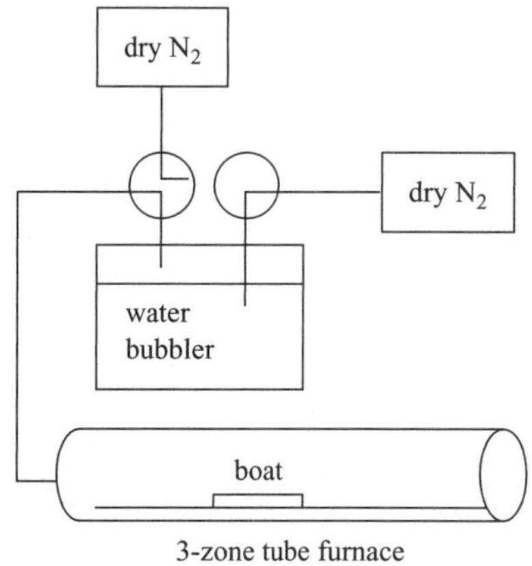

FIGURE 10.12 Diagram of wet oxidation apparatus.

10.4 PRACTICAL WET OXIDATION

The apparatus for performing wet oxidation can be very simple, consisting of a water bubbler, an inert gas source and a tube furnace. It is best to use a quartz tube to avoid introducing contaminant species, such as Na, that can be a problem with ordinary glass at the elevated temperatures required for oxidation (350–525°C). A general system schematic is shown in FIGURE 10.12. For $Al_{0.98}Ga_{0.02}As$ layers thicker than 90 nm, oxidation at 400–440°C using 80–90°C water and an N_2 flow rate of 3 slm in a 4″ diameter three-zone tube furnace works well. It is best to work at a temperature at least two degrees below the boiling point of water, which will depend on the elevation of the laboratory above sea level.

Water vapour is entrained in inert gas (N_2 or Ar) that is bubbled through liquid water. The water temperature will determine the equilibrium vapour pressure of water in the flowing gas and therefore controls the water concentration experienced at the wafer. Both nitrogen and Ar work well as carrier gases as long as they are oxygen-free. Even rather small amounts of oxygen in the gas will stop the wet oxidation process (Section 10.2.1). The bubbler should be of sufficient size that the water temperature within it does not change more than a fraction of a degree during the reaction. Heating the tubing between the bubbler and the inlet to the furnace avoids condensation in transit to the reaction zone.

These reactions are highly temperature dependent, so it is vitally important to maintain a uniform temperature in the reaction zone and to avoid temperature-lag effects as much as possible while loading wafers. It is helpful to use a three-zone furnace for thermal stability in the oxidation region. Generally, the wafer will be situated in the central zone. This allows the flowing gas to pass through a hot zone before reaching the wafer. For AlGaAs, oxidation temperatures are generally between 400 and 475°C. For P-containing materials, higher temperatures are used (>500°C), while Sb-containing materials are oxidised at lower temperatures (325–380°C).

Since reaction initiation can be sensitive to the surface condition, it is most advantageous to perform the oxidation as soon as possible after exposing the layer edge with a mesa etch. Minimal formation of native oxide or its removal (Section 3.2.2) just prior to oxidation is generally desirable if a sample has been etched a long time before oxidising. This is especially important for Sb-containing materials, which can form a number of surface oxides that vary in their reactive chemistry.

While reactant-gas access to the AlGaAs layers is easily provided by mesa etching, an alternative approach involves etching trenches into the wafer in the regions where one wishes oxidation to occur. This has the advantage of providing an essentially planar surface for subsequent metallisations and may prove highly advantageous for reliable fabrication of arrays of devices, such as VCSELs.

Samples are inserted into the reaction zone by sliding in a quartz "boat" holding the wafers. Usually the boat is pulled quickly from the furnace, loaded quickly at the mouth of the furnace and returned to the central zone of the furnace as rapidly as possible. Temperature equilibration is fairly rapid but not instantaneous; even with rapid handling it can take a few minutes to achieve thermal stability. This can manifest itself as a non-zero intercept in depth versus time plots (FIGURE 10.6). If an initial delay is observed under a specific set of reaction conditions, it may be desirable to employ a bypass flow design for the most uniform and repeatable process. Dry nitrogen or Ar is flowed through the tube furnace until the boat and wafers thermally equilibrate. As discussed in Section 10.3.5, the best dry-gas flow rate to use during equilibration may differ from the flow rate of the water-saturated gas. When the desired reaction temperature is attained, the water-saturated gas is switched to pass through the reaction zone. Likewise, switching back to dry gas before unloading will remove reaction time uncertainties. This is less of a problem during unloading than during loading since cooling after unloading is rapid compared to the initial time needed to thermally equilibrate a boat that has been pulled briefly from the furnace for loading.

Flow rates and tube dimensions typically used in wet oxidation result in the development of a parabolic flow-velocity profile across the diameter of the tube. The virtually zero flow rate at the tube wall and the maximum flow at the tube centre produce a non-uniform reaction rate across a wafer due to variation in the effective water concentration. For a particular water temperature, wafer temperature and tube radius, the rate increases as the sample becomes more distant from the first wall, then plateaus across the distance where a certain flow rate is exceeded, and drops again as the flow rate drops while approaching the second wall. This can give uniform oxidation across the central portion of the boat, but samples outside the plateau zone will oxidise irregularly. For example, a wafer of mesa-etched VCSELs that is 1.5″ wide can be uniformly oxidised in a 4″ tube furnace.

This problem can be solved by disrupting or "spoiling" the parabolic flow profile by inserting a baffle system just upstream from the position of the boat. This baffle can be a simple honeycomb structure mounted on the upstream end of the boat. The disruption of the flow profile produces a uniform flow rate across the entire width and length of the boat where the samples are located. This is especially important for trench oxidation; mesa structures produce some spoiling of the flow without an additional baffle.

10.5 APPLICATIONS IN OPTOELECTRONIC DEVICES

Wet oxidation is a true enabling technology for high efficiency VCSELs, so this will be the primary focus of this section. There are both structural considerations and electronic consequences due to defect generation that should be considered by the VCSEL designer. These are discussed below.

10.5.1 Structural issues for oxide VCSELs

Wet oxidation of AlGaAs can be extremely useful for locally altering the refractive index of a heterostructure since the oxide has a much lower refractive index of n = 1.6. The insulating character of the oxide also makes current channelling within a device possible. High-efficiency vertical-cavity surface-emitting lasers (VCSELs) with wall-plug efficiencies in excess of 50% have been made possible by wet oxidation processing. They employ oxide apertures made by incorporating a higher Al-content layer next to the optical cavity than those used in the distributed Bragg reflector (DBR) mirror stacks, as illustrated in FIGURE 10.13. Note that while

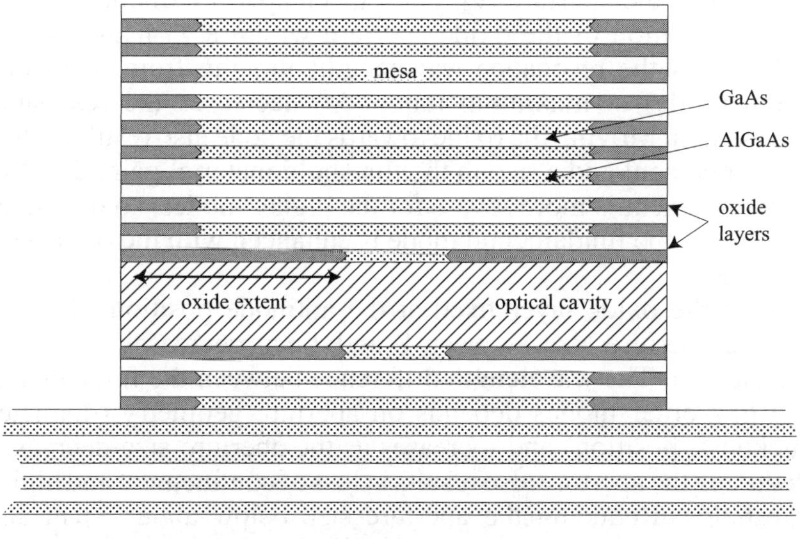

FIGURE 10.13 Oxide aperture VCSEL.

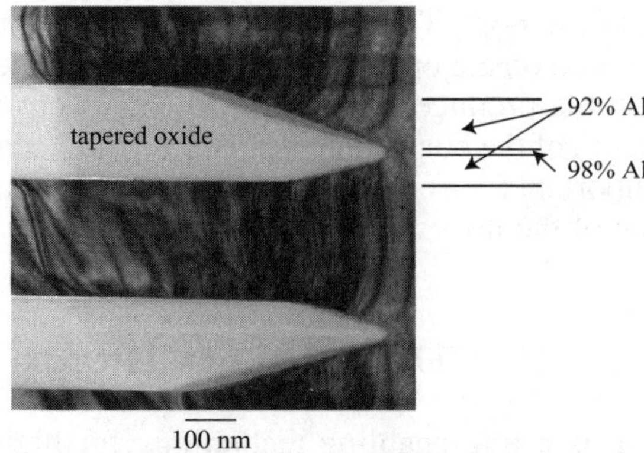

FIGURE 10.14 Tapered oxide aperture from Choquette et al. [8].

the aperture oxide is grown deep into the structure, there is also a lesser depth of oxidation in the lower Al-content layers of the DBR. While general VCSEL design issues will not be discussed here, we will discuss some factors that are specifically related to the as-grown structure that is to be oxidised and the effect of oxidation on some key optical properties.

While it is possible to form an aperture by oxidising an AlGaAs layer of uniform composition, VCSEL performance is improved when the aperture oxide is tapered rather than blunt at its terminus. FIGURE 10.14 [8] shows a tapered aperture that takes advantage of the proximity enhancement effect (Section 10.3.3) to achieve a relatively sharp tip profile with only three AlGaAs layers and only two different Al mole fractions.

The location of the point of the taper within the AlGaAs layer has a strong influence on VCSEL properties, as does the diameter of the aperture [9]. For a given aperture diameter, moving the position of the oxide aperture within a $\lambda/4$ layer in the DBR shifts the resonance wavelength; moving from field minimum (node) to maximum (antinode) decreases the resonance wavelength. Moving the oxide towards the node also results in loss of confinement and a rise in the threshold material gain. The loss of confinement is more severe for the higher modes, so the modal stability of the fundamental mode is enhanced with the oxide near the node.

With the oxide layer fixed at the antinode position, decreasing the aperture size between 8 and 1 μm decreases the lasing wavelength. The wavelength difference between the fundamental and first-order modes depends on aperture geometry (diameter, thickness, position) and increases as the aperture size decreases. Reducing the aperture also results in loss of confinement and higher thresholds. Reducing the aperture size below about 9 μm also reduces the carrier lifetime, as measured by turn-on delay.

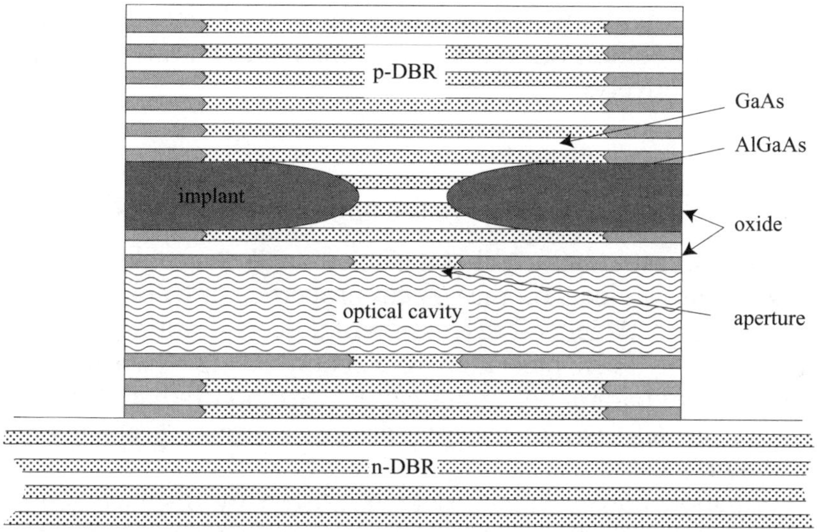

FIGURE 10.15 Double aperture VCSEL: oxide and implanted apertures.

Improved beam quality can be attained with a double-aperture VCSEL, schematically shown in FIGURE 10.15. The combination of both implanted and oxide aperture provides better mode control than either aperture alone and improves single-mode output [10]. In such hybrid structures, the implanted aperture is situated above the oxide aperture. Lattice disorder produced by ion implantation produces current confinement into the unimplanted area. When the implanted aperture is slightly smaller than the oxide aperture, the confined current flow is directed into the centre of the optical cavity defined by the oxide aperture. This preferentially pumps the fundamental mode over the higher-order modes. A thermal lens can also form in the central region due to the implanted aperture. This changes the local refractive index and further enhances the fundamental mode. The final modal distribution in such devices is a result of optical confinement, localised carrier injection, thermal lensing and enhanced loss of higher-order modes through absorption in the implanted region.

The relative sizes of the two apertures determines whether the hybrid device operates more like an implanted VCSEL or an oxide VCSEL. When the apertures are the same size or the oxide aperture is smaller, the wavelength splitting between the fundamental and first-order modes is determined by the oxide aperture. When the oxide aperture becomes larger than the implanted one, the behaviour begins shifting towards that of an implanted VCSEL. When the oxide aperture is much larger, the implant aperture dominates modal behaviour. By combining a smaller implanted aperture with a slightly larger oxide aperture, both optical and electrical confinement can be combined to direct a majority of the power into the fundamental mode. Excellent single-mode performance has been obtained by combining oxide and implanted apertures of 8 and 6 μm, respectively.

10.5.2 Defect-related issues for optoelectronic devices

Because wet oxidation liberates elemental As, the effect of As-generated defects on the optical properties of a device should be considered. A general discussion of defects resulting from wet oxidation has been presented in Section 10.2.2. Two different problems are paramount in optical devices: non-radiative recombination at defects and vacancy-driven intermixing at quantum well interfaces.

Formation of As antisite defects (As_{Ga}) by displacement of Ga or reaction with a Ga vacancy decreases luminescence by facilitating non-radiative recombination. The infamous emission-quenching EL2 is an antisite defect. Quantum wells near an oxidised layer may have their photoluminescence decreased by more than 90%. The decrease is more severe when the quantum wells are located closer to the oxide layer, but quantum wells tens of nanometres away are also affected. The effect is greater if the sample is exposed to wet oxidation conditions for longer times after the oxidation front has passed that particular part of the quantum well. The As generated by the oxidation then has a longer time to diffuse into that part of the quantum well. The time and temperature dependence of this quenching is consistent with the entry into the QW of a defect whose diffusivity is comparable to that of As in GaAs.

Some restoration of luminescence can be achieved by annealing in a hydrogen atmosphere. Hydrogen is known to passivate deep-level defects, such as EL2. Extended hydrogen annealing can restore up to 70% of the pre-oxidation emission intensity, but H can also passivate dopants in the structure, altering the electrical characteristics.

Another way to avoid the diffusing defect problem is to grow a thin barrier layer into the structure. As little as two monolayers of GaP can stop the diffusion of As through an oxidised amorphous-AlGaAs/GaAs structure. However, introduction of such layers is often prohibited by device design constraints for GaAs-based devices. The possibility of including a thin InP layer in InGaAs/InAlAs/InP materials might also be considered.

Luminescence intensity is not degraded by the oxidation of InAlP on GaAs. Since free As is not being generated directly by oxidation of the As-free layer, only very small quantities of As that are generated as the oxidation slowly advances into adjacent

GaAs will be present. These quantities appear too small to produce a significant effect.

The generation of column III vacancies can produce wavelength shifts by changing the effective size of narrow quantum wells through vacancy-promoted intermixing at the interface. This problem is compounded by post-oxidation annealing, which accelerates defect diffusion. This effect can be of the order of several tens of meV near the oxidised layer but even regions remote from the oxide can blue-shift of the order of 10–20 meV.

10.6 APPLICATIONS IN ELECTRONIC GaAs DEVICES

There has long been a desire to have the GaAs equivalent of Si CMOS. However, while the native SiO_2/Si interface is almost "ideal" in terms of possessing a very low density of interface states within the bandgap, the native oxide/GaAs(100) surface has an extremely high density of interface states that preclude CMOS behaviour. Wet oxidation has been explored as a possible route to a useful interface with $<10^{10}$ states/cm^2. In this section, the failure and probable futility of this approach will be explained. The application of wet oxidation to GaAs-on-insulator (GOI) devices will also be discussed.

10.6.1 Problems with wet and dry oxidation for MIS devices

The presence of As at the interface between GaAs and the oxide is a major source of a high density of interface states. Large quantities of As are generated as a dominant product in wet oxidation, so it is unlikely that enough of the interfacial As could ever be removed to reach $<10^{10}$/cm^2 densities. However, even dry oxidation, which does not directly generate As, will not be a benign process. The dominant As-containing product in dry oxidation is As_2O_3. The stability of this compound with respect to GaAs and AlAs is the issue for long-term, low interface-state densities.

Both GaAs and AlAs strongly favour reaction with As_2O_3 to form Ga_2O_3 or Al_2O_3 and elemental As. Thus, the GaAs/oxide surface is thermodynamically unstable. While the ratc of this reaction will be determined by kinetics, the ability to maintain an As-free interface for extended times under device operation, including resistive heating, is unlikely.

Attempts to oxidise AlAs layers on AlGaAs/GaAs or GaAs have generally produced interface-state densities in the low 10^{11} range at best. Slightly higher densities are obtained with wet oxidation than with dry oxidation. A MOSFET using wet-oxidised AlAs on AlGaAs/GaAs has been reported that operated in depletion

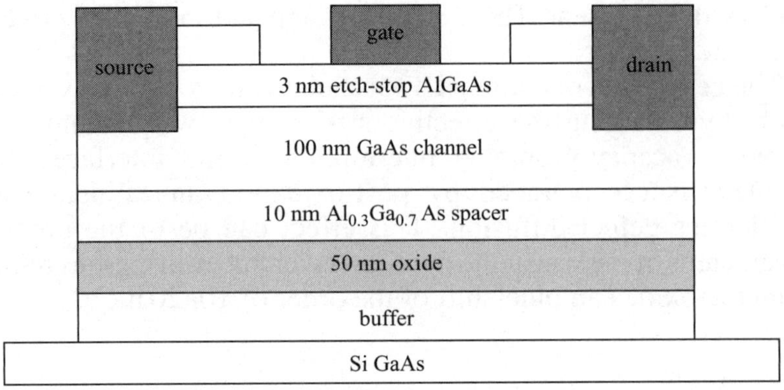

FIGURE 10.16 GaAs-on-insulator (GOI) MESFET.

mode, but the source-to-drain conductance was only one fourth the expected value, possibly due to spatially localised depletion regions around As precipitates.

10.6.2 GaAs-on-insulator applications

Wet oxidation has been explored as a method to electrically isolate a MESFET from the substrate, as illustrated in FIGURE 10.16. As in other wet-oxidised devices, the generation of free As by the reaction and the subsequent generation of other defects by the As must be considered in fabricating the device.

Arsenic precipitates can form in the channel as a result of As interstitials diffusing through an unoxidised spacer layer of AlGaAs [11]. The oxidised devices tend to display reduced mobility and lower sheet charge densities than do unoxidised devices. The reductions are more severe with higher oxidation temperatures and longer oxidation times, both of which favour more diffusion of defects into more distant regions of the device.

10.7 CONCLUSION

While many of the fabrication issues for optoelectronic devices are the same as those for electronic devices, wet oxidation of AlGaAs is a process primarily of utility for optoelectronic devices. By careful design through proper selection of Al mole fractions, layer thicknesses and layer placement, high-power, single-mode VCSELs can be made. While this is an established baseline process for high-efficiency VCSELs, it has several drawbacks that are associated with the generation of free As. Through its role in generating point defects, this elemental As can have an important influence on defect-related materials properties that affect optoelectronic device operation. The electronic problems associated with elemental As severely constrain the use of wet oxidation for electronic device fabrication.

REFERENCES

[1] C.I.H. Ashby, J.P. Sullivan, P.P. Newcomer, N.A. Missert, H.Q. Hou, B.E. Hammons, M.J. Hafich, A.G. Baca [*Appl. Phys. Lett. (USA)* vol.79 (1997) p.2443–5]

[2] C.I.H. Ashby, M.M. Bridges, A.A. Allerman, B.E. Hammons, H.Q. Hou [*Appl. Phys. Lett. (USA)* vol.75 (1999) p.73–5]

[3] K.D. Choquette, K.M. Geib, C.I.H. Ashby, R.D. Twesten, O. Blum, H.Q. Hou, D.M. Follstaedt, B.E. Hammons, D. Mathes, R. Hull [*IEEE Sel. Top. Quant. Electron. (USA)* vol.3 (1997) p.916–26]

[4] M. Ochiai, G.E. Giudice, H. Temkin, J.W. Scott, T.M. Cockerill [*Appl. Phys. Lett. (USA)* vol.68 (1996) p.1898–900]

[5] O. Blum, C.I.H. Ashby, H.Q. Hou [*Appl. Phys. Lett. (USA)* vol.70 (1997) p.2870–2]

[6] O. Blum, K.M. Geib, M.J. Hafich, J.F. Klem, C.I.H. Ashby [*Appl. Phys. Lett. (USA)* vol.68 (1996) p.3129–31]

[7] F.A. Kish et al. [*J. Electron. Mater. (USA)* vol.21 (1992) p.1133–9]

[8] K.D. Choquette, K.M. Geib, H.Q. Hou, D. Mathes, R. Hull [*Proc. 10th Int. Conf. on Semiconducting and Insulating Materials*, Berkeley, CA, USA, 1–5 June 1998 (IEEE, New Jersey, USA, 1998) p.209–13]

[9] P. Bienstman, R. Baets, J. Vukusic, A. Larsson, M.J. Noble, M. Brunner, K. Gulden, P. Debernardi, L. Fratta, G.P. Bava, H. Wenzel, B. Klein, O. Conradi, R. Pregla, S.A. Riyopoulos, J.-F.P. Seurin, and S.L. Chuang, [*IEEE J. Quant. Electron. (USA)* vol.37 (2001) p.1618–31]

[10] E.W. Young, K.D. Choquette, J.-F. Seurin, Shun Lien Chuang, K.M. Geib, A.A. Allerman [*IEEE J. Quantum Electron. (USA)* vol.39 (2003) p.634–9]

[11] P.A. Parikh, P.M. Chavarkar, U.K. Mishra [*IEEE Electron Device Lett. (USA)* vol.18 (1997) p.111–13]

GLOSSARY

AES	Auger electron spectroscopy; a surface analysis technique that probes the chemistry of a surface by measuring the energy of electrons emitted from that surface when it is irradiated with electrons of energy in the range 2–50 keV
AFM	atomic-force microscopy, a surface-analysis technique that measures the position of atoms on the surface using a ceramic or semiconductor tip at the end of a cantilever. As the tip moves over a surface, it contacts the surface continuously or periodically and the cantilever bends as the tip is repelled or attracted to the surface
acceptor	an atom or defect that provides a "hole", e.g. Zn on Ga site in GaAs
activation energy	the energy of a potential barrier that must be crossed for a reaction to proceed
airbridge	a metal interconnect that spans between two contacts through the air rather than crossing over a dielectric
alkane	hydrocarbons with a formula C_nH_{2n+2}. Examples are methane and ethane
alloy	to heat in order to promote a solid-state reaction such as that needed to form many kinds of ohmic contacts
analogue	referring to a circuit or application using an architecture with continuously variable voltages and currents
antisite defect	a type of defect where an atom of one element of a compound is located on a lattice site normally occupied by an atom of a different element of the compound

AR coating	anti-reflection coating, a thin layer of material that changes the reflectivity of a surface; minimum reflection occurs with film thicknesses that are a multiple of 1/4 times the wavelength of light whose reflectance is to be minimised
asher	an etch system used to remove hydrocarbon films from a sample, usually with an oxygen plasma
atomic displacement threshhold	energy required to move an atom off its normal lattice site; about 40 eV for GaAs
β	the DC current gain of a bipolar transistor defined by the ratio of the collector current to the base current
band bending	refers to the curvature of an energy band in proximity to a surface (interface) due to charges at the surface (interface)
bandgap	the energy separation between the valence and conduction bands
bandstructure	the energy states of an ordered solid which are ordered into bands grouped in loose relation to the energy levels of the atoms they were derived from
base	the control electrode of a bipolar transistor; the current from the emitter to the collector is modulated by an applied voltage to the base-emitter junction
BJT	a bipolar junction transistor
blisters	pockets of N_2 gas trapped in photoresist that crosslinked too quickly
blueshift	shift to higher energy
breakdown	describes the condition of a device in a high electric field where the current enters a runaway condition because impact ionisation generates sufficient carriers to be self-sustaining
breakdown walkout	an increase in breakdown voltage during the initial operation of an FET
buffer	a region of semiconductor grown by epitaxy as a transition between the substrate and the active layers of a device
burn-in	an initial operation of a device to stabilise its electrical characteristics and to screen out weak devices
burn-out	the abrupt failure of a device in operation

CAIBE	chemically assisted ion-beam etching
CBM	conduction-band minimum = E_C
collector	the destination of electrons (holes) in a bipolar transistor
conduction band	the energy band of a solid starting with the lowest unoccupied electron state
conduction band offset, ΔE_C	the abrupt energy difference in the conduction band at a heterointerface
critical dimension	smallest feature size
crystallographic etchant	etchant that etches different crystal facets at different rates; also called reaction-rate-controlled or reaction-rate-limited etchant
C–V	capacitance-voltage
CVD	chemical vapour deposition

DBR	distributed Bragg reflector
DC	direct current. The current in a device or circuit that results from applied voltages that do not vary with time
deep-level traps	carrier traps with energies located near the centre of the bandgap
density of states	the number of energy states per unit energy
depletion region	region depleted of free carriers
descum	remove organic residues by reaction with a low-energy oxygen plasma or ozone
developer	a chemical solution that dissolves a lithographically exposed (by light or electrons) polymer
diffusion	carrier movement in a semiconductor in response to concentration gradients
digital	referring to an integrated circuit or application using a binary logic architecture
dilute acid	commonly, an acid diluted to a concentration of 6N (6 normal)
dislocation	an imperfection in a crystal structure resulting from the absence of one or multiple atoms in one or more layers of a crystal
DLTS	deep level transient spectroscopy, a method of probing traps with transient I–V or C–V techniques

donor	an atom that can provide a free electron, e.g. Si on Ga site in GaAs
dopant	an impurity introduced for the purpose of creating free carriers in a semiconductor
dope	to introduce impurities for the purpose of creating free carriers in a semiconductor
drain	the destination of electrons in an FET
drift	carrier movement in a semiconductor in response to an electric field
drain lag	a situation where a drain-current measurement lags the drain-voltage pulse at very short times indicating trapping effects in an FET
ECR	electron cyclotron resonance; used to produce a high-density plasma
EDAX	energy-dispersive analysis of X-rays: same as EDX and EDS
edge bead	resist pile-up at wafer edges during spin coating
EDS	energy-dispersive (X-ray) spectroscopy: same as EDX and EDAX; a method where X-rays are used for microprobe analysis of the elemental composition
EDX	energy-dispersive X-ray analysis: same as EDAX and EDS
effective mass	the mass of an electron (hole) in a solid usually expressed as a fraction of the free-space mass
EL2	an emission-quenching deep-level trap usually ascribed to an antisite defect
electrochemical etching	an etching process where an external power supply is used to provide a voltage to drive the etching process
electroluminescence	a technique that measures light emission in response to applied voltages in a device or semiconductor sample
emf	electromotive force; reduction potential = negative of oxidation potential
EMP	electron-microprobe analysis, a method of elemental analysis by detection of X-ray emission
emitter	the source of electrons (holes) in a bipolar transistor

epi-ready — a substrate ready for epitaxial growth without the need for any external polishing, cleaning or etching

epitaxy — growing a crystal layer on a crystalline substrate so that its crystalline orientation is the same as that of the substrate

equilibrium — the sample has had plenty of time to respond to a perturbation

equivalent circuit model — an electrical model of a device which is constructed of electrical components that represent structural features of the device; equivalent circuit models are not usually accurate physical representations of a device but rather tools for predicting electrical behaviour of a circuit built from devices

etch stop — a layer of material that reacts much more slowly with an etchant than does the layer above it

eutectic — refers to the composition of an alloy that has the lowest melting point of any mixture of the metals

fences — regions of process-altered resist along the upper edges of etched mesas

Fermi level — also called Fermi energy; energy level in a solid at which the Fermi-Dirac distribution function equals 0.5; corresponds to the energy of the highest occupied energy level or state at 0 K

Fermi-level pinning — the position of the Fermi level remains unchanged as carriers move into and out of surface states

FET — field effect transistor

field emission — transport of an electron (hole) across a potential barrier related to its quantum-mechanical probability of tunnelling through the barrier

fluence — total number of particles or photons striking a unit area, i.e., flux × time

flux — number of particles or photons striking a unit area in a unit time

f_{max} — unity power gain in a frequency sweep of a transistor, derived from the unilateral gain which is derived from measured S-parameters of a transistor

frequency dispersion — changes in device parameters as a function of frequency

f_t	unity current gain in a frequency sweep of a transistor, derived from the h_{21} which is derived from measured S-parameters of a transistor
FWHM	full width at half maximum (usually of a spectroscopic peak)

gain	the ratioed increase of a parameter (current, voltage, power, h_{21} etc.) of a transistor
galvanic	relating to direct-current electricity produced chemically
gate	the control electrode of an FET; an applied bias to the gate modulates the current from the source to the drain
gate lag	a situation where a drain current measurement lags the gate voltage pulse at very short times indicating trapping effects in an FET
gate sinking	a reaction of the gate metal with the semiconductor in an FET where the consumed semiconductor allows a sinking of the metal-semiconductor interface and a shift in FET electrical parameters
generation-recombination	the creation of free carriers by the excitation of an electron, leaving behind a hole, and the loss of the free carriers by recombination of an electron and hole
getter	a material that will absorb a targeted substance rendering it unavailable; e.g. a hydrogen getter in a hermetic package prevents hydrogen poisoning in Pt- or Ti-containing FETs
GPS	global positioning satellite
ground state	the lowest energy state of an atom, molecule or solid
Gummel plot	a plot of collector and base currents versus base-emitter voltage for a bipolar transistor

heterointerface	an interface comprising different semiconductors on each side
heterojunction	a pn junction comprising semiconductors with different bandgaps on each side of the junction

heterostructure	a structure with at least two different layered semiconductors, usually with different bandgaps
HBT	a heterojunction bipolar transistor. The emitter is designed with a wider bandgap semiconductor than the base
HDP	high-density plasma; examples are ICP (inductively coupled plasma) and ECR (electron cyclotron resonance) plasmas
HDPE	high-density plasma etching
HEMT	high electron mobility transistor
hermetic	sealed from the ambient, usually in reference to a transistor package
HFET	heterostructure field effect transistor
HV	high vacuum (10^{-2}–10^{-8} Torr)
hot-electron effects	refers to non-ideal behavior due to energetic electrons in an FET
hydrocarbon	organic molecules composed of carbon-carbon and carbon-hydrogen bonds
hydrophilic	attracts water; "wets"
hydrophobic	repels water; does not "wet"
hydrogen poisoning	a situation where hydrogen migrates to a Pt- or Ti-containing gate of an FET and shifts its electrical parameters, usually as a result of enclosure in a hermetic package

IBAE	ion-beam-assisted etching
IBE	ion-beam etching
ICP	inductively coupled plasma; a type of high-density plasma
$I_D(I_G, I_C, I_B,)$	the drain (or gate, or collector, or base) current of a transistor
ideality factor	the coefficient "n" in the equation $I = I_o \exp[V/nkT]$ which relates to the deviation of ideal exponential current increase with applied voltage of a diode or Schottky barrier
image-reversal	method for getting a re-entrant profile with a positive-tone resist
impact ionisation	the process of creating electron-hole pairs when an energetic electron (hole) is accelerated in an electric field and ionises an electron from the valence band

ion implant-ation	the process of introducing impurities into a sample by means of an accelerated ion beam whose energy is chosen in accordance with the maximum desired depth of the impurity
ion milling	the sputter removal of material of a sample by means of a directed ion beam in vacuum
interface states	electronic states specific to the interface between different semiconductors or a metal and a semiconductor
interstitial	refers to a site between atoms in a lattice
IPA	isopropanol; isopropyl alcohol
isotropic etchant	etchant that etches different crystal planes at essentially the same rate, giving nearly equal lateral and vertical etch rates; also called diffusion-controlled or diffusion-limited etchant
I-V	current-voltage

junction temperature	refers to the maximum temperature in a device under a given operating condition, e.g. at the gate of an FET or the pn junction in the emitter of a bipolar transistor or laser diode
junk	the term for a formerly valuable wafer that has been ruined by a serious fabrication error; what we hope this book will help you avoid

knee voltage	the applied voltage (drain for FET or collector for HBT) at which the current reaches saturation
kinetics	refers to the rate at which a reaction occurs

LDMOS	lateral diffused MOS, a type of high-voltage, high-power Si transistor
LDD	lightly-doped drain
LED	light-emitting diode. A semiconductor junction diode that emits light by recombination of electron-hole pairs as current is passed through a pn junction
LEED	low-energy electron diffraction

ligand	a chemical entity bonded to a metal or coordinated to a metal ion in a metal complex ; a hydrocarbon entity such as a methyl group (CH_3) bonded to a transition metal such as Ga for MOCVD source materials
liftoff	the removal of thin films deposited over photoresist by dissolving the resist in a solvent
lithography	a tool such as photolithography or electron-beam lithography that prints patterns on semiconductor and other substrates
loading effects	variation in reaction rates (and etch depths) due to variation in reactant supply across a wafer due to surfaces with different reactivities, e.g., resist versus semiconductor, semiconductor versus platen
loadlock	a separately pumped chamber used to introduce samples into another vacuum chamber without having to vent it; it serves to prevent water and oxygen from contaminating the process chamber

mass spectrometer	an instrument that determines the mass of a gaseous entity in a high-vacuum environment
MBE	molecular beam epitaxy
mercury probe	an instrument that uses mercury droplets as temporary electrical contacts to measure the capacitance-voltage response of a semiconductor
mesa	a flat topped structure with steep sides formed by etching semiconductor away from the edges of the structure, usually to isolate active regions
MESFET	metal-semiconductor field effect transistor
MHEMT	metamorphic HEMT, a HEMT structure (thick enough to relax) such as InAlAs/InGaAs on GaAs that is grown on a substrate with a different lattice constant than the active layers
micromasking	surface contaminants serve as masks and locally retard etching; examples include PR residue, sputtered metal, and non-uniform surface oxides
MIS	metal-insulator-semiconductor
misfit dislocation	a dislocation that results when the crystal lattice constants of two epitaxially grown materials do not match, i.e., display a misfit
mobility	a measure of how an electron (hole) responds to an electric field

MOCVD — metal-organic chemical vapour deposition: a method for depositing a thin film of a material using metal-organic (organometallic) precursors (reactants) as sources for group III and/or group V atoms. Growth is usually epitaxial for semiconductors. Analogous terms: MOVPE, OMVPE

MOS — metal-oxide-semiconductor

MOVPE — metal-organic vapor phase epitaxy: a method of epitaxial semiconductor growth using vapor-phase metal–organic (organometallic) precursors (reactants), e.g., trimethyl gallium. Analogous terms: MOCVD, OMVPE

MTTF — mean time to failure (or median time to failure)

mushroom gate — a gate metal whose cross-section forms a narrow stem at the bottom and widens above the stem

n-type — refers to a semiconductor with a preponderance of electron donor impurities

OEM — original equipment manufacturer

OES — optical-emission spectroscopy

Ohmic contact — a contact with a metal-semiconductor interface that is linear in its I-V characteristic

OMVPE — organometallic vapor phase epitaxy: a method of epitaxial semiconductor growth using vapor phase organometallic precursors (reactants), e.g., trimethyl gallium. Analogous terms: MOCVD, MOVPE

oscilloscope — an instrument that allows the measurement and viewing of fast voltage and current pulses

overcutting — producing angled sidewalls

oxidizing agent — chemical that removed electrons; used to oxidize the surface to form surface oxides; most common is hydrogen peroxide (H_2O_2)

p-type — refers to a semiconductor with a preponderance of electron acceptor, or hole producing, impurities

PA — power amplifier

passivation — refers to a process that either improves the electrical characteristics of an active surface or protects it from further change

PECVD	plasma-enhanced chemical vapour deposition: a method of dielectric deposition using radio-frequency dissociation of gaseous precursors for acceleration of the chemical reactions
pH	negative log(base 10) of the H^+ ion (hydrogen or hydronium ion) concentration
phase-shift mask	a mask capable of producing dimensions smaller than the wavelength of the light of the lithography tool using thin film techniques that alter the phase of the incoming light in select areas of the mask
PHEMT	pseudomorphic HEMT, a HEMT with an InGaAs channel which is strained compared to the GaAs lattice
phonon	lattice vibration; a quantum of thermal energy
photolithography	the patterning of semiconductors or thin films deposited on them using masks, light, and photoresist
photoluminescence (PL)	an optical-analysis technique where a sample is illuminated to form free carriers. Their recombination produces emitted radiation that is detected and analysed by its wavelength and intensity
photoresist	a light-sensitive polymer used in photolithograpy to create patterns in semiconductors
piezoelectric	a crystal property where applied mechanical pressure can produce a voltage and vice versa
pinchoff	a set of bias conditions in an FET where carriers under the gate are completely depleted and no current flows from source to drain
plasma	a gaseous mixture of electrons, ions and neutral atoms and molecules
plasmon	a plasma oscillation in a solid; interaction of plasmons and phonons can be studied using Raman scattering
PL	photoluminescence

PLE — photoluminescence excitation: an optical-analysis technique in which free carriers are excited by light of a single wavelength and the resulting emitted radiation is detected and analysed by its wavelength and intensity. The absorption spectrum of a semiconductor structure can be determined with this technique by varying the wavelength of the excitation light

PMGI — polymethylglutarimide

point defect — localised lattice disruption such as an antisite defect or a vacancy

potential well — a region of minimum (maximum) electrical potential that can confine electrons (holes) to that region

power slump — a unrecoverable drop off in power output of a power amplifier under continual operation

PR — photoresist

probe station — a vibration-isolated apparatus where samples can be electrically measured by means of voltages and currents applied to micro probes that are positioned on the pads of a device on the sample

quad-level mask — a mask consisting of a stack of imaging resist, dielectric, hard-baked resist and lift off layer

quantum-well — a region of minimum (maximum) electrical potential which can confine electrons (holes) to that region and whose dimensions lead to discrete energy levels within the well

quasi-electric field — an electric field arising from the structure of a semiconductor rather than an applied potential

radical — a highly reactive chemical species with an unpaired electron. Examples include alkyl radicals such as methyl radical ($CH_3^{\bullet}$) and ethyl radical ($CH_3\ CH_2^{\bullet}$); and chlorine atoms $Cl^{\bullet}$

rectifying — referring to the property of passing electrical current to an applied bias of only one polarity

refractory metal	a metal with a high melting point; for example, W
RTA	rapid thermal anneal
resist	photoresist
Raman spectroscopy	a spectroscopic technique that measures the inelastic light scattering from a material; peaks appear at the excitation wavelength and at wavelengths shifted from the excitation wavelength by an amount equal to the phonon or vibrational energies of the solid
recessed gate	an FET structure where semiconductor is etched away, with the same mask used for gate deposition prior to the deposition
recombination	the return of an electron to the valence band to combine with a hole
redshift	shift to lower energy
re-entrant profile	a profile that is narrower at the base than at the top of a feature; useful for lift off process masks
reflow	change the sidewall profile of a resist by heating it near or above its glass transition temperature
reticulation	wrinkling of photoresist when heated excessively without proper surface curing
RF	radio frequency
REDR	recombination enhanced defect reaction, a defect creation mechanism in HBTs
RHEED	reflection high-energy electron diffraction
RIBE	reactive-ion-beam etching
RIE	reactive ion etching
R_{th}	the thermal resistance of a device, the proportionality constant that relates the temperature rise in the device to the dissipated power

saponification	hydrolysis of a fat or grease to form soluble products
saturated velocity	the maximum electron (hole) velocity attainable in response to an electric field
Schottky barrier	energy difference between metal Fermi level and the valence band (for p-type) or conduction band (for n-type) at a metal-semiconductor interface

Schottky Contact	a rectifying contact made from a metal-semiconductor interface
self-aligned	an alignment of a feature to an active structure in a device without lithography
semiconductor parameter analyser	an instrument that provides or measures the voltages and currents in a device
sheath	the boundary region between the plasma and the electrodes; the plasma potential is determined by the sheaths; a region of a plasma where electrical neutrality does not prevail
sidewall	the side of an etched feature
SIMS	secondary-ion mass spectroscopy, an analysis technique where a sample is sputtered and the mass of the back-sputtered ions is analysed to identify elements and impurities of a sample
source	the electrode that acts as the origin of electrons in an FET during normal current flow
space-charge region	a region of a doped semiconductor that is depleted of free carriers
S-parameter	scattering parameter. Describes a method of testing where reflected or transmitted signals are measured as an alternative to I-V testing
spectrometer	an instrument used to determine (detect) light intensity as a function of the wavelength of the light
sputtering	the process of material removal by ions accelerated toward the sample in a vacuum; momentum transfer from the ions to the surface ejects surface atoms
STEM	scanning transmission electron microscopy
stepper	a lithography tool that repeats a pattern across many sites of a wafer
STM	scanning tunnelling microscopy
stoichiometry	the ratio of various elements of a chemical compound
stress test	a test that subjects a device beyond its normal operating conditions (bias, temperature, ambient), usually to accelerate a failure mechanism
subcollector	the part of the collector in an HBT that is highly doped for ohmic contact formation
surface states	electronic states specific to the surface of a semiconductor

surface potential	the energy difference between the bottom of the conduction band (or top of the valence band) at the surface and energy of the bottom of the conduction band (or top of the valence band) in the bulk of the semiconductor
surface recombination velocity	S (units of cm/sec); 2D surface analog of $1/\tau$ where τ is the 3D carrier lifetime; essentially the inverse lifetime of a carrier at the surface

TCE	trichloroethane; 1,1,1-trichloroethane
TEM	transmission electron microscopy, a method of high-resolution imaging of solids
tetrahedral	a chemical bonding geometry where four atoms are each bonded to a fifth atom and are equally spaced on the surface of a sphere with the fifth atom at the centre; the characteristic bonding geometry for GaAs and Si
thermionic emission	transport of an electron (hole) across a potential barrier with a probability related to its thermal energy
thermionic field emission	transport of an electron (hole) across a potential barrier with significant probability of either tunnelling or thermionic emission
threshold voltage	the gate voltage of an FET that turns on the current from source to drain
threading dislocation	a structural defect in a semiconductor that extends through the crystal in the irregular shape of a thread; examples are edge, screw and mixed (screw + edge) dislocations
TLM	transmission line method (structure) for ohmic contact testing; refer to Section 6.3.2 for details of the test (structure)
TPD	temperature-programmed desorption; a surface analytical technique where a sample is heated at a controlled rate and the chemical species released at different temperatures are measured by mass spectroscopy

transconductance	the differential of the drain current versus gate voltage in an FET, noted by the symbol g_m
transfer length	refers to the typical distance that electrons (holes) traverse laterally in the semiconductor before crossing the metal semiconductor interface of an ohmic contact
trap	a midgap electronic state, usually localised, that can be filled by electrons (holes)
trenching	deeper etching at feature edges due to reflection of ions from a sidewall that locally increases ion flux on the surface near the feature edge
tunnelling	the passage of a particle across a potential barrier with probability related to the amplitude of its quantum-mechanical wave function on the other side of the barrier
UHV	ultrahigh vacuum (pressures $\leq 10^{-8}$ torr)
undercutting	etching underneath a mask edge
UV	ultraviolet
vacancy	a type of defect consisting of a lattice site in a crystal with no atom residing at that site
vacuum	a pressure below atmospheric pressure: conventional ranges are low = 760 to 10^{-2} torr (overlaps medium range); medium = 1–10^{-2} torr high = 10^{-2} – 10^{-8} torr; ultrahigh $\leq 10^{-8}$ torr
valence band	the energy bands whose electrons are involved in chemical bonding in a solid
valence band offset, ΔE_V	the abrupt energy difference in the valence band at a heterointerface
valence electrons	the electrons in the outermost shells of an atom. These electrons can participate in chemical bonding
VBM	valence band maximum = E_V
VCSEL	vertical-cavity surface-emitting laser

V_{DS} (V_{GS}, V_{CE}, V_{BE}) the drain-source (or gate-source, or collector-emitter, or base-emitter) voltage of a transistor

VPE vapour-phase epitaxy, an older technique for GaAs epitaxial growth with source materials that are created by flowing Cl_2 gas over solid, heated Ga and As sources

WDM wavelength-division multiplexing. A method of sending multiple communication channels along an optical fibre by using (slightly) different wavelengths for each channel

WDS wavelength-dispersive spectrometer, a high-resolution X-ray detector used for electron microprobe analysis

work function the minimum amount of energy required to remove an electron from the surface of a metal; the amount of energy required to take an electron from the Fermi level within a material and remove it to infinity

XPS X-ray photoelectron spectroscopy, a technique where X-rays are absorbed by a sample and emitted electrons are detected. Also called ESCA (electron spectroscopy for chemical analysis)

zinc blende the crystal structure of GaAs with each atom tetrahedrally bonded to four atoms of the opposite type

INDEX

Index

Index